ATOMIC THEORY FOR STUDENTS OF METALLURGY

ATOMIC THEORY
FOR STUDENTS OF METALLURGY

WILLIAM HUME-ROTHERY

*Late Isaac Wolfson Professor of Metallurgy
University of Oxford*

and

B. R. COLES

*Professor of Solid State Physics
Imperial College of Science and Technology*

Monograph and Report Series
No. 3

1969

Published by

THE INSTITUTE OF METALS
17 BELGRAVE SQUARE, LONDON, S.W.1
for the Metals and Metallurgy Trust
of the
Institute of Metals and the Institution of Metallurgists

Book 424
ISBN 0 901462 39 X

First published	.	.	.	.	1946
First (revised) reprint	.	.	.	.	1947
Second (revised) reprint	.	.	.	.	1948
Second (revised) edition	.	.	.	.	1952
Third (revised) reprint	.	.	.	.	1955
Third (revised) edition	.	.	.	.	1960
Fourth (revised) reprint	.	.	.	.	1962
Fifth (revised) reprint	.	.	.	.	1969

Facsimile reprint published in 1988 by

The Institute of Metals
1 Carlton House Terrace
London SW1Y 5DB

and

The Institute of Metals North American Publications Center
Old Post Road Brookfield VT05036 USA

© THE INSTITUTE OF METALS
ALL RIGHTS RESERVED

British Library Cataloguing in Publication Data
Hume-Rothery, William, *1899-1968*
  Atomic theory for students of metallurgy.
  – 3rd ed. rev. and repr.
  1. Metals. Chemical properties
  I. Title II. Coles, B.R. III. Institute of Metals (1985-)
  546′.3
  ISBN 0-901462-39-X

Library of Congress Cataloguing in Publication Data
applied for

First printed by Richard Clay (The Chaucer Press) Ltd,
Bungay, Suffolk

Reprinted in England by Antony Rowe Ltd,
Bumper's Farm, Chippenham, Wilts

FOREWORD TO REPRINT OF THIRD EDITION

At a remove of some forty years, it is appropriate to recall the considerable and far-reaching impact of this work when first published. Ostensibly written to help young metallurgists, it also acquainted established members of the profession with an exciting new world of ideas. Hume-Rothery wisely played down the underlying and supportive theory of mathematical physics. While gaining acceptance as a standard introductory text in Universities, it was also purchased by many young metallurgists who, of necessity, studied on a part-time basis for their professional qualifications. Both full-time and part-time students responded to its easy, almost conversational, style.

In educational terms, the book was very effective in providing metallurgists with new ways of looking at the natural world through working descriptions of difficult concepts such as uncertainty, wave mechanics, energy quantisation, Fermi surfaces and Brillouin zones. It is very unlikely that qualified metallurgists of today require the special introduction to atomic theory which was the main aim of the book. Nevertheless, in spite of this change in the educational background of the profession, the concepts covered are basic to an understanding of the electronic properties of materials such as electrical conductivity, semiconductivity and superconductivity. To the young student meeting these topics for the first time the concepts are still difficult and therefore, although this particular book was introduced forty years ago it still has a part to play in today's educational process.

R E Smallman

CONTENTS

PREFACE

PART I. THE GENERAL BACKGROUND
1. Introduction 1
2. The Bohr Theory of the Hydrogen Atom . . . 15
3. The Uncertainty Principle of Heisenberg . . . 18
4. The Ideas of Wave Mechanics 23
5. Wave Equations and Their Interpretation . . . 33
6. The Representation of the Individual Electron . . 39
7. The Meaning of Velocity 44
8. Potential Boundaries in Wave Mechanics . . . 45

PART II. THE STRUCTURE OF THE FREE ATOM
1. The Electron Groups 48
2. Representation of Electron States 65
3. The s, p, and d States 71

PART III. ASSEMBLIES OF ATOMS
1. The Soft X-Ray Spectra of Solids 85
2. The Elements 99
3. Bonding Energies 109
4. Assemblies of Unlike Atoms 114
5. Atomic Attraction and the Nature of van der Waals Forces 131
6. The Interpretation of the Co-valent Bond and the Nature of Exchange Forces 134
7. Directed Valencies 150
8. Resonance Bonding and the Metallic Linkage . . . 157

PART IV. THE FREE-ELECTRON THEORY OF METALS
1. The Fermi–Dirac Statistics and the Electron Gas . . 166
2. The Models of Wave Mechanics 179
3. Applications of the Free-Electron Theory . . . 184

PART V. THE BRILLOUIN-ZONE THEORY OF METALS
1. The Simple Theory of Brillouin Zones 195
2. Insulators, Semi-Conductors, and Metals . . . 215
3. Electron Theories of Metallic Crystals 232
4. The Experimental Examination of the Electronic Structure of Metals 254

PART VI. ELECTRONS, ATOMS, METALS, AND ALLOYS
1. The Alkali Metals 284
2. Copper, Silver, and Gold 301
3. Some Metals of Higher Valency 329
4. The Transition Elements 345
5. The Rare-Earth Metals and Actinides 375
6. Models for Dilute Alloys 377

PART VII. SOME PHYSICAL PROPERTIES
1. Some Magnetic Properties 381
2. The Electrical Resistivity of Metals and Alloys . . 405
3. Superconductivity 411

APPENDIX A: THE METALLIC ELEMENTS AND THEIR CRYSTAL STRUCTURES . 419
APPENDIX B: INTERATOMIC DISTANCES IN THE CRYSTALS OF THE ELEMENTS 422
NAME INDEX 423
SUBJECT INDEX 425

GREEK ALPHABET

A	α	Alpha	I	ι	Iota	P	ρ	Rho
B	β	Beta	K	κ	Kappa	Σ	σ	Sigma
Γ	γ	Gamma	Λ	λ	Lambda	T	τ	Tau
Δ	δ	Delta	M	μ	Mu	Υ	υ	Upsilon
E	ϵ	Epsilon	N	ν	Nu	Φ	ϕ	Phi
Z	ζ	Zeta	Ξ	ξ	Xi	X	χ	Chi
H	η	Eta	O	o	Omicron	Ψ	ψ	Psi
Θ	θ	Theta	Π	π	Pi	Ω	ω	Omega

PREFACE TO FIFTH (REVISED) REPRINT

DURING the preparation of this revised reprint metallurgy lost one of its most distinguished scientists, and the writer a wise teacher and generous friend, with the death of Professor Hume-Rothery. Among his great gifts was an insight into the changing needs of metallurgy students, and the present book appeared first when wartime experience had shown that physical metallurgists would need to draw upon knowledge of the electronic structures of metals (as well as their crystal structures and phase constitution) to understand both their alloying behaviour and their physical properties. Increasingly the metallurgist was being asked to produce alloys and intermetallic compounds whose applications depended upon their electronic rather than their mechanical properties. Professor Hume-Rothery thus anticipated much of the thinking that has led to undergraduate and postgraduate courses in the Science of Materials. Because of his death the revisions of the later sections of this revised reprint have not had the benefit of his gift for combining in the highest degree clarity of expression and scientific accuracy, but the general pattern of the changes made had been discussed between us.

The principal changes have been brought about by improvements in the confidence that can be given to theoretical calculations of electronic structures, the increasing volume of direct experimental evidence about electronic structures, and the possibility of correlating with these structures many of the macroscopic physical properties. There have therefore been significant additions to Chapters 25 and 26, the addition of short new chapters on rare-earth metals and dilute alloys in Part VI and the creation of a new Part VII on physical properties to include accounts of electrical resistivity and superconductivity, as well as the former chapter on magnetic properties to which a section on the Mössbauer effect has been added. Since this is a revised reprint, not a new edition, no effort has been made to modify the units in which physical quantities are expressed to the SI system.

My thanks are due to many colleagues for helpful discussions, to Mr. John Dunlop for the preparation of the indices, and to Mrs. A. D. Moss for help in the correcting of the proofs.

B. R. COLES

Department of Physics,
 Imperial College of Science and Technology,
  London.

PREFACE TO FOURTH (REVISED) REPRINT

THE present monograph was first published in 1946, and was written in the hope of providing a bridge by means of which the student of metallurgy might be led to an understanding of the general ideas on which the electron theories of metals and alloys are based. At the request of the Publication Committee of the Institute of Metals, the monograph was made reasonably self-contained, with the result that some of the subject matter of Parts I, II, and III has already been treated in books of a fairly elementary standard. The book is written primarily for the Honours Student in Metallurgy, and no attempt is made to deal with mathematical technique. This means that equations have sometimes to be presented dogmatically, and it is hoped that this may be excused in view of the desirability of teaching the student not to be afraid of papers which contain numerous equations whose derivation he cannot hope to follow.

In the fourteen years that have elapsed since the appearance of the first edition, there have been spectacular advances in nuclear physics and in the science of semi-conductors. As these subjects have already been dealt with at different levels by many authors, the present description has been left at an elementary level with comparatively few alterations. The remarkable developments in the science of dislocations and crystal imperfections are not dealt with, because it was felt that this subject would have led too far from the electronic theory of the ideal crystal with which the book is mainly concerned.

In the earlier sections of the book, the chief alteration is a somewhat more detailed account of the theory of the co-valent bond. The work of the last ten years has emphasized the extreme difficulty of producing any really quantitative theory of the electronic structure of metals, except for those of the alkali group. The different lines of approach have been outlined briefly in Chapter XXV, but no attempt has been made to describe the details, many of which can hardly be explained in non-mathematical terms. In view of the increasing recognition that, in spite of the great mathematical skill involved, most of the electronic theory of metals is only a crude approximation, an account has been added in Chapter XXVI of some of the experimental methods for exploring the electronic structure of metallic crystals. In the later chapters most of the text has been rewritten in order to include new work, but the last chapter on magnetic properties has been left at an elementary level.

Preface

For easy reference, the crystal structures of the metallic elements are listed in an Appendix on p. 419.

The author would again like to express his gratitude for the great help he has received from many friends in the preparation of this monograph. The thanks expressed to those who helped with the first edition must be repeated. For the new edition, particular thanks are due to Dr. J. W. Christian, Dr. S. L. Altmann, and Dr. V. Heine, who have read large sections of the manuscript. Professor N. F. Mott, F.R.S., Professor C. A. Coulson, F.R.S., Professor L. Bates, F.R.S., Dr. A. B. Pippard, F.R.S., and Dr. B. V. Rollin have also shown much kindness in discussing some of the theoretical points. The author must also acknowledge the great help received from Mr. N. B. Vaughan (of the Institute of Metals) in connection with the preparation of the manuscript and diagrams, and from Mrs. A. D. Moss in correcting the proofs. It need scarcely be said that the fact that the friends mentioned above have read the manuscript in no way commits them to the views expressed in the monograph.

W. HUME-ROTHERY.

Department of Metallurgy,
Parks Road,
Oxford.

Note: The Part, Chapter, and page numbers in Professor Hume-Rothery's Preface refer to the fourth (revised) reprint.

PART I. THE GENERAL BACKGROUND.

1. GENERAL INTRODUCTION.

THE modern theory of atomic structure is the result of work in chemistry, physics, and mathematics, extending over more than 150 years, and involving very great changes in the underlying ideas. We do not propose to deal here in detail with the historical development of the subject, but it may be of interest to point out some of the more prominent stages by which the present position has been reached.

The atomic theory of John Dalton provided the Science of Chemistry with a sound foundation, and its developments by the great chemists of the nineteenth century culminated in Mendeleev's Periodic Table of the Elements which appeared in 1869. This table, which is reproduced in a modified form on p. 2, showed that a periodic repetition of physical and chemical properties was observed if the elements were arranged in a definite order which, with a very few exceptions, was the order of increasing atomic weight. These developments of what may be called the simple atomic theory of chemistry clearly supported the view that matter consisted ultimately of atoms, which in ordinary chemical reactions were indestructible, but which could combine together to form molecules in accordance with the principles of valency. At the same time the growth of the Science of Electrochemistry, which began with the work of Davy and Faraday, suggested that electricity had some kind of an atomic structure, since the gram-molecular weights of the different ions were associated with small whole-number multiples of a definite electric charge. The periodicity revealed in Mendeleev's table suggested that the atoms themselves had structures, and that on passing down the Periodic Table, the changes in atomic structure involved the building up of a series of stable groups or units, so that each time a group was completed, the building up process began again.

The development of the Science of Chemistry led to the determination of the relative weights of the atoms of the different elements. For many years these were expressed as a scale in which the atomic weight of oxygen was taken as 16·000. Recently, this has been replaced by a scale in which the atomic weight of the carbon isotope ^{12}C is taken to be 12·000. On this scale, the atomic weight of hydrogen is 1·00797 and the atomic weights at present accepted are given in Table I.

In the middle and latter half of the nineteenth century, the development of the kinetic theory of gases showed that many properties of the gaseous state could be explained on the assumption that a gas consisted of an assembly of small particles, the average velocity of which increased with rise of temperature. These " particles " were naturally identified with the molecules of chemistry, and several different lines of approach led to the conclusion that the number of molecules present in the gram-molecular weight of a substance was of the order of 10^{24}. Later and more perfect methods have shown that this

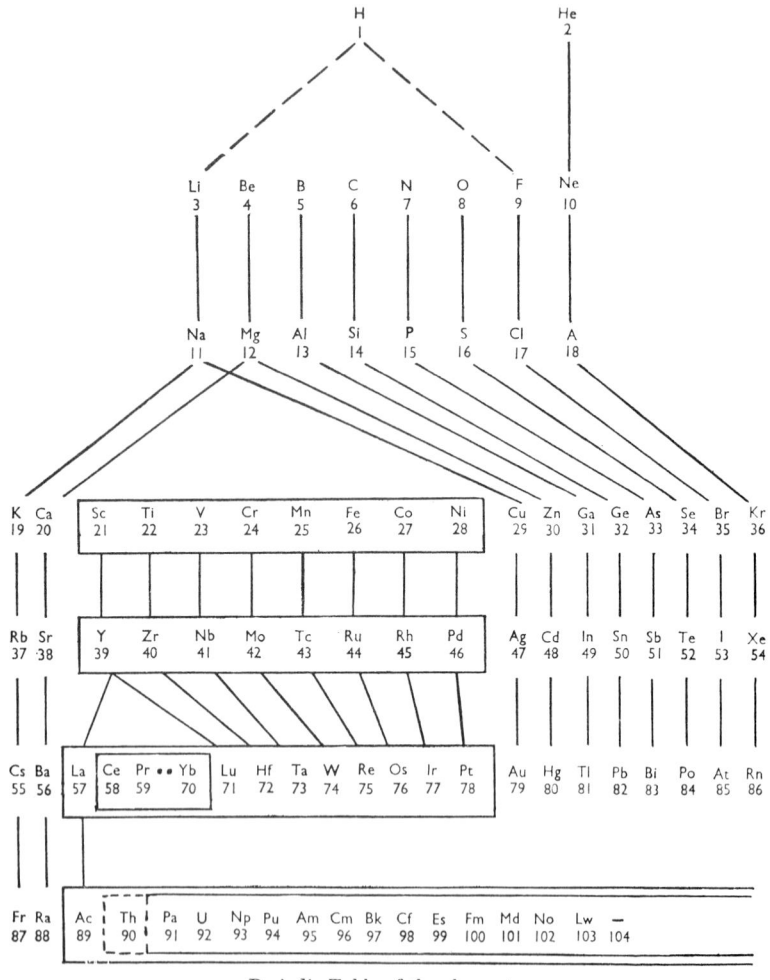

Periodic Table of the elements.

The General Background

TABLE I. *Relative International Atomic Weights, 1965.*
Based on the atomic mass of $^{12}C = 12$.

	Symbol	At. No.	At. Wt.*		Symbol	At. No.	At. Wt.*
Actinium	Ac	89	[227]	Mercury	Hg	80	200·59
Aluminium	Al	13	26·9815	Molybdenum	Mo	42	95·94
Americium	Am	95	[243]	Neodymium	Nd	60	144·24
Antimony	Sb	51	121·75	Neon	Ne	10	20·183
Argon	A	18	39·948	Neptunium	Np	93	[237]
Arsenic	As	33	74·9216	Nickel	Ni	28	58·71
Astatine	At	85	[210]	Niobium	Nb		
Barium	Ba	56	137·34	(Columbium)	(Cb)	41	92·906
Berkelium	Bk	97	[249]	Nitrogen	N	7	14·0067
Beryllium	Be	4	9·0122	Nobelium	No	102	[254]
Bismuth	Bi	83	208·980	Osmium	Os	76	190·2
Boron	B	5	10·811†	Oxygen	O	8	15·9994
Bromine	Br	35	79·904	Palladium	Pd	46	106·4
Cadmium	Cd	48	112·40	Phosphorus	P	15	30·9738
Caesium	Cs	55	132·905	Platinum	Pt	78	195·09
Calcium	Ca	20	40·08	Plutonium	Pu	94	[242]
Californium	Cf	98	[251]	Polonium	Po	84	[210]
Carbon	C	6	12·01115	Potassium	K	19	39·102
Cerium	Ce	58	140·12	Praseodymium	Pr	59	140·907
Chlorine	Cl	17	35·453	Promethium	Pm	61	[147]
Chromium	Cr	24	51·996	Protactinium	Pa	91	[231]
Cobalt	Co	27	58·9332	Radium	Ra	88	[226]
Copper	Cu	29	63·546	Radon	Rn	86	[222]
Curium	Cm	96	[247]	Rhenium	Re	75	186·2
Dysprosium	Dy	66	162·50	Rhodium	Rh	45	102·905
Einsteinium	Es	99	[254]	Rubidium	Rb	37	85·47
Erbium	Er	68	167·26	Ruthenium	Ru	44	101·07
Europium	Eu	63	151·96	Samarium	Sm	62	150·35
Fermium	Fm	100	[253]	Scandium	Sc	21	44·956
Fluorine	F	9	18·9984	Selenium	Se	34	78·96
Francium	Fr	87	[223]	Silicon	Si	14	28·086
Gadolinium	Gd	64	157·25	Silver	Ag	47	107·868
Gallium	Ga	31	69·72	Sodium	Na	11	22·9898
Germanium	Ge	32	72·59	Strontium	Sr	38	87·62
Gold	Au	79	196·967	Sulphur	S	16	32·064†
Hafnium	Hf	72	178·49	Tantalum	Ta	73	180·948
Helium	He	2	4·0026	Technetium	Tc	43	[99]
Holmium	Ho	67	164·930	Tellurium	Te	52	127·60
Hydrogen	H	1	1·00797	Terbium	Tb	65	158·924
Indium	In	49	114·82	Thallium	Tl	81	204·37
Iodine	I	53	126·9044	Thorium	Th	90	232·038
Iridium	Ir	77	192·2	Thulium	Tm	69	168·934
Iron	Fe	26	55·847	Tin	Sn	50	118·69
Krypton	Kr	36	83·80	Titanium	Ti	22	47·90
Lanthanum	La	57	138·91	Tungsten	W	74	183·85
Lawrencium	Lw	103	—	Uranium	U	92	238·03
Lead	Pb	82	207·19	Vanadium	V	23	50·942
Lithium	Li	3	6·939	Xenon	Xe	54	131·30
Lutetium	Lu	71	174·97	Ytterbium	Yb	70	173·04
Magnesium	Mg	12	24·312	Yttrium	Y	39	88·905
Manganese	Mn	25	54·9380	Zinc	Zn	30	65·37
Mendelevium	Md	101	[256]	Zirconium	Zr	40	91·22

* A value given in brackets denotes the mass number of the isotope of longest known half-life, which is not necessarily the most important isotope in atomic-energy work.
† Because of natural variations in the relative abundance of their isotopes, the atomic weights of boron and sulphur have a range of ± 0·003.

number known as *Avogadro's number*, is equal to $6·0225 \times 10^{23}$ when the ^{12}C atomic weights are used (see p. 14 for the relation between physical and chemical scales). In the case of carbon monoxide, for example, the atomic weights are C = 12·011 and O = 15·999, and consequently 28·010 g of carbon monoxide contain $6·0225 \times 10^{23}$ molecules of CO.

The determination of Avogadro's number led to an estimation of the actual as distinct from the relative weights of the atoms of the different elements, and this knowledge, when combined with the known densities of substances, suggested that the distances between atoms in solids were of the order 10^{-7}–10^{-8} cm. The subsequent development (1913–20) of crystal analysis by X-ray diffraction methods confirmed this experimentally, and in so far as the interatomic distances in crystals are a measure of the sizes of the atoms concerned, we now know that the values lie between 10^{-7} and 10^{-8} cm, and the interatomic distances in solid metals have been determined to a degree of accuracy which is often as high as 1 part in 50 000.

The periodicity shown in Mendeleev's table suggested clearly that atoms were not analogous to very minute billiard balls, but that they possessed definite inner structures. The modern theory of atomic structure is founded essentially on the experimental work at the end of the nineteenth century which led to the discovery of the electron. In this work it was shown that under appropriate conditions, such as electrical discharges in gases, high temperatures (thermionic emission), and strong electric fields, many substances could be made to emit " rays " which on passing through electric or magnetic fields behaved as though they consisted of very minute particles carrying charges of negative electricity. These experiments showed that although the velocities of the emitted electrons varied greatly with the experimental conditions, the mass m and charge e of the electrons were always the same, irrespective of the substance from which the electrons were produced. The values accepted at present for these constants are: $m = 9·1091 \times 10^{-28}$ g; and $e = 4·8030 \times 10^{-10}$ electrostatic units. The mass of the electron is thus about $\frac{1}{1840}$ that of a hydrogen atom, whilst its electric charge is the same as that on a univalent negative ion.*
The development of the methods of mass spectroscopy by J. J. Thomson and F. W. Aston enabled the masses and charges of electrons and ions

* The charge on one gram-molecular weight of a univalent ion is called a Faraday and is equal to 96487 absolute coulombs (using the ^{12}C atomic weights). The charge on one single ion is therefore $96487 \div 6·0225 \times 10^{23} = 1·60210 \times 10^{-19}$ coulombs. Since 1 coulomb = $2·99793 \times 10^9$ electrostatic units, the charge on one ion is equal to $1·60210 \times 2·99793 \times 10^{-10} = 4·8030 \times 10^{-10}$ e.s.u., which is the value given above for the electronic charge. These figures refer to the new value for the velocity of light (p. 12).

to be studied in detail, and this work showed that whilst both positive and negative ions existed, only negative electricity was associated with a mass as small as that of an electron. The work in nuclear physics referred to below has led to the discovery of positive electrons, or *positrons*, in which a charge of $+e$ is associated with the same mass as that of an electron. This does not, however, disturb the earlier conclusion that, under ordinary experimental conditions, only negative electricity is associated with a mass as small as that of an electron. This led naturally to the idea of an atom as consisting of a relatively heavy, positively charged body to which was attached a number of light, negatively charged electrons, and from this it was an easy step to the concept of a nuclear atom with negatively charged electrons revolving round a positively charged core. This concept of a nuclear atom received direct experimental confirmation from the investigations of Rutherford on the deflection of α-particles* when passing through matter, and his work led to the conclusion that an atom consisted of a minute positively charged nucleus, surrounded by a sufficient number of electrons to keep the atom as a whole neutral. The dimensions of the nucleus are of the order of 10^{-12} cm, and are thus very small compared with the distances between atoms in molecules or crystals (10^{-7}–10^{-8} cm), though the positively charged nucleus contains almost the whole of the mass of the atom. The zone occupied by the surrounding electrons is of the order of 10^{-7}–10^{-8} cm, and is thus of the same magnitude as the interatomic distances in crystals The positive charge on the nucleus is equal to $+Ze$, where Z, the so-called *atomic number*, gives the position of the element in the Periodic Table. Since the atom as a whole is neutral, an atom of atomic number Z is surrounded by Z electrons, so that an atom of, say, copper of atomic number 29 contains a nucleus of charge $+29e$, surrounded by 29 electrons.

The development of the idea of a nuclear atom was accompanied by increasing evidence (1910–30) that under some conditions the nuclei of atoms were not indestructible. The first discoveries were in connection with the phenomenon of radioactivity, where it was found that some of the heaviest elements underwent spontaneous disintegration into new elements, the process being accompanied by emission of positively charged α-particles, negatively charged electrons (β-particles), or very penetrating radiation (γ-rays). These investigations showed that certain elements, notably lead, could exist in forms which had different atomic weights, but were chemically indistinguishable. These

* α-particles or α-rays are produced in some radioactive transformations and are doubly charged He^{++} ions, *i.e.*, helium atoms which have lost two electrons.

discoveries, in the hands of Soddy, led to the concept of isotopes, this name being given to cases where an element could be shown to exist in the form of atoms which were of different atomic weights, but which had the same positive nuclear charge, so that they occupied the same position in the Periodic Table and were chemically indistinguishable.* Subsequent work showed that nearly all the elements met with on the earth are really mixtures of isotopes; thus chlorine, of atomic weight 35·453, consists of a mixture of isotopes of atomic weights 35 and 37.

The conclusion reached was, therefore, that the characteristics of the atoms of elements are determined not by their atomic weights, but by the positive charges on their nuclei. The order of the elements in the Periodic Table is thus the order of increasing positive charge on the nucleus, each step in the table corresponding to an increased nuclear charge of $+ e$. This order was first established conclusively by the X-ray spectroscopic investigations of Moseley (1912), whose work showed that in the few cases where the order of the elements in the Periodic Table was not the order of increasing atomic weight, the regular increase in nuclear charge was maintained. Thus, as can be seen from the table on p. 2, the order of the elements in the middle of the First Long Period is iron $\rightarrow$ cobalt $\rightarrow$ nickel, for which the nuclear charges are $+ 26e$, $+ 27e$, and $+ 28e$, respectively, whilst the atomic weights are 55·85, 58·93, and 58·71, so that cobalt and nickel are out of order as regards atomic weights, but in order as regards the nuclear charges.

Much of the later experimental work has been concerned with the structure of the nucleus, and has led to the methods for the artificial breaking up of atomic nuclei which have culminated in the various atomic-energy projects. This work has shown that whereas the positively charged hydrogen ion, or *proton*, is a fundamental particle, the nuclei of the heavier elements have structures which under suitable conditions may be broken up. From the Table of Atomic Weights on p. 3 it will be seen that the atomic weight of an element is of the order of twice its atomic number. The earlier theories of nuclear structure tried to account for this by assuming that in an isotope of atomic weight N, the nuclei contained N protons and roughly $N/2$ electrons, so that the resultant nuclear charge or atomic number was of the order of $N/2$. This view led to many difficulties, which were not overcome until 1932, when Chadwick discovered a new kind of particle which is now called a *neutron*. This has the same mass as the proton but no

* Some chemical properties (*e.g.*, rates of reaction involving diffusion) may depend slightly on the weights of the atoms, but the differences are usually very small.

The General Background

electric charge, and it may be represented by the symbol n or $_0n^1$ to indicate that the charge is zero, and the mass is 1 on the hydrogen scale. Neutrons were originally discovered by allowing α-particles from polonium to collide with beryllium, but they have since been produced in many other ways, and are now regarded as a universal constituent of all the heavier atoms. The discovery of the neutron meant that if there were some way of holding protons and neutrons together, it was no longer necessary to assume that electrons were present in atomic nuclei, but that an atom of atomic number Z and atomic weight W might contain Z protons and $(W - Z)$ neutrons, where W was approximately equal to $2Z$. This view received support from the fact that when the isotopes of the different elements were compared, it was found that nuclei of even mass number* were very much more common than those of odd mass number. This clearly suggested that neutrons and protons went together in pairs in atomic nuclei. The fact that, in the elements of higher atomic number, the atomic weights become increasingly greater than twice the atomic numbers is readily explained from this point of view. A neutron is not charged electrically, and consequently there is no repulsion between the protons and neutrons in a given nucleus. There is, however, a repulsion between two protons, and this becomes so great that atoms of the elements of higher atomic number would be unstable if the proportion of neutrons : protons did not increase. The increase in the relative proportion of neutrons serves to dilute the protons, and so enables a stable structure to be formed.

As explained later (p. 23), electrons possess properties of both particles and waves. Their wave-like properties mean that they undergo diffraction by the regular arrangements of atoms in crystals, and electron diffraction is one of the methods of determining crystal structure. Neutrons also possess wave-like characteristics and neutron diffraction can be used for crystal-structure determination. It is important to note that in refined work, such as the determination of the exact positions of atoms in complicated structures, X-ray, electron, and neutron diffraction do not involve exactly the same process. X-ray diffraction effects are determined by the distribution of electron density around the atom; neutron diffraction involves the nuclei; while electron diffraction involves the potential distribution around the nuclei.

The forces that hold the neutrons and protons together in a nucleus are not electromagnetic, but are forces of a new kind. Their origin and

---

* To a first approximation the atomic weights of individual isotopes are whole-number multiples of that of hydrogen, but this relation is not exact because the large amount of energy evolved when protons and neutrons combine results in a slight loss of mass according to the theory of relativity. The mass number of an isotope is the nearest whole number to its atomic weight on the hydrogen scale.

related problems associated with various groups of unstable elementary particles, of which some are mediators of the nuclear forces, lie outside the scope of the present book.*

The reader may note that the term *nucleon* is often used to describe a *proton* or *neutron*, while *pion* and *muon* refer to different kinds of *meson* which are particles of mass intermediate between those of an electron and a nucleon. *Hyperons* or heavy *mesons* are unstable particles with masses somewhat greater than those of nucleons.

In general, the set of anti-particles which, at extremely high energies, can be produced along with corresponding particles of ordinary matter are of little concern outside a nuclear-physics laboratory. The first of these to be discovered, and the most easily produced, is the anti-electron or *positron*, which has the same mass as an electron but a positive charge of $+e$. The positron can be used as a probe for the electronic structure of solids, since the positron and an electron will annihilate one another in a manner depending on the momentum of the latter.

In most problems of metallurgy we are concerned with the atom as a whole, and not with the nucleus, but we may point out here that the nucleus has a magnetic moment whose magnitude is described on p. 263. These magnetic moments are important in connection with Nuclear Magnetic Resonance (p. 262), which is a valuable method of investigating metals and alloys.

With the advance in large-scale atomic-energy work, it has been found possible to split up most of the heavier atoms, with the production of isotopes of many of the lighter elements, and many of these isotopes are radioactive and undergo one or more transformations with the emission of characteristic β-rays (electrons), α-particles (He^{++} nuclei), or γ-rays (X-rays of very short wave-length). γ-ray emission or absorption by certain nuclei (especially ^{57}Fe) in solids can be of extremely sharply defined energy—*i.e.*, occurs over a very narrow range of frequencies—and can therefore be a sensitive measure of the atomic environment, in what is known as the Mössbauer effect (p. 402). These radioactive isotopes are now finding increasing use in metallurgical research on problems such as those of diffusion. In this kind of application, minute quantities of a radioactive isotope of, say, silver, are electroplated on to an alloy, and after an appropriate heat-treatment, the penetration of the radioactive element can be studied by measurement of the radioactivity of the specimen at different depths. The same method

---

* Readers interested may consult:
D. Halliday, " Introductory Nuclear Physics ". **1960**: New York and London (John Wiley).
D. H. Frisch and A. M. Thorndyke, " Elementary Particles ". **1964**: Princeton, N.J. (D. Van Nostrand).

The General Background

may be used for the study of self-diffusion, by electroplating a radioactive isotope on to the ordinary non-radioactive form of the same element. The work in nuclear physics has also shown that in some cases it is possible to add a neutron to the nucleus of a heavy atom. The resulting atom is generally unstable and if it emits a β-ray (electron) from its nucleus, it has an atomic number and a mass number each one unit greater than the element which absorbed the neutron at the beginning of the process. In this way new elements lying beyond uranium in the Periodic Table have been prepared. Thus, the uranium isotope 238 can absorb a neutron, with the formation of a new isotope of uranium 239 which is unstable and undergoes a β-ray transformation, with the formation of the new element neptunium of atomic number 93 and mass number 239. The atoms of neptunium are again unstable and suffer a β-ray transformation with the formation of the new element plutonium of atomic number 94 and mass number 239. The resulting plutonium is unstable and undergoes a slow α-ray decomposition in which the mass number is reduced by 4 owing to the emission of the He^{++} nucleus, while the atomic number is reduced by 2 because the emitted He^{++} nucleus removes two positive charges from the heavy nucleus. The α-ray decomposition of plutonium thus results in the formation of uranium 235, and the final result of the sequence of changes is to convert uranium 238 into uranium 235. The principles described above are quite general, and in all β-ray emission processes the atomic number increases by 1; in α-ray emission processes the atomic number decreases by 2 and the mass number by 4; whilst in neutron transformations the atomic number is unaltered and the mass number increases or decreases by 1 according to whether the neutron is absorbed or emitted.

The new *transuranic* elements, with their chemical symbols, are:

Name	Neptunium	Plutonium	Americium	Curium	Berkelium	Californium
Symbol	Np	Pu	Am	Cm	Bk	Cf
Atomic No.	93	94	95	96	97	98

Name	Einsteinium	Fermium	Mendelevium	Nobelium	Lawrencium	—
Symbol	Es	Fm	Md	No	Lw	—
Atomic No.	99	100	101	102	103	104

From the above survey it will be evident that the general concept of a nuclear atom is the result of a long series of experiments of a widely varying nature, which indicate that the theory of atomic structure must involve the behaviour of electrons in the fields of force of a minute positively charged nucleus, and at distances from the nucleus which do not exceed about 10^{-7} cm. Before describing the theory of the subject, we may consider briefly the units of measurement which are usually employed in connection with the structure of the atom.

The unit of length in most general use for atomic phenomena is the Ångstrom unit (Å or A), which is equal to 10^{-8} cm. In the earlier X-ray crystallography, use was made of a so-called " Ångstrom unit " which was based on the Siegbahn scale of X units, an X unit being so defined that the (200) spacing of calcite at 18° C is equal to 3029·45 X units. This definition made an X unit as nearly as possible equal to 10^{-3} Å on the basis of the then known value of Avogadro's number. Later work showed that a slight error was made and that the crystallographic or kX unit was equal to 1·00202 Å. As this is slightly greater than the Å unit, values in kX units require multiplying by 1·00202 in order to give values in true Ångstroms. The difference between kX and Å units was recognized increasingly in the period 1940–50. In all early crystallographic papers results are really in kX units, even though the symbol Å or A is used. From 1945 onwards, there was a steadily increasing tendency to give results in true Ångstroms, denoted A, but the reader should be warned that, in some countries, the symbol " A " was used for kX units until quite recently, and the resulting confusion is unfortunate. At present, most authors give results in true Ångstroms, but some prefer to use kX units until the conversion factor is known with greater certainty.

For some purposes it is convenient to use the so-called " atomic unit of length," which is equal to 0·5292 Å. As we shall see later (p. 16), the Bohr theory assumed that the one electron of the hydrogen atom could travel round the nucleus in a number of definite orbits or stationary states, and the atomic unit of length is the radius of the first Bohr orbit, and is equal to:

$$h^2/4\pi^2 me^2$$

where m and e are, respectively, the mass and charge of the electron, and h is Planck's Constant of Action, which is referred to later and which equals $6·6256 \times 10^{-27}$ erg-second. The atomic unit of length is employed because its use avoids the necessity for working out terms of the form $h^2/4\pi^2 me^2$ which occur in some calculations.

For the measurement of energy, the results can be expressed either as the energy change per atom or as the energy change per gram atomic weight. When we are concerned with the individual atom or electron, the most convenient unit of energy is often the energy change involved when one electron falls through a potential difference of 1 volt. This

The General Background 11

quantity of energy is called an *electron volt*, and is equal to $1{\cdot}60210 \times 10^{-12}$ erg. Avogadro's number states that there are $6{\cdot}0225 \times 10^{23}$ atoms in the gram atomic weight of an element, and hence an energy change of one electron volt per atom is equivalent to $1{\cdot}60210 \times 10^{-12} \times 6{\cdot}0225 \times 10^{23} = 9{\cdot}6486 \times 10^{11}$ ergs per gram atomic weight. Now 1 *joule* is equal to 10^7 ergs, and 1 *calorie* is equal to $4{\cdot}184$ joules. An energy change of one electron volt per atom is thus equivalent to:

$$\frac{9{\cdot}6486 \times 10^{11}}{4{\cdot}184 \times 10^7} = 23{\cdot}06 \times 10^3 \text{ cal/g-atom}$$
$$= 23{\cdot}06 \text{ kcal/g-atom}.$$

If, therefore, an electron in an atom or a molecule undergoes an energy change of 1 electron volt, the energy change involved is equivalent to that of a heat of transformation or reaction of $23{\cdot}06$ kcal per g-atom, and this conversion factor should be borne in mind when trying to visualize the magnitude involved when results are expressed in electron volts per atom. The latent heats of fusion of most metals lie in the range $0{\cdot}5$–$10{\cdot}0$ kcal per gram atomic weight, so that a change of $\frac{1}{10}$ of an electron volt per atom is of the same order as a latent heat of fusion.

The energy changes involved in atomic transformations are also sometimes expressed in terms of wave numbers or frequencies. It is well known that the phenomena of interference effects led to the development of the wave theory of light, but that when absorption or emission processes occur, the facts can only be explained by means of the quantum theory. This theory was first advanced by Planck (1901) in order to explain the distribution of energy among the different wave-lengths of radiation emitted by hot bodies. According to the quantum theory (see p. 24), energy in the form of radiation can only be emitted in units or quanta of magnitude $E = h\nu$, where E is the energy, ν the frequency of the radiation, and h is Planck's constant referred to above. A beam of radiation of frequency ν is thus the result of an immense number of atomic processes each involving an energy change equal to $h\nu$. Conversely, if an atom gives rise to emission, as the result of an energy change E, it expels one quantum of radiation of frequency E/h. It can therefore be understood that in many cases where atomic processes result in the emission or absorption of radiation, it is convenient to express energy in terms of frequency by means of the relation $E = h\nu$. The velocity of light is usually denoted by the symbol c, and until recently was thought to be equal to $2{\cdot}99776 \times 10^{10}$ cm/sec. Since velocity is equal to the product of wave-length and frequency, it follows that light of frequency ν has a wave-length λ equal to c/ν, whilst the *wave-number*, n, or number of waves/cm* is equal to ν/c. Work in the National Physical Laboratory† showed that the value of c was greater than the previously accepted figure, and the new generally agreed value is $2\cdot997925 \times 10^{10}$ cm/sec. From the relation $E = h\nu$, it follows that if an atom undergoes a transition in which it emits one quantum of radiation of frequency ν, the energy change involved is $6\cdot6256 \times 10^{-27} \nu$ ergs; and since the wave-number n equals ν/c, the energy of a quantum of radiation of wave-number n is $6\cdot6256 \times 10^{-27} \times c \times n = 1\cdot9863 \times 10^{-16} \times n$ ergs/atom, and to 4-figure accuracy this is the same as the value given by the old physical constants (see Note below). It is possible to use wave-numbers as a measure of energy, and a combination of the above factors will show that an energy change of y electron volts/atom is equivalent to a wave-number $n = 8\cdot066 \times 10^3 \times y$, and an energy change of wave-number n is equivalent to $1\cdot240 \times 10^{-4} \times n$ electron volts per atom. The wave-lengths of visible light are of the order of 3000–7000 Å, so that an atomic transition which gives rise to visible light involves an energy change of the order of 30 000–15 000 wave-numbers per cm or *ca.* 3·5–1·8 electron volts per atom. The wavelengths of X-rays are of the order of 1 Å, so that the atomic transitions involve energy changes about 1000 times as great as those which give rise to optical spectra.

Fig. I.1, which is taken from H. W. Thompson's " A Course of Chemical Spectroscopy " (Cambridge University Press), enables a comparison to be made between the above different units of energy over a certain range. The reader will also encounter energies measured in *Rydberg Units*, where 1 Rydberg = 13·605 eV. This quantity of energy is equal to the first ionization potential of hydrogen, *i.e.*, the work required to remove an electron from a hydrogen atom in its normal or lowest energy state.

It is also useful to remember that in the equation for a perfect gas, $pv = R\theta$, the quantity $R\theta$ represents the energy of a gram-molecule of the gas at the absolute temperature θ; the constant R has the value 1·9872 cal/deg. The corresponding equation referred to a single molecule is $pv = k\theta$, where k is the so-called Boltzmann's constant, and equals $1\cdot38055 \times 10^{-16}$ erg/deg. At room temperature θ is equal to approximately 300, so that $k\theta$ is of the order of 4×10^{-14} ergs, *i.e.*, of the order of 3×10^{-2} eV. This should be remembered when equations are

---

* It has, unfortunately, become common practice to use the symbol ν both for the frequency and for the wave-number which we have denoted by the symbol n, and the reader must be careful to distinguish between cases where ν means the real frequency and those where it means the wave-number.

† J. W. M. Du Mond and E. R. Cohen, *Rev. Modern Physics*, 1953, **25**, 691.

encountered with terms of the form $e^{-\frac{E}{k\theta}}$, where E is the energy associated with some process.

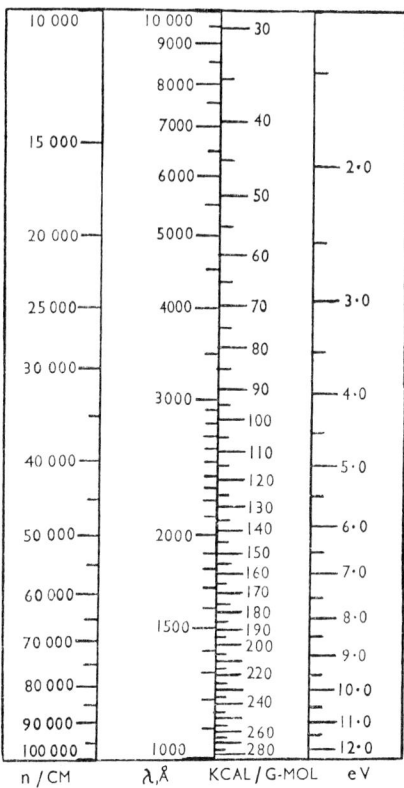

[*Courtesy Cambridge University Press.*]

FIG. I.1. In this figure the right-hand scale is a logarithmic scale of electron volts. The next scale is the corresponding scale of kcal/g-molecule. On p. 11 we have explained that an energy change of 1 eV/molecule is equivalent to 23·06 kcal/g molecule, and it will be seen that 10 eV on the right-hand scale of Fig. I.1 corresponds with 230·6 kcal/g-molecule on the adjacent scale. When an atom or molecule emits a quantum of radiation of frequency ν, the change in energy, E, of the atom is given by the relation:

$$E = h\nu = hc/\lambda,$$

where λ is the wave-length and c is the velocity of light. The third scale from the right shows the wave-lengths of the radiation corresponding to the energy changes of the first scale of eV/molecule or of the second scale of kcal/g molecule. Visible light has wave-lengths of the order of 3000–7000 Å, and hence corresponds with energy changes of the order of 4–2 eV. The left-hand scale is the corresponding scale of wave numbers, *i.e.*, number of waves/cm.

Note on Physical Constants.

In the last 30 years slight changes have occurred in the accepted values of many of the physical constants used in atomic theory. The first of these changes arose from the discovery of errors in the oil-drop method of determining the electronic charge, e, and also in some methods of determining the ratio h/e. The effects of these were discussed by R. T. Birge* in 1941, and for some years a confusing position arose in which increasing use was made of Birge's values, although they had not been officially recognized, and were not found in the majority of textbooks.

The matter has since been fully discussed several times, and the values given here are taken from the review by Cohen and Du Mond.† These values are based on the new physical scale of atomic weights in which the isotope ^{12}C is assigned an atomic weight of exactly 12·000. If atomic weights are used in which atmospheric oxygen is assigned an atomic weight of exactly 16, the values given for F, N, k, and R, and for the energy in kilocalories per gram molecule for one absolute electron volt per molecule, require dividing by 1·00093. The physical scale is preferred because *ordinary* oxygen is a mixture of isotopes, whose proportions may vary slightly.

Physical Constants
(Cohen and Du Mond, 1965)

Velocity of light: old value $c = 2·99776 \times 10^{10}$ cm/sec
new value $c = 2·997925 \times 10^{10}$ cm/sec
1 Faraday, $F = 9·64871$ absolute coulombs per gram equiv.
Avogadro's number, $N = 6·02253 \times 10^{23}$
Electronic charge, $e = 4·8030 \times 10^{-10}$ absolute e.s.u.
Planck's constant, $h = 6·6256 \times 10^{-27}$ erg sec
Electronic mass, $m = 9·1091 \times 10^{-28}$ g
Boltzmann's constant (see p. 12), $k = 1·38055 \times 10^{-16}$ erg/deg
Gas constant per gram molecule, $R = 8·31434$ erg/deg
$= 1·9872$ thermochemical cal/deg
Energy in ergs of one absolute electron volt (see p. 11) $= 1·60210 \times 10^{-12}$ erg
Energy in kilocalories per gram molecule for one absolute electron volt per molecule (see p. 11) $= 23·06$ kcal
Magnetic moment of one Bohr magneton $= 0·92732 \times 10^{-20}$ erg/gauss
Wavelength associated with one absolute volt (see p. 12) $= 1·240 \times 10^{-8}$ cm
Wave-number associated with one absolute volt (see p. 12) $n = 8066$ cm/absolute volt
1 Ångstrom unit $= 10^{-8}$ cm
1 Crystal Ångstrom $= 1$ kX unit $= 1·00202 \times 10^{-8}$ cm
Atomic unit of length, $a = 0·5292 \times 10^{-8}$ cm

---

* *Rev. Modern Physics*, 1941, **13**, 233; another paper by the same author appears in " Reports on Progress in Physics ", Vol. VIII. **1941:** London (The Physical Society).

† E. R. Cohen and J. W. M. Du Mond, *Rev. Modern Physics*, 1965, **37**, 537.

The General Background

2. THE BOHR THEORY OF THE HYDROGEN ATOM.

In developing the theory of atomic structure, attention was naturally concentrated first on the hydrogen atom, since this consists of a single electron in motion round a nucleus with a single positive charge of $+e$. In its simplest form, the first theory of the hydrogen atom assumed that the electron described a circular orbit round the nucleus, so that the inward attraction of the nucleus was balanced by the outward centrifugal force. Although the idea of a precise mechanical orbit has now been completely discarded, many of the general conclusions of this model are still valid and must be appreciated.

An electron describing a circular orbit round a positively charged nucleus of charge $+e$, is attracted by a force equal to e^2/r^2. The electron possesses both potential and kinetic energies, and the potential energy, W, must be measured relatively to some arbitrary zero. For this purpose the zero of potential energy is assumed to be that of an electron at rest at an infinitely large distance from the nucleus. In a circular orbit of radius r, the potential energy is then *negative*, and equals $-e^2/r$.* The kinetic energy equals $\frac{1}{2}mu^2$, where u is the velocity. Since the centrifugal force, mu^2/r, is to be balanced by the attraction, we have:

$$\frac{mu^2}{r} = \frac{e^2}{r^2} \quad \ldots \ldots \quad \text{I (1)}$$

so that the kinetic energy, K, is given by:

$$K = \tfrac{1}{2}mu^2 = \tfrac{1}{2}\frac{e^2}{r} \quad \ldots \ldots \quad \text{I (2)}$$

and the total energy, E, is therefore:

E = potential energy + kinetic energy =

$$-\frac{e^2}{r} + \tfrac{1}{2}\frac{e^2}{r} = -\tfrac{1}{2}\frac{e^2}{r} \quad \ldots \ldots \quad \text{I (3)}$$

The total energy is thus always negative† and becomes increasingly negative as the orbit becomes smaller. It is thus necessary to add energy if the electron is to be pushed out into a larger orbit. The

---

* The negative sign of the potential energy is the result of the choice of zero and should give rise to no difficulty. If, for example, we consider a hill and a valley, and assume the potential energy of a stone in the valley to be zero, its potential energy on the top of the hill will be positive. But this procedure is clearly arbitrary, and we might equally well call the potential energy on the top of the hill zero, and in this case the potential energy in the valley would be negative.

† In some treatments it is customary to use some symbol, say Y, for the *negative* total energy, so that $Y = +\tfrac{1}{2}e^2/r$. It is this convention which often causes confusion as to whether large or small orbits have the greater total energy.

kinetic energy equals $\frac{1}{2}e^2/r$, and hence the kinetic energy of an electron becomes smaller as the orbit becomes larger, and on comparing a large and a small orbit, we see that the larger orbit has the greater total energy and the greater potential energy, but the smaller kinetic energy. In so far as it is justifiable to think of " orbits " (see p. 43), these considerations apply to all the later theories.

The defect of the above classical theory was that it did not explain the reason either for the formation of sharp spectral lines, or for the stability of the atom. An electron moving in an orbit as described above would, on the basis of classical mechanics, continually approach the nucleus, with a steady emission of radiation of gradually changing frequency. The first step towards the solution of this problem was made by Bohr, whose original quantum theory of the atom involves two main assumptions:

(1) The electrons are assumed to revolve round the nucleus in definite orbits called *Stationary States*, and emission or absorption of radiation takes place by an electron jumping from one stationary state to another. When the electron jumps from an orbit of energy E_1 to one of energy E_2, the atom emits or absorbs a quantum of radiation of frequency ν, given by the Einstein relation:

$$E_1 - E_2 = h\nu \quad \ldots \quad \text{I (4)}$$

where h is the fundamental constant of Planck and equals $6\cdot 626 \times 10^{-27}$ erg-second. Emission of radiation takes place if E_1 is greater than E_2, and absorption occurs if the atom gains energy during the transition.

(2) The only stationary states which are stable are assumed to be those for which the angular momentum is an integral multiple of $h/2\pi$. In this way an electronic orbit or state is associated with a whole number, the so-called *quantum number*. An electron can, for example, revolve in an orbit for which the angular momentum is $\frac{h}{2\pi}, \frac{2h}{2\pi}, \frac{3h}{2\pi} \ldots$ or in general $\frac{nh}{2\pi}$, where n is the quantum number of the orbit. In a simple circular orbit, the angular momentum equals mur, so that the conditions for quantization may be written:

$$mur = \frac{nh}{2\pi} \quad \ldots \quad \text{I (5)}$$

This condition is an arbitrary postulate of the theory, and when combined with the relations given above leads to the expression:

$$E = -\frac{2\pi^2 me^4}{h^2} \times \frac{1}{n^2} \quad \ldots \quad \text{I (6)}$$

The General Background

The negative energies of the orbits are thus proportional to the inverse squares of whole numbers. From equation I (4), the frequency of the quantum of energy emitted when an electron jumps from an orbit of energy E_1 to one of energy E_2 is equal to $(E_1 - E_2)/h$. Since the wave number is equal to ν/c, where c is the velocity of light, the wave number associated with a transition from an orbit of quantum number n_1 to one of quantum number n_2 will be:

$$\frac{2\pi^2 me^4}{ch^3}\left(\frac{1}{n_1^2} - \frac{1}{n_2^2}\right) = R\left(\frac{1}{n_1^2} - \frac{1}{n_2^2}\right) \quad . \quad . \quad \text{I (7)}$$

where
$$R = \frac{2\pi^2 me^4}{ch^3}$$

This agrees with the fact that some of the lines of the hydrogen spectrum can be classified in series whose wave numbers involve the differences of the reciprocals of the squares of whole numbers. The arbitrary assumptions of the theory were, of course, chosen to explain this fact, and the really striking point about the theory is not so much that it predicts the above types of series, as that the numerical value of the *Rydberg Constant*, R, is almost exactly that required by experiment, and an exact agreement is obtained when the theory is elaborated so that the electron is regarded as revolving, not with the nucleus as the centre of revolution, but about the centre of gravity of the nucleus and the electron together, this being the more correct assumption.

The circular orbit is the simplest example of the more general type of orbital motion, which is that of an ellipse with the nucleus at one of its foci. The Bohr theory was extended to the case of elliptical orbits with the additional complication of the variation of the mass of the electron with velocity, in accordance with the Theory of Relativity.* In these theories, each electron state is characterized by two quantum numbers n and k. The principal quantum number n is analogous to that of the theory of circular orbits and is a measure of the energy of the orbit. The secondary quantum number k is a measure of the angular momentum of the orbit which is equal to $kh/2\pi$. In this way a more complete theory of the hydrogen atom was obtained, and some progress was also made in the interpretation of the spectra of more complex atoms, particularly

---

* According to the Theory of Relativity, the energy of a material particle at rest is equal to $m_0 c^2$, where m_0 is the rest mass of the particle and c the velocity of light. The mass, m, of a particle moving with velocity u, is given by the relation $m = m_0 \bigg/ \sqrt{1 - \frac{u^2}{c^2}}$, so that the mass varies with the velocity, but the effect is only appreciable at very high velocities. The total energy of a particle moving with velocity u is equal to $mc^2 + V$, where V is the potential energy, and if u is small compared with c, the total energy is equal to $m_0 c^2 + \frac{1}{2} m_0 u^2 + V$. Here $m_0 c^2$ is called the mass energy of the particle, and $\frac{1}{2} m_0 u^2$ is the kinetic energy.

those of the alkali metals, whose atoms contain a relatively loosely bound electron (see p. 59) which was regarded as revolving round the " core " formed by the nucleus and the inner electrons. Bohr was also the first to produce a scheme of electronic structures of the atoms of the elements, involving groups and sub-groups of electrons with different values of n and k.

In spite of its great success, the Bohr theory was not completely satisfactory, for in the first place facts were found with which it did not agree, and, secondly, it involved a logical inconsistency, in that when an electron was moving in an orbit, it was assumed to obey the laws of classical mechanics, though these were subjected to the apparently quite arbitrary restrictions of the quantum hypothesis. The solution to this difficulty was found, not by making further arbitrary assumptions, but by recasting the laws of mechanics into a new mathematical framework, and we may approach this difficult subject by considering first some points in connection with the Uncertainty Principle of Heisenberg.

SUGGESTIONS FOR FURTHER READING.

For a detailed account of the Bohr theory, the reader may consult:

E. N. da C. Andrade, " The Structure of the Atom." **1927**: London (G. Bell and Son). Although old, this book is most useful.

M. Born, " Atomic Physics " (translated by J. Dougall and revised in collaboration with R. J. Blin-Stoyle), 6th edn. **1957**: London and Glasgow (Blackie and Son).

F. K. Richtmyer, E. H. Kennard, and T. Lauritsen, " Introduction to Modern Physics," 5th edn. **1955**: New York and London (McGraw-Hill).

3. THE UNCERTAINTY PRINCIPLE OF HEISENBERG.

If the previous pages are examined critically, it will be seen that they involve the assumption that quantities such as position, velocity, momentum, &c., may be ascribed to an electron at any instant, and that these quantities have the physical meanings associated with them in the classical mechanics. From the modern viewpoint, it is meaningless to speak of a physical quantity unless we can describe a method which, if not yet actually realizable, will at least in principle enable us to measure the quantity concerned. If we consider the ordinary methods of measuring quantities such as position, velocity, &c., it will be seen that they all involve the assumption that we can ignore most of the alteration produced by the act of measurement. If, for example, we measure the length of a body, we must compare it with a standard scale. If this is done by making contact between one body and another, we are uncertain as to the exact deformation produced by the actual

The General Background

contact. If we seek to overcome this difficulty by using a beam of light to compare the object with the scale, the indicating beam of light exerts a pressure, and we are again uncertain as to the exact deformation caused. With ordinary objects these effects are so small as to be negligible, but when we come to consider electrons, the position is very different.

The usual illustration of the Heisenberg Principle is that of an attempt to measure the simultaneous position and velocity of an electron. To do this we might in principle* use two microscopes and a beam of light, and so observe the positions of the electron at two instants of time and hence deduce its velocity. The resolving power of a microscope is limited, and is given by the relation $\Delta x = \dfrac{\lambda}{\sin \theta}$, where λ is the wave-length of the light employed, and 2θ the aperture of the objective. The accurate measurement of position therefore involves the use of light of a very short wave-length and of an objective of high aperture, the objective receiving light from all directions included within the angle 2θ. The work of Compton has, however, shown that when light interacts with an electron, the latter undergoes a recoil, the process being known as the *Compton Effect*. This process is a typical quantum phenomenon in which a ray of light of frequency ν behaves as though it consisted of a stream of particles or *photons* each of energy $h\nu$, where h is Planck's constant. The observation of the electron requires interaction with at least one photon, and in this case if we imagine the photon to travel along the x-axis and to be scattered through an angle α, the electron acquires a momentum whose component in the x direction is $\dfrac{h\nu}{c}(1 - \cos \alpha)$, where c is the velocity of light. The observation of the first position of the electron will therefore alter the velocity of the electron by an amount which increases with ν (*i.e.*, as the wave-length becomes smaller) and which cannot be estimated exactly, because the objective of high aperture receives light from the whole range of the angle 2θ. We see, therefore, that the very factors which favour the accurate measurement of the position produce the greatest uncertainty in the velocity and *vice versa*. No method can be devised which avoids this difficulty, and the Uncertainty Principle of Heisenberg states that if Δx is the uncertainty in the position and Δp the uncertainty in the momentum, then the product $\Delta x \Delta p$ has a minimum value of h, where h is Planck's Constant of Action.

This principle is not confined to the above case. If, for example,

---
* The experiment is a purely hypothetical one.

a particle exists with an energy E for a time t we find that the conditions necessary to measure the energy with the greatest accuracy are those which produce the greatest uncertainty in the time, and again $\Delta E \Delta t \not< h$. This implies that if we are dealing with an electron in a stationary state the latter will only have an exact energy if the time is completely indeterminate. If we deal with a restricted period of time, then there will be a slight uncertainty in the energy. With periods of time of the order of one second, the effect is negligible, because of the small value of the constant h, and it is for this reason that we usually obtain sharp spectral lines, whose frequencies are connected by the relation

$$E_1 - E_2 = h\nu$$

In some cases an atom may go through a series of changes, in which it passes through a number of energy states E_1, E_2, E_3 ... In such a case it may happen that one of the states has a very brief existence, and if so its energy is slightly uncertain, in accordance with the Heisenberg Principle, and a diffuse spectral line results.* This effect has, in fact, to be considered in the study of the soft X-ray spectra of solids, which has thrown great light on the nature of metals. Similarly, if we try to measure the angular momentum, M, of a particle revolving round a fixed point, and the angle ω made with some fixed reference line by the line joining the particle to the fixed point, the uncertainties are given by $\Delta M \Delta \omega \not< h$. It will be noted that in all the above cases, the product of the two quantities connected by the Uncertainty Principle has the dimensions† ML^2T^{-1}, i.e., the dimensions of *action* (see p. 28). It will be noted that the Principle refers to the product of the two uncertainties and refers only to the minimum value of this product.

In considering the meaning of the Uncertainty Principle we may adopt two points of view. On the one hand, we may argue that our methods of measurement are at fault, and that if these could be perfected we could measure the real momentum and position at the same time. This clearly leads to difficulties about the meaning of reality, since we are practically compelled to say that by the " real position and momentum of an electron at a given instant " we mean something which, by the very nature of things, we cannot measure. Apart from this we should still be left with the problem of how the electron would behave in the hypothetical world where momentum and position, energy and time,

---

* There are, of course, many other factors which may produce diffuse spectral lines.

† Position has dimensions L, and mass M, and velocity L/T, so momentum multiplied by position has dimensions ML^2T^{-1}. The technical expression is that two quantities which are canonically conjugate cannot be measured exactly at the same instant.

The General Background 21

&c., could be measured simultaneously. Without going into details, it may be said that this line of approach has failed. The alternative is to accept the Uncertainty Principle as embodying a real characteristic of the physical world, which had escaped our notice in large-scale phenomena, because the magnitudes involved are very much larger than Planck's constant h, but which is of supreme importance in connection with the behaviour of particles of such small mass as an electron. It is from this point of view that very great progress has been made, though examination will show that a very strange world is being visualized. Suppose, for example, that we consider an electron contained in a one-dimensional box of length L. Then the Uncertainty Principle implies that the particle cannot be at rest, for if it were at rest its velocity would be exactly zero, and if this were so the uncertainty in its position would be infinite, and it would not be confined to the length L. Even at the absolute zero of temperature, we are thus led to expect some motion, although the effect is only appreciable when the length L is of atomic dimensions. Heisenberg's Principle then states that the uncertainty in momentum must be at least h/L, and we cannot consider changes of momentum of less than this, since to do so would imply that the uncertainty in momentum was less than h/L, and the electron would therefore not be confined to the length L. These considerations clearly lead to great difficulties if we try to describe small-scale events in terms of our usual ideas of space and time, since we have become so accustomed to imagining that each of the properties, position, velocity, kinetic energy, momentum, time, &c., can be defined exactly, without reference to the others, that we find it difficult to form a mental image of the restrictions imposed by the Uncertainty Principle. In the methods developed by Heisenberg and Dirac, all attempts to describe phenomena in terms of our usual concepts of space and time are given up, and are replaced by systems of mathematics, which deal only with the quantities we actually observe, and express these in suitable mathematical forms. Although quantitatively satisfactory, these methods are difficult to visualize, and for the non-mathematician it is more easy to consider the alternative line of approach, the so-called *wave mechanics* of Schrödinger, since this expresses some of the results in a way to which we are already accustomed. The methods of Dirac, Heisenberg, and Schrödinger are mathematically equivalent, as of course they must be if each is to lead to the correct answer, but some of the results of wave mechanics can be visualized, although care is necessary to avoid using the wave analogy unjustifiably.

Confusion may be caused by the different ways in which the

22 Atomic Theory

Uncertainty Principle is expressed. For simplicity we may consider the simultaneous position and momentum of an electron in a one-dimensional problem. If the momentum is measured to a certain degree of accuracy, the Principle implies that the position cannot be specified exactly, but involves some uncertainty. There is a probability of finding the electron over a small region, but this region cannot be defined uniquely by sharp boundaries; the region in which there is a probability of finding the electron merges asymptotically into the regions where the probability is vanishingly small. Frenkel (" Wave Mechanics: Elementary ") uses the expression " effective uncertainty " to denote the region outside which the probability of finding the electron is negligible. It is in this sense that we have used the expression " uncertainty " in the above description, and following some authors we have written the Uncertainty Principle in the form $\Delta x \Delta p \not< h$. It will be appreciated that the effective uncertainty cannot be defined uniquely, and some authors therefore write the Principle in the form $\Delta x \Delta p \sim h$ (i.e., $\Delta x \Delta p$ is of the order of h).

In wave mechanics the small region in which there is an appreciable probability of finding the electron is called a wave packet (see I.6). In the more advanced developments of the subject, the relative probabilities of finding the electron at different places within the wave packet, i.e., the relative probabilities at different places within the region of effective uncertainty, are analysed. It is then possible to define lengths smaller than the effective uncertainty (i.e., smaller than the length of the wave packet) in such a way that the probability of finding the electron within the length concerned is related by a mean or a root mean square relation with the mean value of the x co-ordinate over the whole region of the wave packet. Some writers use the symbol Δx to denote these lengths, and refer to them as uncertainties, in which case the Principle takes the form $\Delta x \, \Delta p \geqslant \dfrac{h}{2\pi}$, or $\Delta x \, \Delta p \geqslant \dfrac{h}{4\pi}$, according to the exact definition used.* In these cases the term Δx is precisely defined, but it is perhaps unfortunate to call it an uncertainty in position, because as Δx is shorter than the length of the wave packet, there will be a definite probability of finding the electron outside the region of uncertainty as thus defined. If the word " uncertainty " is used in its ordinary sense, it undoubtedly implies the " effective uncertainty " in the sense used by Frenkel.

In view of the difficulties in connection with the definition of " uncertainty in position," it is probably better to write the Principle in the

---

* The reader should note that the symbol $\hbar$ is often used to denote $\dfrac{h}{2\pi}$.

The General Background

form $\Delta x \, \Delta p \sim h$, and to note that the numerical factor by which h is multiplied depends on the definition of uncertainty which is used. Similar considerations apply to uncertainties in momentum, time, energy, &c.

4. The Ideas of Wave Mechanics.

In the preceding sections we have described the general failure of ordinary mechanics to deal with problems of atomic structure, and we have seen how Heisenberg's Principle suggests that when we consider events on the atomic and electronic scale we can no longer regard properties such as position, time, velocity, momentum, kinetic energy, &c., as being defined exactly and independently without reference to one another. If we ignore the historical development of the subject and consider the experimental evidence alone, we may say that the solution to these difficulties lies in the fact that electrons have been shown to give rise to diffraction and interference effects. This discovery was made first by Davisson and Germer (1927), and by G. P. Thomson (1928), who showed that diffraction patterns could be observed and recorded photographically when beams of electrons were passed through crystals. This remarkable discovery showed that electrons possessed the properties of waves as well as of particles, and a beam of electrons of velocity u was found to behave as though it were associated with waves, of wave-length λ, given by the relation

$$\lambda = \frac{h}{p},$$

where h is Planck's constant, and $p =$ momentum.

For electrons whose velocity is much smaller than the velocity of light, c, the momentum p may be put equal to mu, and the relation reduces to:

$$\lambda = \frac{h}{mu}.$$

For abbreviation we shall always use this simplified form, and this relation is the basis of the methods of investigating crystal structure by electron-diffraction techniques; considered experimentally it is the foundation of the whole of the modern theory of atomic structure and of the properties of metals.

We see, therefore, that the electron which was at first thought to possess only the properties of a charged particle, exhibits also characteristics which are usually associated with waves, and it is well known that light shows a similar dual character. In this respect the development of optical theory may be divided into three well-marked stages. The older science of *Geometrical Optics* dealt with the passage of rays of light from one point to another, and led to the discovery of the laws of reflection and refraction. The fact that light rays travel in straight lines in a uniform medium naturally suggested that light consisted of streams of particles or corpuscles, and a corpuscular theory of light was advanced by Sir Isaac Newton. The discovery of interference phenomena led to the development of the wave theory of light by Huygens (1690) and others. This theory accounted satisfactorily for the large-scale phenomena of geometrical optics, and provided such a beautiful explanation of so many additional effects (interference, dispersion, &c.) that by the end of the nineteenth century, the corpuscular theory of light was almost completely discarded, and the wave theory was universally accepted and was often thought to imply that light consisted of waves in a definite medium, the so-called *ether*.

The wave theory, however, led to great difficulties when the quantum phenomena were discovered. In this class of phenomenon—the photoelectric effect and the Compton effect are examples—the evidence seemed overwhelming that light of frequency ν gave up its energy in definite quanta equal to $h\nu$, where h is Planck's constant. This behaviour seemed impossible to reconcile with any picture in which the energy was associated with the whole of an expanding wave-surface, and clearly suggested the existence of light particles or *photons* with kinetic energy $h\nu$. There was thus an apparent conflict between the wave and particle properties of light, and the position was made more confused by the fact that all experimental attempts to detect the hypothetical " ether " resulted in complete failure.

This conflict between the two theories of light was solved by a compromise which may be summarized briefly as follows:

(1) A light ray is to be regarded as a stream of moving particles or photons, light of frequency ν being composed of photons of kinetic energy $h\nu$. In this sense the old corpuscular theory is correct, and whenever we observe light we observe one or more whole photons.

(2) When we wish to calculate the intensity of light, *i.e.*, the density of photons or the probability of finding a photon at a particular place, we use the equations of the wave theory with the assumption that the square of the amplitude of the waves is proportional to the light intensity. In an interference experiment, for example, the dark areas of low intensity are the regions where the wave equations indicate a small amplitude, and this is to be interpreted as a *small probability* of finding a photon at the place concerned.

The General Background

In this way the whole of the mathematical framework of the wave theory is retained, but we no longer imagine the existence of a vibrating medium. Light does not involve any actual physical vibration, and although as a matter of convenience we still speak of light waves, wave-lengths, frequencies, &c., what we really mean is that the intensities of light can be calculated by wave equations, with the interpretation that the square of the amplitude is proportional to the probability of finding a photon at the place concerned.

(3) The interpretation of the wave theory in terms of probabilities applies not only to beams of light consisting of streams of photons, but also to the behaviour of an individual photon, and this implies that there is some degree of uncertainty in the behaviour of the individual photon. If, for example, we pass rays of light through a diffraction apparatus, we may obtain on a screen a series of bright and dark rings, the former being at the places where the amplitude of the wave equation is large and the latter where it is small. If now we pass one single photon through the same apparatus, we cannot predict where the photon will travel. All that we can say is that there are different probabilities of its travelling in different directions, these probabilities being greatest in the directions corresponding to the light rings observed when a stream of photons is passed through the apparatus. There is thus a certain indeterminancy in the behaviour of the individual photon, analogous to the behaviour of the electron according to Heisenberg's Principle. If, for example, we pass light through an aperture, then the smaller the aperture the greater is the diffraction effect, so that the more the aperture restricts the cross-section of the beam the greater is the uncertainty of the direction of the individual photon after passing through the aperture. Effects such as these can readily be described in terms of probabilities, but it is, of course, quite impossible to describe them in terms of either the older geometrical optics or in terms of the simple particles of the early corpuscular theory. The use of wave equations with a probability interpretation does not enable us to " understand what is happening," but it places us in a position in which we can describe and predict experimental results in terms of the appropriate probabilities, and can include both the wave and particle aspects of light in one theory. It will be noticed that this increased power is gained at the cost of an admission that we cannot always follow the individual photon in detail along its path. In the experiment referred to above, for example, nothing enables us to predict which direction the individual photon will take after passing through the diffraction apparatus. We may calculate exactly the relative probabilities of the photon moving in different directions after passing through the

apparatus, but we then reach the limit of our powers, and the rest is indeterminate and unpredictable.

We have seen that electrons give rise to diffraction effects as though they were associated with waves, and it is therefore natural to enquire whether the wave-like and the particle aspects of an electron can be united in one single theory, as has been done for the case of a photon. Clearly, so far as the simple diffraction effects are concerned, the problem can be approached in the same way, and we may develop a " probability wave theory " of electrons on the assumption that the electron is a particle, and that the amplitude of a wave equation is a measure of the probability of finding an electron at the place concerned. The question then arises as to whether the wave-like characteristics of electrons may affect their behaviour in other ways, and here an examination of the relation $\lambda = \dfrac{h}{mu}$ is instructive.

We know from experiment that an electron is accelerated or retarded by an electric field, with a corresponding change in its momentum, mu. From the relation $\lambda = \dfrac{h}{mu}$ it follows that the wave-length associated with the electron is inversely proportional to its momentum, and we can see, therefore, that as the electron accelerates, the wavelength will change. We have, thus, a condition of affairs in which as an electron moves through a field of varying potential, the associated wave-length is continually changing. Speaking generally, it may be said that if the changes in potential are appreciable only over distances which are large compared with the associated wave-length, the electron moves more or less as though it were a " particle " obeying the laws of " ordinary mechanics." If, however, the changes in potential are appreciable over distances of the order of a wave-length, some completely new phenomena are met with, and these cannot be explained in terms of the older mechanics. In the following pages we shall describe the general methods by which mechanics has been re-moulded so as to take into account these wave-like characteristics of the electron, this branch of mechanics being known as " Wave Mechanics." General considerations of the magnitudes involved in the relation $\lambda = \dfrac{h}{mu}$ show that the wave-lengths associated with electrons in atoms are of the same order as the " sizes " of the atoms, or of the distances between atoms in metallic crystals. Under these conditions the wave-like characteristics of the electrons produce many strange effects, and some of these are directly responsible for the properties of metals. Many of these effects can be understood only in terms of the new mechanics, and if the

The General Background

metallurgist wishes to understand the behaviour of electrons in metals, it is essential for him to understand the ideas which underlie wave mechanics. In the following pages we shall describe the main steps or lines of argument by means of which the mathematicians passed from the old to the new mechanics, but we shall present the results dogmatically and not concern ourselves with mathematical technique. We have adopted a rather formal method of approach, not merely because of its historical and logical interest, but because it serves to emphasize the correct way of regarding the wave-like characteristics of electrons. The reader who does not wish to follow the argument may note that the method adopted is to re-write the laws of mechanics in a form which is mathematically similar to certain equations of optical theory. In this way a problem of the motion of a beam of electrons in a varying field of force is made mathematically similar to a problem

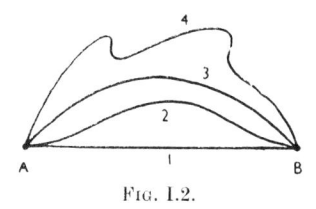

FIG. I.2.

of the passage of a ray of light in a medium of varying refractive index. This involves the setting up of a differential wave equation, the interpretation of which is discussed in Section I. 5.

In the preceding sections we have referred to the circular orbits of classical mechanics. These are merely one particular example of the more general problem of the motion of a particle in a variable field of force, and we shall assume that the reader is familiar with the elementary equations of dynamics by means of which we can calculate the motion of a particle under different conditions of attraction, repulsion, &c. These different equations may be expressed in more generalized forms, one of which is the *Principle of Least Action* of de Maupertuis (1698–1759), and this may be understood in the following way. We may suppose that we wish to shoot a particle of given total energy E from a point A to a point B (Fig. I.2), subject to the condition that potential energy, W, and kinetic energy, K, may be interchanged, but that the total energy $E = W + K$, remains constant. Then clearly the path which the particle will follow depends on the field of force in the space between A and B. In the absence of a field, the particle will travel in a straight line from A to B. In the presence of a uniform field such as the gravitational field of the earth, we should have to " aim " the particle upwards from A, and the path followed would be a parabola. The Principle of Least Action then states that if for all conceivable paths from A to B, we evaluate the integral $\int_A^B 2Kdt$, where t is the time, then this integral will have a minimum value* for the actual mechanical path. That is to say that, if in Fig. I.2 we evaluated the above integral for a series of different paths 1, 2, 3, 4 . . ., then in the absence of a field of force, the minimum value of the integral would be obtained for the straight-line path No. 1, whilst the correct parabola would be predicted for the gravitational problem, and similarly with more complicated fields. For simple problems this principle has no advantage over the usual equations of motion, but for complicated problems it is sometimes more convenient. The integral $\int_A^B 2Kdt$ defines the quantity known as " action," and this has the dimensions of ML^2T^{-1}. It will be noted also that the value of the integral depends only on the space co-ordinates of A and B, and not on the time. The minimum value of this integral for the actual mechanical path may be called S.

The above remarks refer to the trajectory of a particle on passing from one point to another, and we may now consider the analogous problem of the passage of a beam of light from one point A to another B. In this case we know that the path of the beam of light depends on the refractive index, and hence on the velocity† of light in the space between the two points. If the refractive index is uniform, light travels in a straight line from A to B, just as in the mechanical problem the particle travelled in a straight line from A to B in the absence of a field of force. If the refractive index is not uniform between A and B, the light travels in a path which is bent or curved, and which depends on the exact variation of the refractive index. The science of Geometrical Optics deals with this kind of problem, and the well-known laws of reflection and refraction enable us to deduce the path of a beam of light when the variation of refractive index and the positions of reflecting surfaces are known in the space surrounding A and B. Just as the Laws of Mechanics could be generalized in the form of the Principle of Least Action, so the Laws of Geometrical Optics may be generalized in a form known as *Fermat's Principle of Least Time*. This states that if in a problem where the variation of refractive index is known we consider a large number of

* Strictly speaking we should refer to a stationary value rather than a minimum value, but for a simplified description, the latter may be accepted.
† The velocities of light in different media are inversely proportional to the refractive indices of the different media.

The General Background

conceivable optical paths for a beam of light travelling from A to B, then the actual optical path which the light follows is characterized by a minimum* value of the integral:

$$\int_A^B \frac{ds}{v},$$

where v is the velocity of light at the place ds, and ds is an element of length of the path. This implies that of all conceivable paths the light travels by the path which enables it to reach B as quickly as possible, and this principle enables the laws of reflection and refraction to be predicted.

It will be seen, therefore, that there is a certain resemblance between the laws governing the trajectories of particles in the mechanical problem and the paths of beams of light in the optical problem, since in each case the actual path followed is characterized by the minimum value of an integral. Further, in a general way it can be seen that the variation of the refractive index in the optical problem plays the same part as the field of force or potential in the mechanical problem. This correspondence can be expressed in a precise mathematical form, so that the problem of the motion of a particle in a field of force becomes mathematically similar to the problem of the path of a ray of light in a medium of variable refractive index. These resemblances between the equations of mechanics and geometrical optics were noted many years ago by Hamilton, and were long thought to be nothing more than an interesting mathematical coincidence; it was only with the work of de Broglie and Schrödinger that it was realized that a supremely important principle lay concealed.

So far we have described the correspondence between the mechanical problem of the trajectory of a particle in a field of force and the optical problem of the path of a ray of light in a medium of varying refractive index, and we have discussed this in terms of the older geometrical optics. It is well known that the laws of geometrical optics can be expressed in terms of the wave theory of light, and all the equations for the paths of rays under conditions of reflection, refraction, &c., can be expressed in terms of the wave theory. For simple problems of geometrical optics, no advantage is gained by using the wave theory, but when we deal with problems in which the refraction index varies appreciably over distances of the order of one wave-length, the wave theory predicts a large number of new effects (dispersion, diffraction, &c.) which were quite inexplicable in terms of the older theory of rays. We see, therefore, that the change from the older geometrical optics

* See footnote to p. 28.

to the wave theory did not affect what may be called large-scale phenomena (paths of rays, reflection, &c.), but accounted for a number of small-scale phenomena which could not otherwise be understood. Since the change from geometrical optics to wave theory does not affect the large-scale phenomena, the correspondence between the equations of mechanics and those of geometrical optics to which we have referred above can be expressed as a correspondence between the equations of mechanics and those of the wave theory of light. In this way it is possible to rewrite the equations of mechanics in forms which are mathematically similar to those of a wave theory. If this is done, the new equations naturally contain terms which play the same part that wave-lengths, frequencies, &c., would play if the equations actually referred to a wave motion. The mathematical correspondence is so complete that for many purposes it is justifiable and helpful to visualize the equations as though they were equations of a wave motion. In this way we may think of a moving particle as being associated with wave-lengths, frequencies, &c., but we must remember that this is only a convenient way of visualizing the equations, and does not imply the existence of any physical waves.

As stated above, it is possible to rewrite the equations of mechanics in forms which are mathematically similar to those of a wave theory. So long as we are dealing with large-scale problems, this procedure has no advantage, and it was at first looked upon as being little more than clever mathematical jugglery. The genius of de Broglie and Schrödinger lay in the fact that they were prepared to carry out the transformation of the equations of mechanics to those of a wave theory, and at the same time to hold fast to the correspondence between the effect of a field of force in the mechanical problem and of a varying refractive index in the optical problem, and to carry this correspondence to its logical conclusion. In the optical problem a variation of refractive index which is appreciable over the distance of a wave-length produces effects which cannot be explained in terms of the older theory of rays. Hence it was suggested that in the mechanical problem a variation of force over very small distances might produce effects which could not be explained in terms of the older mechanics. But these effects might be expressed in terms of the new equations, because these transformed equations were those of a wave theory, and a wave theory was able to deal with the effects observed when the refractive index varied appreciably over a distance of the order of a wave-length. In this way by rewriting the equations of mechanics in a form similar to that of a wave theory, and by retaining the correspondence between force and variation of refractive index, it was possible to build up a new mechanics, the so-

The General Background

called "wave mechanics," which predicted new effects when the force varied greatly over very small distances and was thus directly applicable to the problem of the motion of an electron in the field of an atom or in the periodic field of a crystal lattice. It need scarcely be said that in its early stages this development of mechanics was regarded as little but mathematical speculation, but the whole position was transformed when it was discovered that electrons showed diffraction effects as though they were somehow associated with waves. This discovery showed that the rewriting of the laws of mechanics in the form of equations of a wave theory was not just an example of mathematical ingenuity, but had revealed a real characteristic of the physical world. It is highly important for the student to realize that wave mechanics arose before the discovery of the wave-like properties of electrons, and as the result of pure mathematical reasoning which showed how the equations of mechanics could be rewritten in the form of a wave theory.

The student who wishes for further details should consult the references given on p. 39. For the general reader it is sufficient to note that wave mechanics came into being as the result of the line of argument described above. When the equations of mechanics are transformed into those of a wave theory, there are naturally relations between the terms of the wave equations and the dynamical properties such as momentum, energy, &c. In our comparison of the optical and mechanical problems we have dealt with the similarity between the laws governing the paths of particles and of rays of light, and have not considered the individual photons. Our first description should therefore refer to beams of electrons, rather than to single electrons or other particles.

Historically it was first pointed out by de Broglie that if waves were associated with particles, then the Principle of Relativity indicated that:

$$\frac{\text{Energy}}{\text{Frequency}} = \frac{\text{Momentum}}{\text{Wave number}} = \text{Constant} \quad . \quad . \quad . \quad \text{I (8)}$$

The constant was then intuitively identified with Planck's constant, h, and since the above relations were known to hold in the case of light (energy $h\nu$, momentum $h\nu/c$) it was suggested that they might also apply to particles. The second of these relations was then confirmed experimentally by the electron-diffraction experiments of G. P. Thomson and of Davisson and Germer, which showed that electrons of velocity u were diffracted by a crystal as though they possessed a wave-length equal to $\frac{h}{mu}$.* Considered experimentally, this relation between wave-

---

* As explained on p. 23, this relation requires modifications if the velocity u is very great.

length and momentum is the foundation of the whole subject. It could therefore be concluded that the first of the above relations would also be satisfied, and the arguments of de Broglie showed that for particles, $h\nu$ would be equal to the relativity energy E_m given by:

$$E_m = \frac{mc^2}{\sqrt{1 - \frac{u^2}{c^2}}} + W \quad \ldots \quad \text{I (9)}$$

which for ordinary problems where the velocities are small compared with that of light is the same as:

$$h\nu = E_m \simeq mc^2 + \tfrac{1}{2}mu^2 + W \quad \ldots \quad \text{I (10)}$$

The energy E_m thus includes the energy mc^2 associated with the mass of the particle. It may be thought curious that we should suddenly change from the ordinary total energy $E = \tfrac{1}{2}mu^2 + W$, to the total energy E_m, but it must be remembered that the potential energy, W, has always to be measured from some arbitrary zero, so that the addition of the constant term mc^2 is merely equivalent to a shift in the zero from which the potential energy is measured, and consequently it does not affect the problems which we have been discussing. As will be appreciated from what follows the absolute value of ν is never required,* and the physical properties dealt with involve either the differences between two frequencies or the differential of the frequency, for which the constant term mc^2 is not involved.† We have, therefore, the two fundamental relations:

$$E_m = h\nu \quad \ldots \quad \text{I (11)}$$

$$\lambda = \frac{h}{mu} \quad \ldots \quad \text{I (12)}$$

In the wave mechanics developed by Schrödinger, the essential assumption made was that the waves were of a sine form, and it could then be shown that the above two relations (11) and (12) were those necessary to make the mathematics self-consistent. That is to say, it could be shown that if the hypothetical waves were of a sine form, in the equations of which the time t entered as a term $2\pi\nu t$, then the mathematical scheme became consistent if the frequency ν were directly proportional to the energy.‡ In some books the reader will find the

---

* Thus N. F. Mott ("An Outline of Wave Mechanics") writes $\nu = E/h$, and for the problems with which he is concerned the change makes no difference although the frequencies are enormously smaller than those given by $\nu = E_m/h$. This may serve to emphasise the purely symbolic and mathematical nature of the hypothetical waves.

† The reader should be warned that in some books the same symbol is used indiscriminately for E and E_m, and much confusion is caused in this way.

‡ The general equation of a sine wave is of the form $\sin 2\pi(\nu t - x/\lambda + a)$.

The General Background

relation (11) deduced in this way as the relation necessary to produce a consistent scheme, and in the same way the relation (12), which has a direct experimental basis, fits into the general scheme. We have, therefore, now carried our discussion to a point at which the mechanical problem has reached the second stage of the optical theory (p. 24), because the hypothetical waves have been associated with frequencies and wave-lengths which agree with the observed interference phenomena.

From the above relations it will be seen that the velocity of the hypothetical waves is given by:

$$U = \nu\lambda = \frac{E_m}{mu} \simeq \frac{mc^2 + \tfrac{1}{2}mu^2 + W}{mu} \qquad . \quad . \quad \text{I (13)}$$

In most ordinary problems, the term mc^2 is so much the greatest that U is of the order of c^2/u, and since u can never be greater than c, this implies that U must always be greater than c. This serves to emphasize* that the waves are merely mathematical functions, and are not of any material nature, because nothing material can travel faster than light. The velocity U is called the phase-velocity of the waves, and gives the velocity with which the hypothetical waves travel. U is thus greater than c, and consequently U cannot be the velocity of the actual electron, since this cannot exceed c. We shall consider later what function gives the velocity of the electron itself, and for the present we may say that the argument leads to the conclusion that moving electrons can be treated mathematically as though they were associated with waves of phase-velocity c^2/u, where u is the velocity of the electron.

We have seen how the laws that express the motion of streams of electrons can be manipulated mathematically into a form which resembles that of the optical theory, and we have now to see how the electron density, i.e., probability of finding an electron at a particular place, is to be calculated, and, as in the case of the optical theory, the interpretation is that the amplitude of the hypothetical waves is a measure of the electron density at the place concerned.

5. Wave Equations and Their Interpretation.

We have now to proceed to the final stage of setting up a wave equation and interpreting it in terms of electrons. A beam of electrons of uniform velocity will correspond to a monochromatic ray of our hypotheti-

---

* We have reproduced above the description which the reader will usually find in the text-books. A little consideration will show that since the potential energy W has to be measured from some arbitrary zero, the frequency ν cannot be uniquely defined, because W in equation I (10) may have any arbitrary constant added to it. On the other hand, the difference between two frequencies and the differential of the frequency do not involve the constant term.

cal mechanical waves, although the velocity u of the electrons is totally different from the velocity $\left(\sim \dfrac{c^2}{u}\right)$ of the waves; we shall consider later what function of the waves corresponds with the actual velocity of the electrons. The hypothetical "something which vibrates" is denoted by Ψ, and in accordance with the general analogy with the theory of light, the amplitude of the hypothetical Ψ vibrations is a measure of the electron density at the place concerned. A large value of the amplitude of the Ψ vibrations corresponds with a high electron density and a large probability of finding an electron at the place concerned. In expressing this precisely, a difference exists between the optical theory and the wave mechanics. In the electromagnetic theory of light, the state of the vibrating medium is described by two vectors,* E and H, and the intensity of illumination, or the density of photons, is proportional to $(E^2 + H^2)$, and this intensity fluctuates with the time, because E and H are taken to describe waves in phase with one another. In the same way† we may describe the hypothetical Ψ vibrations by two functions f and g, so that the electron density is proportional to the value of $(f^2 + g^2)$ at the place concerned. The wave-mechanics theory is then developed on the assumption that at a given place $(f^2 + g^2)$, and hence the electron density, is independent of time.‡ This implies that f and g must be one-quarter of a period apart. In the very simple case of a plane wave, for example:

if $\qquad f = a \cos 2\pi(\nu t - x/\lambda) \quad . \quad . \quad . \quad . \quad . \quad$ I (14)

then $\qquad g = a \sin 2\pi(\nu t - x/\lambda).$

So that $\quad f^2 + g^2 = a^2 \left[\cos^2 2\pi(\nu t - x/\lambda) + \sin^2 2\pi(\nu t - x/\lambda)\right] = a^2.$

In this way $(f^2 + g^2)$ is independent of time, and in this very simple case is also independent of the space-co-ordinate x, although in the more general case $(f^2 + g^2)$ varies from place to place.

The method universally employed is to replace the pair of real functions, f and g, by the single complex function:

$$\Psi = f + ig \quad . \quad . \quad . \quad . \quad . \quad \text{I (15)}$$

---

* A vector is a quantity which has both magnitude and direction, and a scalar a quantity involving magnitude only. If the radius of a circle revolves like the hands of a clock, the length of the radius is a scalar quantity independent of direction.

† This treatment is taken from "An Outline of Wave Mechanics," by N. F. Mott (Cambridge University Press), with the permission of the author and publisher.

‡ The theory developed in this way because when wave mechanics was invented it was necessary to assume wave functions to describe the phenomena. The essential requirement was a measure of the intensity of electrons in a beam, and it was therefore convenient to adopt a function which made the intensity of a beam independent of the time at any one place.

The General Background

where $i = \sqrt{-1}$. The conjugate complex of Ψ which is denoted Ψ^* is equal to $f - ig$, and the electron density is proportional to:

$$f^2 + g^2 = (f + ig)(f - ig) = \Psi\Psi^* \quad . \quad . \quad . \quad \text{I (16)}$$

The reader who is not familiar with the use of complex numbers may content himself with noting that the method used is merely a mathematical device for expressing the assumption that the hypothetical Ψ waves are the result of two vibratory processes which are one-quarter of a period apart, and so unite to give a resultant which is independent of time at any one place. The assumption often leads the beginner into difficulties, because there is a natural tendency to think of a wave process as involving a fluctuation with time, but as can be seen from the above example of sine and cosine functions, it is quite possible for two vibratory processes to unite so as to produce a resultant which is independent of the time. In the symbolism of this method, the electron density is proportional to the value of $\Psi\Psi^*$ at the place concerned, and the number of electrons in an element of volume ΔV is proportional to $\Psi\Psi^*\Delta V$. It is helpful to remember that $\Psi\Psi^*$ is real, even though Ψ be complex, and the electron density is always proportional to $\Psi\Psi^*$. In some problems it is possible to express the electron density in other ways, and to understand this the details given in the Appendix (p. 38) must be appreciated.

For a stationary state in which the energy is fixed and equal to $h\nu$, the general form of the function Ψ may be written

$$\Psi(x, y, z, t) = \psi(x, y, z)e^{2\pi i \nu t} \quad . \quad . \quad . \quad \text{I (17)}$$

where $\psi(xyz)$ is sometimes called the *space factor* or the *amplitude factor*, and $e^{2\pi i \nu t}$ is the *time factor*. This is merely another way of expressing the assumption that the electron density is independent of time, since

$$e^{2\pi i \nu t} = \cos 2\pi\nu t + i \sin 2\pi\nu t \quad . \quad . \quad . \quad \text{I (18)}$$

which is of the form $(f + ig)$ used in the previous description. The scalar value of $e^{2\pi i \nu t}$ is thus unity, though this is not to be taken as implying that the vibratory process is independent of time; rather is it imagined to be the result of two vibrating processes, each of which fluctuates with time, but which compound so that their resultant, although in the general case varying from place to place, is independent of time at any one place. In equation I (17), ψ is thus the amplitude of Ψ, and Ψ and ψ have the same scalar value, so that as long as we are not concerned with the time, Ψ and ψ are often interchangeable, since they have the same scalar value,† and obey the same differential

---

† It is this which leads to the unsatisfactory position in which symbols for Ψ and ψ are often interchanged, or are not distinguished, so that the reader may find the same symbol with different meanings on adjacent pages of the same book.

equations (see below), and the electron density is thus proportional to $\psi\psi^*$.

We have seen that the general problem of the motion of a particle in a variable field of force is analogous to the motion of light rays in a medium of variable refractive index. In general the variation of the potential energy as a function of the space co-ordinates will be known, and the wave equation expressing the magnitude of Ψ first proposed by Schrödinger is of the form:

$$\nabla^2\Psi + \frac{8\pi^2 m}{h^2}(E - W)\Psi = 0 \quad . \quad . \quad . \quad \text{I (19)}$$

Here W is the potential energy and E the total energy, whilst the symbol $\nabla^2\Psi$ stands for $\left(\frac{\partial^2\Psi}{\partial x^2} + \frac{\partial^2\Psi}{\partial y^2} + \frac{\partial^2\Psi}{\partial z^2}\right)$ and is sometimes written $\Delta\Psi$ by Continental writers. This equation is known as the First Schrödinger Equation, or the time-free equation, for ordinary (non-relativity) mechanics, and it represents a family of surfaces, each corresponding to one value of E. (As explained in the Appendix (p. 38) the amplitude ψ obeys an equation of exactly the same form. In the present book we are not concerned with the mathematical development of these equations, or of the further equation described in the Appendix, but the reader will meet the equations so frequently that the meaning of the symbols should be understood.) This equation amounts to the replacement of the total energy of classical mechanics (the Hamiltonian) by a set of mathematical instructions. The classical Hamiltonian was

$$-\frac{1}{2m}(p_x^2 + p_y^2 + p_z^2) + W(x, y, z) = E$$

where p_x is the x component of the momentum.

The mathematical instructions are:

"Operate on the wave function with the mathematical processes $-\frac{\hbar^2}{2m}\left(\frac{\partial^2}{\partial x^2} + \frac{\partial^2}{\partial y^2} + \frac{\partial^2}{\partial z^2}\right)\Psi + W(x, y, z)$. Ψ and the result is the product of E and Ψ." Here $\hbar$ stands for $h/2\pi$. For this reason the Schrödinger equation is sometimes written

$$\mathscr{H}\Psi = E\Psi$$

where $\mathscr{H}$ is the Hamiltonian operator so defined.

Other measurable variables, such as, for example, the z component of the orbital angular momentum of an electron in an atom, are also replaced by operators.

In the general type of problem with which we are concerned, the variation of W is known (thus, in the problem of the hydrogen atom

The General Background

the potential energy is equal to $-\frac{e^2}{r}$) and the methods of wave mechanics consist in finding suitable solutions to the appropriate wave equation, and then interpreting these on the assumption that $\Psi\Psi^*$ is proportional to the electron density at the place considered. As we have already explained, so far as the paths traversed are concerned, the effect of the wave mechanics is to reduce the problem of the path of a beam of electrons in a field of varying potential to a problem of rays of light in a medium of varying refractive index. Examination of the equations then shows that with a particle of total energy, E, and potential energy, W, the effect of varying W is equivalent to the effect on a light ray produced by a varying refractive index, μ, given by:

$$\mu = \left(1 - \frac{W}{E}\right)^{\frac{1}{2}}$$

In general, wave equations of the above type have large numbers of solutions, and the correct solution has to be chosen so that it represents something which is physically possible. Such solutions may be called acceptable solutions, and in most simple problems they result in the condition that ψ shall be everywhere finite and single-valued, and that both ψ and $\frac{d\psi}{dx}$ shall be continuous. In most cases where the solution represents a stationary state, the conditions to be satisfied result in the wave-length, λ, being simply related to some length which the problem involves. Thus, if we consider an electron contained in a one-dimensional box of length L, acceptable solutions are obtained only for which the wave-lengths are equal to $2L/n$, where n is a whole number.* This in its turn implies that only certain energies can give rise to acceptable solutions, and these energies are naturally interpreted as being those of the stationary states of the older quantum theories, and the whole-number relations which the wave-lengths must obey are naturally regarded as connected with the older quantum numbers. In this way it may be said that quantum numbers and stationary energy states appear in wave mechanics, not as arbitrary postulates imposed on one system of mechanics, but as the inevitable conditions which must be satisfied if the equations are to yield acceptable solutions. It is thus

---

* If a stretched wire is set vibrating, a steady vibration (*i.e.*, a stationary state of vibration) can be set up only for wave-lengths which are definite fractions of the length of the wire. The wave-lengths of the various possible steady vibrations are then related by whole numbers. This is a physical vibration, but the mathematics which express the whole number relations are roughly analogous to those which introduce the whole numbers into the above problem of electrons. As previously emphasized, this does not mean that the electron is a physical wave.

of particular fascination to see how wave mechanics, which by its very nature deals with continuous quantities, results in the introduction of the whole numbers required to explain the quantum phenomena.

APPENDIX TO SECTION I.5.

The relations implied by the use of complex quantities may be understood by the usual diagram in which the real and imaginary quantities are plotted at right angles, *i.e.*, one-quarter of a revolution apart, just as the sine and cosine terms in I (14) are one-quarter of a revolution apart. In this case, in Fig. I.3, points on

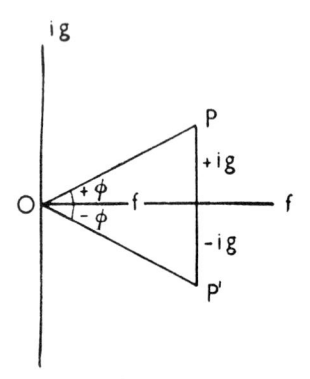

FIG. I.3.

the horizontal axis represent the real numbers $(f + i0)$, whilst those on the vertical axis represent the purely imaginary numbers $(0 + ig)$. The vectors $\overrightarrow{OP}$ and $\overrightarrow{OP'}$ represent the complex numbers $\Psi = (f + ig)$ and $\Psi^* = (f - ig)$, respectively. The length OP equals $\sqrt{f^2 + g^2}$ and is a real number, called the modulus of Ψ, and is written $|\Psi|$. The angle ϕ is called the angle† of Ψ, and to multiply two complex numbers the angles are added and the moduli are multiplied. It will be seen, therefore, that on multiplying $(f + ig)$ by $(f - ig)$ the product will be a point on the horizontal axis, *i.e.*, a real number, of magnitude $f^2 + g^2$, so that ‡ $\Psi\Psi^* = |\Psi|^2$, and in the problems with which we are concerned the electron density may be written as proportional to $|\Psi|^2$.

We have explained above the general form of the function Ψ as given by equation I (18), and have seen how this contains both an amplitude factor $\psi(x, y, z)$ and a time factor $e^{2\pi i\nu t}$. If the time factor is chosen so that Ψ is of the form given by equation I (17), then this equation represents a standing wave if ψ is real, and a progressive wave system if ψ has a suitable complex form. It is common practice to omit the time factor in equation I (17), and the reader will often encounter equations, such as $\psi = e^{ikx}$, described as representing travelling waves, and in such cases it is understood that the time factor is to be added.

As explained on p. 36, the electron density is proportional to $\psi\psi^*$, but in many problems concerning stationary states, ψ is real and $\psi\psi^*$ is identical with ψ^2. In

---

† This angle is often called the amplitude of the complex number, but we have avoided the use of this expression in order to prevent possible confusion with the amplitude of the Ψ waves.

‡ This is sometimes expressed by saying that the scalar value of Ψ is $\sqrt{f^2 + g^2}$. This is correct, but should not lead one to write Ψ^2 instead of $|\Psi|^2$.

The General Background

such problems it is possible to draw a graph showing how ψ varies with the distance in a certain direction, and from this graph the electron density can be read off by squaring† the values of ψ. It is only in some problems concerning stationary states that this is possible, and the reader should note that $\psi\psi^*$ and ψ^2 are not always interchangeable, and that the electron density is always proportional to $\psi\psi^*$.

The use of imaginary quantities in the above way is merely a mathematical device which reduces some of the equations to a form with which the mathematician is familiar, and when expressed in simple terms the above treatment merely assumes that the hypothetical Ψ vibrations are the result of two vibratory processes, which are one-quarter period apart, and so combine to give a resultant independent of time at any one place.

As stated above, the first Schrödinger equation, I (19), represents a family of surfaces each corresponding to one value of E. If Ψ is assumed to be of the form given by I (17), the time factor may be cancelled out, and the amplitude equation

$$\nabla^2 \psi + \frac{8\pi^2 m}{h^2}(E - W)\psi = 0 \quad \ldots \quad \text{I (20)}$$

is obtained, so that as stated before, both Ψ and ψ obey the same differential equation. If the time factor has the same form as in I (18), the energy E may be eliminated from equation I (20) to yield the Second Schrödinger Equation:

$$\nabla^2 \Psi - \frac{8\pi^2 m}{h^2} W\Psi + \frac{4\pi m i}{h} \cdot \frac{\partial \Psi}{\partial t} = 0 \quad \ldots \quad \text{I (21)}$$

This is the more fundamental equation, and can be used for problems where W is an explicit function of the time. Those who wish for further information on the derivation or use of the equations should consult the following books:

Elementary.

H. T. Flint, "Wave Mechanics." 5th edn. **1946**: London (Methuen & Co., Ltd.).
W. Heitler, "Elementary Wave Mechanics," 2nd edn. **1956**: London (Oxford University Press).
C. N. Hinshelwood, "The Structure of Physical Chemistry." **1951**: Oxford (Clarendon Press).
P. T. Matthews, "Introduction to Quantum Mechanics." **1965**: London and New York (McGraw-Hill).
L. Pauling and E. B. Wilson, Jr., "Introduction to Quantum Mechanics." **1935**: New York (McGraw-Hill).

More Advanced.

J. D. Landau and E. M. Lifshitz, "Quantum Mechanics." **1958**: Reading, Mass. (Addison Wesley).
A. Messiah, "Quantum Mechanics." Vols. I and II (translated by G. M. Temmer and J. Potter. **1958**: Amsterdam (N. Holland Publishing Co.): New York and London (Interscience Publishers).

6. THE REPRESENTATION OF THE INDIVIDUAL ELECTRON.

In the preceding description we have dealt with the rays or beams consisting of a continuous stream of electrons, and the wave function was interpreted so that $\Psi\Psi^*\Delta V$ was a measure of the probability of finding an electron in the particular element ΔV concerned. We have

† Confusion has arisen in many of the books owing to the fact that the description says that the electron density is proportional to $\psi\psi^*$, whilst the examples dealt with are stationary states of the above type, with the result that the beginner is puzzled by the change from $\psi\psi^*$ to ψ^2.

next to consider what happens when we try to identify the individual electrons in the stream, and here the interpretation offered is that the total Ψ is regarded as the sum of the effects due to the individual electrons, so that $\Psi\Psi^*\Delta V$ is also proportional to the probability of finding any one particular electron in the volume ΔV. The constant of proportionality is obtained by a process known as *normalization* of the wave function. If, for example, we deal with a problem in which one electron is contained in a volume V, then the integral $\int \Psi\Psi^* dV$ over the whole volume V, must equal one, since the probability that the electron is somewhere in the volume V is clearly unity.

This interpretation of the wave equation as giving the probabilities for one single electron implies naturally that we can no longer regard an electron as being localized at a given point, since all we can know are the probabilities of its being in various small volumes. Similarly, the motion of the electron can no longer be described by expressing the space co-ordinates (xyz) of the electron as a function of the time, but has to be considered in terms of the whole wave function which involves x, y, z, and t. From practical experience we know that what we observe as the electron is confined to a relatively small region of space, and if we are in any way to localize the electron, we must find some function of the hypothetical Ψ waves which has an appreciable amplitude only over a small region. We may understand this by considering how we could isolate a small portion from an infinite monochromatic wave. To do this we might arrange an aperture provided with a shutter which could be opened for a very short time, thus allowing a small group of waves to pass through. In this case, if we are dealing with actual material waves, we find that the action of the shutter slightly disturbs the head and the tail of the wave group, whilst the sides of the aperture disturb the sides of the wave group. The effect of this is to produce a wave group which is not homogeneous, but may be regarded as composed of a large number of slightly different frequencies. The result of this is that the individual waves stream forward with the phase-velocity U, but that owing to their slightly different frequencies they interfere with one another, and so have a negligible amplitude, except over a small region where they reinforce one another to produce a small wave group, as shown in Fig. I.4. The centre of the region in which the amplitude is appreciable moves forward with a velocity u, which is very much smaller than U, and which is given by the equation for the group velocity:

$$u = \frac{d\nu}{d\left(\frac{1}{\lambda}\right)}$$

The General Background

Since $U \gg u$, the individual waves which reinforce one another to produce the wave packet are continually changing and may be regarded as streaming on through the packet, but the centre of the region in which

PHASE VELOCITY OF WAVES
= $U \gg u$

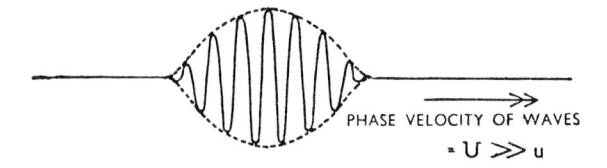

Fig. I.4. Velocity of centre of wave packet = group velocity $u = \dfrac{d\nu}{d(1/\lambda)}$.

there is an appreciable amplitude moves on with the much slower group velocity u. Further, as the wave packet moves on, it tends to spread out, and this spreading out is more rapid the shorter the group of waves cut out by the shutter, since it will be appreciated that the shorter the wave group, the relatively greater is the disturbing effect produced by the shutter. In the case of the Ψ waves we are dealing with nothing material, but are to interpret the hypothetical waves so that their amplitude is a measure of the probability of finding an electron at the place concerned. If the shutter is opened for such a short time that only one electron passes through, the wave packet of the hypothetical Ψ waves will refer to the one single electron. The wave packet of Ψ waves is thus not to be interpreted as the electron itself, but as representing a limited region of space in which there is an appreciable probability of finding the electron. We have explained that the Ψ waves are assumed to be the result of two vibratory processes, one-quarter of a period apart, which reinforce one another so that the resultant amplitude does not oscillate with the time, and we should therefore think of the electron wave packet as the envelope shown by the dotted line in Fig. I.4, and this envelope moves along with the velocity u. That this region of space moves on with the velocity of the electron is thus understandable,*

---

* We can readily see that if for free electrons

$$\nu = \frac{E_m}{h} = \frac{mc^2 + \tfrac{1}{2}mu^2}{h}$$

and

$$\frac{1}{\lambda} = \frac{mu}{h}$$

then

$$\frac{d\nu}{d\left(\dfrac{1}{\lambda}\right)} = u.$$

but the electron may be found at any point within the region represented by the wave packet, the probability being greatest in the region of the packet where the amplitude is greatest.

We have next to consider how we may interpret the spreading out of the wave group referred to above, and it will readily be seen that this is equivalent to an uncertainty in the velocity of the electron. As the wave group spreads out the amplitude diminishes, and instead of a large probability of finding the electron in a very small region of space, we have a probability which is smaller on the average, but extends over a larger region of space, and since there is only one electron this is equivalent to an uncertainty in its velocity. We have seen that the wave-group spreads out more rapidly, the smaller it is at the beginning, and from this it can be seen that the " probability wave-packet " concept expresses the exact characteristics of the Heisenberg Uncertainty Principle. A very small wave packet means that the electron's position is located precisely, because the region in which there is a probability of finding it is very restricted, and under these conditions the probability wave packet spreads rapidly, and the velocity is correspondingly uncertain. If we are asked what process is concerned, we can only reply that if we used any kind of shutter device in order to cut one electron out of a beam, then the movement of the shutter very near to the electron would produce fields slightly disturbing the latter, and the experiment is in fact only an attempt to measure the exact position of the electron, using a shutter instead of the beam of light described on p. 19. If we diminish the size of the aperture, and so diminish the uncertainty in the initial position of the electron in directions at right angles to the beam, we then produce diffraction effects which are equivalent to an uncertainty in the velocity in these directions, since there is now a greater probability of the electron being found moving in a direction different from that of the original beam. It will be seen, therefore, that the probability wave-packet concept has the exact characteristics of the Heisenberg Uncertainty Principle in so far as exact description of the position leads to an uncertainty in the velocity and *vice versa*, and it is easy to show that $\Delta x \times \Delta p \sim h$.

The above description will show that when we are dealing with the motion of an electron over distances which are very large compared with the wave-length, we may to some extent follow the electron in its path by considering the motion of the probability wave packet, although there is always some uncertainty in the simultaneous position and velocity, energy and time, &c. But this simple visualization breaks down when we consider the motion of the electron over distances comparable with the wave-length. The probability wave packet by its

The General Background 43

very nature extends over several wave-lengths, but the probabilities indicated by its varying amplitude refer to the one single electron, and this, therefore, cannot be visualized as moving about within the length of the wave packet. Over this small distance all that can be described is the probability of finding the electron in a given small region, and all attempts to follow the electron about in space must be given up.

The theory, which is confirmed by experiment, indicates that the "sizes of atoms" (*i.e.*, the regions in which there is an appreciable probability of finding electrons) are of the same order as the wave-lengths of the electrons in the atoms in their normal states, and from the point of view of wave mechanics it is meaningless to speak of electrons "moving in orbits," or to try to visualize them as particles "running round" orbits of the circular, elliptical, or rosette types. All that we can obtain is a wave function from which we can evaluate $\Psi\Psi^*$ at different points round the nucleus, and in this way we can visualize the electron in an atom as giving rise to a "probability cloud," or "electron cloud" of varying density, a high density of the electron cloud corresponding to a high probability of finding the electron at the place concerned. It was here that the early wave mechanics was in error, since it regarded the electron as *being* a wave running round and round an elliptical orbit, whereas the probability interpretation is the correct one, and requires a complete surrender of attempts to follow the electron round the orbit. It is unfortunate that the terms "orbit" and "orbital" from the older theories are often retained in order to describe the electron-cloud patterns of the new theories, because in many cases the concept of an orbit is misleading.

In atoms with more than one electron, the total electron cloud is formed by the summation of the probability clouds of the individual electrons, and some of the details of these electron-cloud patterns will be considered in later chapters. These electron-cloud patterns constitute our "pictures" of the atoms concerned, but the electron-cloud pattern of a given energy state must be accepted as a whole, and no attempt must be made to visualize the electron as moving about within the pattern. If it is asked why this should be so, the answer is that if we speak of an electron moving about within a given energy state, we imply that we can observe it at successive instants, but in any actual experiment the first observation would disturb the electron and remove it to a different energy state, so that two observations of the same state are impossible. These difficulties of describing events on an atomic scale are merely expressions of the Heisenberg Uncertainty Principle, and the probability interpretation of wave mechanics is the nearest approach we can make to a visualization of electrons over distances of the same order as the size of an atom.

In spite of the fact that an electron in an atom is no longer to be thought of as a particle running round an orbit, an electron state of an atom corresponds with a quite definite angular momentum, which in the hydrogen atom is equal to $\frac{h}{2\pi}\sqrt{l(l+1)}$, where l is a whole number.

If it is asked what is meant by angular momentum when we must no longer imagine a particle revolving in an orbit, the answer is that the electron state has associated with it a property which in processes such as collisions behaves analogously to the angular momentum of classical mechanics. If $l = 0$, the angular momentum is thus zero, but this is not to be taken as indicating that the electron is stationary, but rather that its motion is as likely to be in one direction as another, and consequently gives rise to no resultant angular momentum.

7. THE MEANING OF VELOCITY.

We have seen above how the Heisenberg Principle leads to a degree of uncertainty in the velocity of the electron. Apart from this, the use of wave mechanics introduces a new phenomenon in connection with the motion of a particle in a field of force, which is quite absent from classical mechanics, and which requires a careful definition of exactly what is meant by the term velocity. For simplicity we may deal with motion in one dimension, and in this case the motion of an electron in a field of varying potential is analogous to the passage of a ray of light in a straight line through a medium in which the refractive index varies from place to place, but is constant in any one plane perpendicular to the ray. In the optical problem we know that a certain fraction of the light is transmitted and a certain fraction reflected, the proportion of the reflected light becoming smaller as the wave-length is smaller. Expressed in terms of photons, this means that we cannot say what will happen to an individual photon, but at each place there is a certain probability that its velocity will be reversed in direction. The wave mechanics leads to an analogous effect in the case of an electron moving in a varying potential field, and at each place, although it is impossible to say what will happen to the individual electron, there is a certain probability that the velocity of the electron may be reversed, without change of magnitude. If the wave function Ψ is known, then at any one point $\Psi\Psi^*\Delta V$ is a measure of the probability of finding an electron in the volume ΔV, and a corresponding function can be found which

The General Background

gives the probability of a reversal in direction of the velocity. We have therefore to distinguish carefully between the mean velocity v associated with an electron represented by a given wave function and the root mean square velocity u. If, for example an electron state is obtained by the superposition of two travelling waves of equal amplitude but moving in opposite directions, we shall have a standing wave for which the mean velocity v is zero, since there is an equal probability of finding the electron moving in either direction, but the root mean square velocity will, of course, have a definite value. In the Brillouin zone theories of electrons in a crystal, this point is of great importance. Here the potential varies greatly over very small distances, and a given factor may affect v and u quite differently; a quite misleading impression may therefore be gained unless it is realized that the mean velocity v takes into account the probability of the reversal of direction of the velocity.

8. Potential Boundaries in Wave Mechanics.

The wave equations in Section I.5 involve the terms $(E - W)$, where E is the total energy (kinetic plus potential) and W is the potential energy. In the problems of the older mechanics, the kinetic energy could only be zero or positive, and so $(E - W)$ was always positive, and regions where $(E - W)$, or the kinetic energy, would be negative were forbidden regions into which the particle concerned could not penetrate. If, for example, a ball were thrown vertically into the air with a given kinetic energy, it would gain potential energy and lose kinetic energy until it came to rest, and the region beyond this point was one into which it could not enter unless additional energy were supplied. In wave mechanics, on the other hand, there is no sharp boundary between permissible and forbidden regions, and a particle has a probability of penetrating into the region where $(E - W)$ is negative, although the probability rapidly diminishes as this region is entered. In the problem of a ball thrown into the air, this effect would be entirely negligible, owing to the small value of Planck's constant, but in problems concerning electrons, the effect often produces interesting results. The wave equations of Schrödinger can be applied to both the region where $(E - W)$ is positive and the region where $(E - W)$ is negative, and the ψ functions in the two regions fit smoothly on to one another.

We may illustrate this by considering an electron contained in a one-dimensional potential box, the potential being constant inside the box, but rising sharply at the walls whose positions are shown by dotted lines in Fig. I.5. Then, according to the old mechanics, an electron

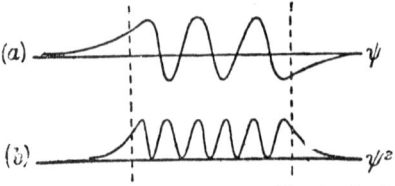

FIG. I.5. The upper diagram shows one of the possible ψ patterns for an electron enclosed in a one-dimensional box, whose length, L, is bounded by the dotted vertical lines. The ψ pattern chosen is the stationary state for which the wave-length is equal to $2L/5$. The probability of finding the electron at different places is proportional to ψ^2, and the lower curve shows the curve for ψ^2.

whose kinetic energy was less than the potential step at the boundary was confined to the box, and there was a zero probability of finding it beyond the boundaries. We have already explained that the stationary states of an electron in a one-dimensional box are characterized by wave-lengths such that the length of the box is an integral number of half wave-lengths. Fig. I.5(a) shows a hypothetical ψ pattern with two and a half wave-lengths in the length L. The meaning of this pattern is that if, as in Fig. I.5(b), we square the value of ψ at any point, then ψ^2 is proportional to the probability of finding the electron at the point concerned. In this case the solution of the equation shows that in the forbidden region, ψ falls off exponentially, *i.e.*, according to a relation of the type e^{-kx}, where k is very large, so that ψ, and hence ψ^2 and the probability of finding the electron, diminish very rapidly as the forbidden region is entered. This figure shows how ψ falls off exponentially as the forbidden area is entered, and the reader will encounter numerous diagrams in which ψ patterns have exponential tails of this kind, instead of the sharp boundary which would exist in classical mechanics. The rapid fall in the exponential curve means that as the forbidden region is entered the probability of finding an electron very soon becomes negligible, and if the potential wall is thick there is little difference between the new and old concepts. But if the thickness of the potential wall is of the same order as the wave-length, very interesting effects are met with, and an electron may penetrate a thin potential wall which it could not surmount according to the older view,* this effect being of importance in some problems of electronic emission from metals with surface films and in some electronic devices.

* A similar " tunnelling " process must be invoked in discussions of the escape of α-particles from radioactive nuclei.

The General Background

The second curve in Fig. I.5 shows the probability of finding the electron in different places across the box.† It will be seen that there are places where this probability is zero, and the question is often asked as to how, if this is so, the electron can travel from one part of the box to another. The answer to this question is that the above " probability pattern " refers to a stationary state of the electron in the box, but if we speak of the electron moving about in the box we imply that we can make successive observations on the electron, and so follow its path. Actually, however, the first observation will disturb the electron, and so remove it from the stationary state concerned. We have thus to accept the probability pattern as giving all that can be known of the electron in the one stationary state. If we wish to consider the electron as travelling about inside the box, we shall have to regard it as a wave packet, smaller than the box itself, and in this case the wave packet will be built up of a group of states, whose probability patterns will be superimposed. This difficulty is continually met with in dealing with electron-cloud patterns or probability patterns of stationary states, and we have to remember that the probability pattern of a given energy state always implies that the time is completely indeterminate, so that no attempt must be made to follow the electron about within the pattern of the state.

† The problem is one of a stationary state in which $\psi\psi^*$ is equivalent to ψ^2 (see p. 39), and the second curve is obtained by squaring the ordinates in the first curve.

PART II. THE STRUCTURE OF THE FREE ATOM.

1. THE ELECTRON GROUPS.

As we have explained in the preceding sections it is not possible to give any precise mechanical picture of the motion of the electrons round the positively charged nucleus of an atom. If an atom contains more than one electron, each individual electron moves in the field resulting from the attraction of the nucleus and the repulsion of the other electrons, all of which are in motion. Wave mechanics then gives us for the system as a whole, a number of possible wave functions, and these have to be interpreted on the assumption that $\Psi\Psi^*$ is a measure of the electron density at the place concerned. From this point of view the spatial distribution of the electrons in an atom may be represented by a series of diagrams, each giving the total electron density as a function of the distance from the nucleus in a particular direction. The complete series of diagrams then gives us what may be termed a " probability pattern," representing the electron density in the space surrounding the nucleus, and it is the shape of this pattern which constitutes our " picture " of the atom concerned. To a fairly close approximation we may assume each electron to move in a potential field which represents the attraction of the nucleus and the average of the instantaneous repulsions of all the other electrons. Wave mechanics then gives us for each electron a number of possible wave functions, and in the simplest theory the total electron density at a given place is proportional to the sum of the values of $\Psi\Psi^*$ for each of the electrons in the atom. In the more complete theories account is taken of the way in which the different electrons interfere with each other's motion.

The electrons in an atom can occupy different energy states, and each electron state in a given atom gives rise to a definite electron cloud pattern, and is characterized by a definite energy, and the first thing to be grasped is the division of the electrons among the main energy groups or quantum shells. These groups are characterized by quantum numbers, and the state of an electron in an atom is described by four quantum numbers, n, l, m_l, and m_s. The precise significance of these is described later (p. 61), and for the moment we may note that, speaking in a general way, the quantum number n is a measure of the energy of the electron in the state concerned, and l is a measure of its angular momentum; l may have any value from 0 to $(n-1)$, but a state where $l = 0$ is not to be regarded as one in which the electron is at

The Structure of the Free Atom 49

rest, but rather as one in which the motion does not give rise to an angular momentum, and cannot be visualized in terms of " orbits." In classical mechanics there is an " orbit " corresponding to $l = 0$, namely, oscillation of an electron along a straight line passing through the nucleus, but this kind of motion is clearly impossible, and in the new mechanics the state $l = 0$ should be regarded as one in which the motion is as likely to be in one direction as in another.

The quantum number m_l is a measure of the component of the angular momentum in a particular direction, which may be taken to be that of an applied magnetic field; m_l may have any value from $+l$ to $-l$, including 0. As explained later (p. 61), it is really meaningless to speak of the component of the angular momentum in a perfectly free atom, since the term " component " by its very nature implies that there exists some axis with reference to which measurements can be made. The policy usually adopted is to define the z axis as the direction of an applied magnetic field, and then to imagine the field to be reduced until it is vanishingly small. By a free atom we mean, therefore, an atom in which the direction is defined by a very weak magnetic field. In free atoms, electron states with the same n and l, but with different values of m_l, have the same energy and are said to be *degenerate*, this term being used to describe any case in which different wave functions lead to the same energy. When the atom is placed in a weak magnetic field, the degeneracy is removed, and each value of m_l gives rise to a slightly different energy.

The fourth quantum number, m_s, is due to the electron spin (see p. 63). In a magnetic field, the spinning electron can take up either of two directions, and this effect is represented by the spin quantum number m_s, which can assume the two values $\pm \frac{1}{2}$, according to the direction of the spin. In general, the difference between the energies of electrons with a given value of n, l, and m_l, and with $m_s = +\frac{1}{2}$, or $m_s = -\frac{1}{2}$, is very small, but it increases with the atomic number.

In order to avoid the continual printing of expressions such as $(n = 2, l = 1, m = 0)$, it is customary to use the small letters s, p, d, and f to describe electron states for which $l = 0, 1, 2$, and 3, respectively, whilst a large figure before the letter indicates the value of n. The symbol $3d$, for example, stands for a state in which $n = 3$, $l = 2$. If the symbol is enclosed in a bracket with a small superscript number outside, this indicates the number of electrons in the sub-group concerned. Thus the structure of the normal atom of boron may be written $(1s)^2(2s)^2(2p)^1$, which implies that the atom contains two electrons in the $(1s)$ state, two in the $(2s)$ state, and one in the $(2p)$ state.

The *state of an electron* in an atom is thus described by the four quantum numbers n, l, m_l, and m_s, and in a given atom each electron state has a definite energy. The energy of the atom as a whole can thus assume a number of different values, according to the way in which its electrons are arranged in the different possible electronic states. Each of these arrangements may be said to give rise to a *state of the atom*, and the most stable state is that for which the energy is a minimum. This may be referred to as the *normal state* of the atom concerned, and the states of higher energy content are called *excited states*. It will be noted that the term " state " is used to describe both the condition of an individual electron in an atom, and the condition of the atom as a whole. In general, the context makes clear the sense in which the word is being used.

It might at first be thought that the normal state of every atom would be such that all the electrons occupied the lowest quantum state. Actually, however, this is not the case, and the facts are summarized by what is known as the *Pauli Exclusion Principle*. According to this Principle no two electrons in an atom can be in exactly the same state as defined by all four quantum numbers. An atom may, for example, have one electron in the state characterized by $n = 1$, $l = 0$, $m_l = 0$, and $m_s = +\frac{1}{2}$, and one electron in the state characterized by $n = 1$, $l = 0$, $m_l = 0$, and $m_s = -\frac{1}{2}$, but only one electron can be in a given state as defined by all four quantum numbers. Since l varies from 0 to $(n-1)$, and m_l varies from $+l$ to $-l$, including zero, this implies that the maximum possible numbers of electrons with principal quantum numbers $n = 1, 2, 3, 4, \ldots n$, are 2, 8, 18, 32, $\ldots 2n^2$, respectively. This Principle was first put forward empirically, and was later shown to be, if not a logical outcome of the wave mechanics, at any rate in complete agreement with its requirements.*

The Pauli Principle is of very great importance because it applies to all structures in which electrons occupy stationary states characterized by quantum numbers. In all such cases there cannot be more than one electron in each state as defined by all the quantum numbers, including that of the spin. In some treatments it is customary to omit the spin quantum number, and to say that each electron state can contain not more than two electrons; in this case, if a state is fully occupied it will contain two electrons, one of each spin. The difference between these two definitions depends simply on whether the

* The Pauli Principle is necessitated by wave mechanics if the additional assumption is made that the wave function of the whole system of electrons is completely anti-symmetric in the co-ordinates of the electrons, *i.e.*, on interchanging the co-ordinates of any two electrons the wave function must change its sign. For this purpose the co-ordinates are taken to be the three space co-ordinates x, y, and z, and the spin is regarded as a fourth co-ordinate of value $\pm \frac{1}{2}$. (See p. 138.)

The Structure of the Free Atom

spin quantum number is taken into account in defining an electron state.

The Pauli Principle implies, therefore, that in atoms with several electrons, only two of these can be in the (1s) state, and the remainder must be in higher states. The classification of electrons into groups was first made by Bohr, and later extended by Stoner and Main-Smith. For our purpose it is usually sufficient to deal with the main groups and sub-groups described by the first two quantum numbers, and the later division into further sub-groups will not be considered in detail. Table II shows the electronic structures of the free atoms in their normal states, and it must be emphasized that these refer to free atoms, and not to atoms in a liquid or solid.

The characteristics of Table II have been described so often that we shall only refer to them briefly here, and for full details the reader should consult a book such as Pauling's " The Nature of the Chemical Bond."

The atomic number of an element which gives its order in the table is equal to the number of electrons in the atom, with the result that the atom of an element of atomic number Z has a nucleus with a charge of $+ Ze$ surrounded by Z electrons. Hydrogen and helium in their normal states have one and two electrons respectively in the 1-quantum group. By Pauli's Principle this group can contain not more than two electrons which must be of opposite spins, and in the next element, lithium, the third electron enters the 2-quantum shell, and it is this electron which gives rise to the metallic properties. The passage from lithium to neon results in the filling of this shell to its maximum of 8 electrons. During this process, the electrons first enter the (2s) sub-group, and beryllium has two (2s) electrons with opposite spins. On passing from boron to neon, the electrons fill up the (2p) sub-group until this contains six electrons, which is the maximum allowed by Pauli's Principle, since the 2-quantum shell then contains 8 (*i.e.*, 2×2^2) electrons. During the filling up of the p-states, the electrons tend as far as possible not to pair with one another, but to occupy different sub-sub-groups, keeping their spins parallel. Thus, in nitrogen the three p-electrons occupy the three sub-sub-groups ($2p, m_l = 0$), ($2p, m_l = + 1$), and ($2p, m_l = - 1$), and the three electron spins are parallel. In the atom of oxygen there is one more p-electron, and this must pair with one of the above three electrons, leaving the other two unpaired. The tendency for the electrons to enter different sub-sub-groups so as to keep their spins parallel as long as this is possible, is known as *Hund's Principle of Maximum Multiplicity*, and it applies also to the filling of the d and f sub-groups.

The 2-quantum shell is completely filled for neon with atomic

TABLE II. *Atomic Structures* (1).

Element and Atomic Number.	Principal and Secondary Quantum Numbers.									
$n =$	1	2		3			4			
$l =$	0	0	1	0	1	2	0	1	2	3
1 H	1									
2 He	2									
3 Li	2	1								
4 Be	2	2								
5 B	2	2	1							
6 C	2	2	2							
7 N	2	2	3							
8 O	2	2	4							
9 F	2	2	5							
10 Ne	2	2	6							
11 Na	2	2	6	1						
12 Mg	2	2	6	2						
13 Al	2	2	6	2	1					
14 Si	2	2	6	2	2					
15 P	2	2	6	2	3					
16 S	2	2	6	2	4					
17 Cl	2	2	6	2	5					
18 A	2	2	6	2	6					
19 K	2	2	6	2	6		1			
20 Ca	2	2	6	2	6		2			
21 Sc	2	2	6	2	6	1	2			
22 Ti	2	2	6	2	6	2	2			
23 V	2	2	6	2	6	3	2			
24 Cr	2	2	6	2	6	5	1			
25 Mn	2	2	6	2	6	5	2			
26 Fe	2	2	6	2	6	6	2			
27 Co	2	2	6	2	6	7	2			
28 Ni	2	2	6	2	6	8	2			
29 Cu	2	2	6	2	6	10	1			
30 Zn	2	2	6	2	6	10	2			
31 Ga	2	2	6	2	6	10	2	1		
32 Ge	2	2	6	2	6	10	2	2		
33 As	2	2	6	2	6	10	2	3		
34 Se	2	2	6	2	6	10	2	4		
35 Br	2	2	6	2	6	10	2	5		
36 Kr	2	2	6	2	6	10	2	6		

TABLE II (cont.). *Atomic Structures* (2).

Element and Atomic Number.	Principal and Secondary Quantum Numbers.										
$n =$	1	2	3	4				5			6
$l =$	—	—	—	0	1	2	3	0	1	2	0
37 Rb	2	8	18	2	6			1			
38 Sr	2	8	18	2	6			2			
39 Y	2	8	18	2	6	1		2			
40 Zr	2	8	18	2	6	2		2			
41 Nb	2	8	18	2	6	4		1			
42 Mo	2	8	18	2	6	5		1			
43 Tc	2	8	18	2	6	5		2			
44 Ru	2	8	18	2	6	7		1			
45 Rh	2	8	18	2	6	8		1			
46 Pd	2	8	18	2	6	10		—			
47 Ag	2	8	18	2	6	10		1			
48 Cd	2	8	18	2	6	10		2			
49 In	2	8	18	2	6	10		2	1		
50 Sn	2	8	18	2	6	10		2	2		
51 Sb	2	8	18	2	6	10		2	3		
52 Te	2	8	18	2	6	10		2	4		
53 I	2	8	18	2	6	10		2	5		
54 Xe	2	8	18	2	6	10		2	6		
55 Cs	2	8	18	2	6	10		2	6		1
56 Ba	2	8	18	2	6	10		2	6		2
57 La	2	8	18	2	6	10		2	6	1	2
58 Ce	2	8	18	2	6	10	1	2	6	1	2
59 Pr	2	8	18	2	6	10	3	2	6		2
60 Nd	2	8	18	2	6	10	4	2	6		2
61 Pm	2	8	18	2	6	10	5	2	6		2
62 Sm	2	8	18	2	6	10	6	2	6		2
63 Eu	2	8	18	2	6	10	7	2	6		2
64 Gd	2	8	18	2	6	10	7	2	6	1	2
65 Tb	2	8	18	2	6	10	9	2	6		2
66 Dy	2	8	18	2	6	10	10	2	6		2
67 Ho	2	8	18	2	6	10	11	2	6		2
68 Er	2	8	18	2	6	10	12	2	6		2
69 Tm	2	8	18	2	6	10	13	2	6		2
70 Yb	2	8	18	2	6	10	14	2	6		2
71 Lu	2	8	18	2	6	10	14	2	6	1	2
72 Hf	2	8	18	2	6	10	14	2	6	2	2

TABLE II (cont.). Atomic Structures (3).

Element and Atomic Number.	$n =$	1	2	3	4				5				6			7
	$l =$	—	—	—	0	1	2	3	0	1	2	3	0	1	2	0
73 Ta		2	8	18	32				2	6	3		2			
74 W		2	8	18	32				2	6	4		2			
75 Re		2	8	18	32				2	6	5		2			
76 Os		2	8	18	32				2	6	6		2			
77 Ir		2	8	18	32				2	6	7		2			
78 Pt		2	8	18	32				2	6	9		1			
79 Au		2	8	18	32				2	6	10		1			
80 Hg		2	8	18	32				2	6	10		2			
81 Tl		2	8	18	32				2	6	10		2	1		
82 Pb		2	8	18	32				2	6	10		2	2		
83 Bi		2	8	18	32				2	6	10		2	3		
84 Po		2	8	18	32				2	6	10		2	4		
85 At		2	8	18	32				2	6	10		2	5		
86 Rn		2	8	18	32				2	6	10		2	6		
87 Fr		2	8	18	32				2	6	10		2	6		1
88 Ra		2	8	18	32				2	6	10		2	6		2
89 Ac		2	8	18	32				2	6	10		2	6	1	2
90 Th		2	8	18	32				2	6	10		2	6	2	2
91 Pa		2	8	18	32				2	6	10		2	6	1	2
92 U		2	8	18	32				2	6	10	2	2	6	1	2
93 Np		2	8	18	32				2	6	10	4	2	6	1	2
94 Pu		2	8	18	32				2	6	10	5	2	6	1	2

The exact electronic configuration of the later elements is uncertain, but according to Katz and Seaborg* the most probable arrangements of the outer electrons are:

95 Am	$(5f)^7(7s)^2$
96 Cm	$(5f)^7(6d)^1(7s)^2$
97 Bk	$(5f)^8(6d)^1(7s)^2$
98 Cf	$(5f)^{10}(7s)^2$
99 Es	$(5f)^{11}(7s)^2$
100 Fm	$(5f)^{12}(7s)^2$
101 Md	$(5f)^{13}(7s)^2$
102 No	$(5f)^{14}(7s)^2$
103 Lw	$(5f)^{14}(6d)^1(7s)^2$
104 —	$(5f)^{14}(6d)^2(7s)^2$

* J. J. Katz and G. T. Seaborg, "The Chemistry of the Actinide Elements." **1957**: London (Methuen).
Element 104 has only recently been prepared and has not been named at the moment of writing.

number 10, and this element is an inert gas because the outer group of 8 electrons is very stable. On passing to sodium with atomic number 11, the last electron has therefore to enter the 3-quantum shell, and sodium is thus a univalent metal. On passing from sodium to argon, the process is analogous to that occurring between lithium and neon,

The Structure of the Free Atom

and the 3-quantum shell is built up to a group of 8. According to Pauli's Principle the 3-quantum shell can contain a maximum of 18 electrons, but the group of 8 has a provisional stability, with the result that argon is an inert gas, and on passing to the next element potassium, the new electron enters the 4-quantum shell, and potassium is a univalent metal. From potassium to calcium the process is analogous to that occurring between sodium and magnesium, but on passing to scandium the additional electron does not enter the 4-quantum or valency shell, but enters the 3-quantum shell, and, as can be seen from Table II, the passage from scandium to copper results in the expansion of the 3-quantum shell from a group of 8 electrons to one of 18. These elements are known as the Transition Elements, and are all of variable valency, and this is to be taken as implying that at this stage in the Periodic Table, the energies of the electrons in the $(3d)$ and $(4s)$ states are not very different. Similar transition processes occur in the later Periods, although, as will be seen from Table II, the exact details of the process are not always the same. It must again be emphasized that the electronic structures given in Table II refer to the free atoms, and not to the atoms in a solid or liquid. Thus, in the case of nickel the outer electronic arrangement in the free atom is $(3d)^8(4s)^2$, but in solid nickel there are on the average 9·4 electrons per atom in $(3d)$ states, and only 0·6 per atom in $(4s)$ states. There has been an unfortunate tendency in some metallurgical papers to assume that the finer details of Table II apply to the atoms in a solid metal, whereas this is not so.

By Pauli's Principle, the 4-quantum shell can contain 32 electrons, and although the group of 18 which is completed at palladium has a provisional stability, it expands later into a complete group of 32. This process takes place in the group of elements known as the Rare Earths or *lanthanides* which lie between lanthanum and hafnium in the Periodic Table, and is immediately followed by the building up of the 5-quantum shell into a group of 18. By Pauli's Principle, the 5-quantum shell can contain a maximum of $2 \times 5^2 = 50$ electrons, and so the group of 18 electrons, although of provisional stability, is able to undergo a further expansion. It is now known that the building up of the $(5f)$ sub-group of electrons (*i.e.*, the 14 electrons in states for which $n = 5$, $l = 3$) is taking place in the new trans-uranic elements which have resulted from the work on atomic energy. As will be seen from Table II, the filling of the $(4f)$ sub-group begins at cerium, which is the fourth element of the Period. In the next Period the filling up of the $(5f)$ sub-group begins at the fifth element (protactinium) of the Period, and the following elements are sometimes called *actinides* by analogy with the lanthanides. The details of the filling of the $(4f)$ and $(5f)$ sub-

E

groups are not the same. In the actinides there is more variable valency than in the lanthanides, and the energies of the (5f) and (6d) electrons are more nearly equal than are those of the (4f) and (5d) electrons in the lanthanides.

In the first three Long Periods the elements copper, silver, and gold are univalent elements with one valency electron outside a group of 18. On passing from copper to krypton, silver to xenon, and gold to radon, the changes in electronic structure are similar to those occurring on passing from lithium to neon. In each case a sub-group of two s-electrons is first formed, and then the sub-group of six p-electrons is built up. One point of great importance is the stability of the sub-group of two s-electrons, which increases with the number of the Period. Aluminium does not give rise to a stable Al^+ ion, but, in the Third Long Period, the stability of the $(6s)^2$ sub-group is so great that thallium forms univalent thallous salts which are more stable than the trivalent thallic salts, and similarly lead forms divalent salts in which the $(6s)^2$ sub-group persists unchanged. The existence of this stable $(6s)^2$ sub-group is of considerable interest in the study of alloys, because its stability is so great that it persists unchanged in some alloys, although in other alloys the sub-group breaks down. With indium and tin, there is a considerable tendency for the $(5s)^2$ sub-group to persist as a stable sub-group, but with gallium and germanium the stability of the $(4s)^2$ sub-group is not so great.

The energies necessary to remove one, two . . . &c., electrons from an atom are known as the first, second . . . &c., *ionization potentials* of an atom, and it is instructive to see how these vary for the elements of the First Short Period. The data are summarized in Table III, and it will be seen that the ionization potential for the removal of the first electron from a free atom of lithium is only 5·37 V, whilst the removal of the second electron, which involves the breaking up of the completed $(1s)^2$ group requires 75·28 V. Part of this increase is due to the fact that the second electron is being removed from a positive ion instead of from a neutral atom, but as will be appreciated from what follows the greater part of the increase is due to the stability of the completed $(1s)^2$ group. The first ionization potential of beryllium is 9·28 V, and is thus greater than that of lithium. This is the result of the increased nuclear charge in the case of beryllium. A valency electron of beryllium is not subject to the full attraction of a nucleus of charge $+ 4e$, because part of the attraction is neutralized or screened by the other electrons of the atom. This screening action is due mostly to the electrons of the underlying quantum shells, *i.e.*, the (1s) electrons in the case of beryllium, but also to a slight extent to the electrons of the same group as the one considered.

The Structure of the Free Atom

The second ionization potential of beryllium is 18·12 V, and is thus roughly double that of the first ionization potential. This increase is due almost entirely to the fact that the electron is being removed from a positive ion instead of from a neutral atom. The third ionization potential of beryllium is no less than 153·1 V, and this enormous increase is due to the fact that the removal of the third electron involves the breaking up of the stable $(1s)^2$ group. Examination of the data for boron and carbon will show that there is a similar large increase at the stage which involves the breaking up of the $(1s)^2$ group.

TABLE III.

Element.	Ionization Potential, volts.	Transition.	Resulting Electronic Structure.
He	1st 24·465	He $\longrightarrow$ He$^+$	$(1s)^1$
Li	1st 5·37 2nd 75·28	Li $\longrightarrow$ Li$^+$ Li$^+$ $\longrightarrow$ Li^{++}	$(1s)^2$ $(1s)^1$
Be	1st 9·28 2nd 18·12 3rd 153·1	Be $\longrightarrow$ Be$^+$ Be$^+$ $\longrightarrow$ Be^{++} Be^{++} $\longrightarrow$ Be^{+++}	$(1s)^2(2s)^1$ $(1s)^2$ $(1s)^1$
B	1st 8·28 2nd 25·0 3rd 37·75 4th 258·1	B $\longrightarrow$ B$^+$ B$^+$ $\longrightarrow$ B^{++} B^{++} $\longrightarrow$ B^{+++} B^{+++} $\longrightarrow$ B^{++++}	$(1s)^2(2s)^2$ $(1s)^2(2s)^1$ $(1s)^2$ $(1s)^1$
C	1st 11·217 2nd 24·27 3rd 47·65 4th 64·22 5th 389·9	C $\longrightarrow$ C$^+$ C$^+$ $\longrightarrow$ C^{++} C^{++} $\longrightarrow$ C^{+++} C^{+++} $\longrightarrow$ C^{++++} C^{++++} $\longrightarrow$ C^{+++++}	$(1s)^2(2s)^2(2p)^1$ $(1s)^2(2s)^2$ $(1s)^2(2s)^1$ $(1s)^2$ $(1s)^1$

These figures are taken from Bacher and Goudsmit: "Atomic Energy States" (McGraw-Hill Publishing Co., Ltd.).

Reference to Table III will show that the first ionization potential of boron is less than that of beryllium, in spite of the fact that the atomic number of boron is the greater. This is because the $(2s)$ sub-group can contain not more than two electrons, and the third electron must therefore be in a $(2p)$ state. The second ionization potential of boron is greater than that of beryllium, since both involve the removal of a $(2s)$ electron, and as the screening effect of one $(2s)$ electron upon another is relatively slight, the increased nuclear charge in boron attracts each of the $(2s)$ electrons in B$^+$ more strongly than the nucleus attracts the single $(2s)$ electron in Be$^+$.

The data for carbon illustrate the same general principle. The first ionization potential of carbon is greater than that of boron on account of the greater nuclear charge. The second ionization potentials involve the removal of a (2p) electron from C$^+$, and of a (2s) electron from B$^+$, and consequently the energy required is less for carbon than for boron. The third ionization potentials, on the other hand, involve the removal of (2s) electrons from both C^{++} and B^{++}, and the increased nuclear charge makes itself felt, and the energy required is greater for carbon than for boron. Increasing nuclear charge thus leads to a general increase in the firmness with which electrons are bound to the atoms, although this increase is not smooth and continuous, but is modified by the extent to which the nuclear charge is screened by the different electrons. In some theoretical treatments a screening constant S is used, of such a nature that if Ze is the nuclear charge, the effective charge for the electron considered is $(Z - S)e$. It is to be noted that the screening constant S is not an absolute quantity applicable to all problems. If, for example, a term of the nature $(Z - S)e$ is used to calculate the ionization potential of the electron concerned, the value of S required to give the correct result may be different from that needed when a term of the form $(Z - S)e$ is used to calculate some other property.

It must be emphasized that the above figures refer to the free atoms, and are not a measure of the work required to remove an electron from an atom in a solid or liquid metal. The general principle, however, still applies in the sense that increased nuclear charge leads to a firmer binding, modified by the screening effect.

The ionization potential is the work required to remove an electron from a neutral atom. The *electron affinity* is the work required to remove an electron from a negative ion, and if it is a positive quantity the negative ion is stable compared with the neutral atom. Thus, the negative F$^-$ ion is stable compared with the neutral F atom, and work is required to remove an electron from the F$^-$ ion.

As will be seen from Table II, the 2-quantum shell of electrons is completely filled for neon with atomic number 10, and this element is therefore an inert gas. On passing to the next element, sodium, the additional electron enters the (3s) sub-group of the third quantum shell. The first ionization potentials of these atoms are as follows:

Atom.	1st Ionization Potential.
O	13·55
F	18·6
Ne	21·47
Na	5·12

They illustrate clearly how the firmness of the binding of the electrons is affected. From oxygen to neon there is a steady increase in the ionization potential as the stable $(2s)^2(2p)^6$ group of 8 electrons is

The Structure of the Free Atom

built up. On passing to sodium there is a marked fall in the 1st ionization potential, which now corresponds with the removal of a (3s) electron. The single (3s) valency electron of sodium is thus bound comparatively weakly, and we again have a metallic element. A similar change in the first ionization potential occurs between each inert gas and the following alkali metal. In every case the first ionization potential of an alkaline earth metal is greater than that of the preceding alkali metal, and that of a Group IIB metal is greater than that of the preceding Group IB metal, although the differences are not so marked as those shown above for lithium and beryllium, because the relative change in the screening constants between successive elements tends to become less with increasing atomic number. Thus, the first ionization potentials of potassium and calcium are 4·32 and 6·09 V, respectively, and those of copper and zinc are 7·68 and 9·36 V, respectively. For the transition elements relations between the ionization potentials are naturally more complicated.

The structures of the free atoms are conveniently summarized in Table II, and when it is desired to go into more exact details the energies of the different electrons in an atom may be represented on a scale, so that the distance between two points is a measure of the difference in the energy of electrons in the two states concerned, and is consequently proportional to the frequency of the light emitted when an electron drops from the state of higher to the state of lower energy. In any one atom, for a given value of n, the energies increase in the order s, p, d, f, i.e. with increasing l. For each of the s, p, d, f states, the energy increases with n. In any one free atom, an ns electron always has a higher energy than an $(n-1)p$ electron, so that as far as the s and p electrons are concerned, the energies of the electrons in any one atom increase regularly in the order $1s, 2s, 2p, 3s, 3p, \ldots$ For the d electrons the relations are more complicated. We have seen how, in the transition elements, increasing atomic number results in the electrons entering the $(n-1)d$ shell, instead of the np shell. Expressed in terms of energies, this implies that for these elements the energy of an $(n-1)d$ electron is less than that of an np electron, and this is always so when the electrons in the outer shells of an atom are considered. This point is shown clearly in Fig. II.1, which gives an approximate representation of the energies of the different electron shells as a function of the atomic number.* In the elements of higher atomic number, where the 3-

---

* In Fig. II.1, which is due to Herzberg, the $3d$ level is shown at a constant distance above the $4s$ level for all the transition elements of the First Long Period. Actually, in passing along the series there is a general tendency for the $3d$ level to fall relatively to the $4s$ level, and the two are too close to be distinguished.

60   *Atomic Theory*

quantum shell is further down in the atom, this position is reversed, and, as can be seen from the crossing of the lines in Fig. II.1, the 3s, 3p, and 3d electrons all have lower energies than any of the 4-quantum subgroups. Similar relations apply to the other transition elements. In

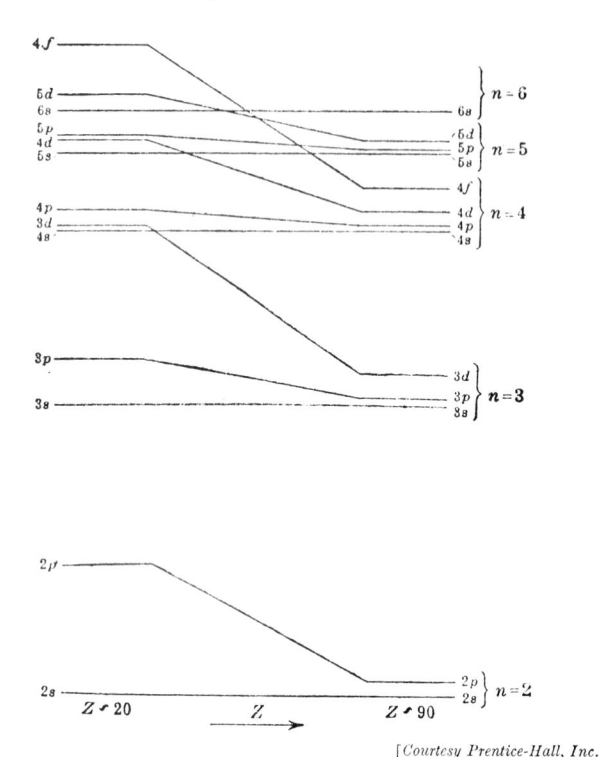

FIG. II.1 Rough representation of the relative energies of the shells for different nuclear charge Z (to explain the filling up of inner shells). The elements potassium, rubidium, and caesium at the beginning of the Long Periods have atomic numbers 19, 37, and 55, respectively. The figure is highly diagrammatic, and over-simplified, and is intended only to show the relative order of the shells. In an accurate diagram the lines would not be horizontal or straight lines, but the relative order would be maintained; the energy decreases on passing from the top to the bottom of the diagram, and the 1s states lie too low down to be shown.

Fig. II.1 the rare earth elements correspond to the position at which the energies of electrons in the 4f sub-group have fallen below those of the 6s and 5d sub-groups, so that with increasing atomic number electrons enter the 4f group until this is completed. In the same way the actinides are in the position at which the energies of the electrons in

The Structure of the Free Atom

the 5f sub-group have fallen below those of the 7s and 6d sub-groups, so that with increasing atomic number electrons begin to enter the 5f sub-group.*

Note on the Exact Meaning of Quantum Numbers.

The quantum number n of the present theories replaces the n of the Bohr theory, and even though we drop the idea of an orbit, n retains some of the older characteristics. In the hydrogen atom the negative energy still varies inversely as n^2, so that energy has to be given to an electron to excite it from a state of lower to one of higher n. From the older viewpoint a large value of n implied a large orbit, and from the present point of view it implies that the electron-cloud pattern extends over a large region of space.

The quantum number l is a measure of the angular momentum of the electron apart from its spin, and this is called the *orbital angular momentum*, because it corresponds with the angular momentum of the orbit in the old sense. The orbital angular momentum of a hydrogen-like atom is of magnitude $\frac{h}{2\pi}\sqrt{l(l+1)}$, as compared with the postulate of the Bohr theory that the angular momentum was of magnitude $\frac{kh}{2\pi}$. The quantum number l is numerically equal to $(k-1)$ of the Bohr theory, and the occurrence of terms of the type $\sqrt{l(l+1)}$ has led to a confusing position in which the symbol l, sometimes modified by an asterisk, is often printed with the understanding that it must be replaced by $\sqrt{l(l+1)}$ for purposes of calculation.

The third quantum number m_l is a measure of the component of the orbital angular momentum in a particular direction, and to understand its meaning we have to consider what is implied when we speak of the orientation of an atom. In the early stages of atomic theory, where an atom was regarded as a hard sphere, the orientation of an atom had little meaning, but in present theories the electron clouds are not of uniform density, and are not necessarily of spherical symmetry. When we speak of the shape or orientation of an atom we have to imagine some axis with reference to which a measurement can be made. In a sense, therefore, it is meaningless to speak of the orientation of a perfectly free atom, since the orientation only acquires a meaning when an experiment has been made with reference to some axis. It is convenient to define the z axis as the direction of an applied magnetic field, and then to imagine that the field is gradually reduced to zero, so that when we speak of the orientation of a free atom we really mean an atom as defined by a very weak magnetic field. This point is important because in some cases it determines the exact mathematical treatment to be used.

We have already explained that the Indeterminancy Principle prevents us from knowing simultaneously the angular momentum of a particle revolving round a fixed point and the angle ω made with some fixed reference line by the line joining the particle to the fixed point. The theory then works out in such a way that we may know simultaneously the magnitude of the total angular momentum M, which in a hydrogen-like atom is equal to $\frac{h}{2\pi}\sqrt{l(l+1)}$, and the component of the orbital angular momentum along the z axis, which is equal to $\frac{h}{2\pi}m_l$. We may not, however, know the components in the directions of the x and y axes. The third quantum number m_l then indicates that

---

* The line for the 5f sub-group is not shown in Fig. II.1.

in a hydrogen-like atom, the component of the orbital angular momentum along the z axis is equal to $\frac{h}{2\pi}m_l$. The quantum number m_l is sometimes called the magnetic quantum number. The orbital angular momentum $\frac{h}{2\pi}\sqrt{l(l+1)}$ is associated with a magnetic* moment whose component along the z axis is equal to:

$$\frac{e}{2mc}\cdot\frac{h}{2\pi}m_l = \frac{eh}{4\pi mc}\cdot m_l.$$

This unit of magnetic moment $\frac{eh}{4\pi mc}$ is called the *Bohr Magneton*, and is equal to 0.928×10^{-20} erg/gauss.

The reader will sometimes find this subject described in conjunction with a diagram of the type shown in Fig. II.2, with a statement that the angular

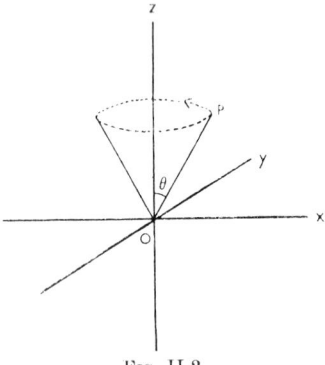

FIG. II.2.

momentum is in the direction OP, which precesses round the z axis, the angle θ being defined by $\cos\theta = \frac{m_l}{\sqrt{l(l+1)}}$. This method of representation is to be avoided, because it easily gives the impression that at any instant the angular momentum can be identified with a particular direction OP, whereas as we have explained above, we can never know simultaneously the magnitude of the total angular momentum and of the three components parallel to the x, y, and z axes, respectively. This kind of diagram has survived as the result of attempts to describe the conclusions of the new theories in terms of the old ideas of electrons in definite orbits, but these nearly always lead to inconsistencies at some point. Thus, the electron-cloud pattern of the $2p$ state with $m_l = 1$ has the form shown in Fig. II.13 (p. 77), and the density is greatest in the xy plane. The natural interpretation of this from the older viewpoint would be an orbit in the xy plane, so that the motion was confined to this plane. In this case the angular momentum would be perpendicular to the xy plane, *i.e.*, parallel to the z axis, and the total angular momentum and its

---

* The symbol m in the term $\frac{e}{2mc}$ is, of course, the mass of the electron and not the quantum number m.

The Structure of the Free Atom

component parallel to the z axis would be the same, whereas they should be in the ratio $\sqrt{l(l+1)}:m_l$. From Fig. II.13 we can see that although the electron cloud is densest in the xy plane, it is not confined to this plane, and this may be interpreted as implying that the motion of the electron is not confined to the xy plane, but includes also motion in other directions. The motion in these other directions gives an average component of the angular momentum along the z axis, although this average component is clearly less than the total angular momentum resulting from these motions.

Fig. II.2 is thus misleading in suggesting that the total angular momentum can ever be associated with a definite direction OP, but the total orbital angular momentum is to be regarded as rotating round the z axis, and positive and negative values of m_l imply rotation in opposite directions. It is this rotation which is concerned in the so-called *Zeeman Effect* in which the application of a magnetic field causes a splitting of the spectral lines.

The more detailed study of atomic spectra showed that all the facts could not be accounted for by means of the three quantum numbers n, l, and m_l. This difficulty was overcome by Uhlenbeck and Goudsmit, who assumed that the electron has a rotatory motion or spin such that the angular momentum of the spinning electron is of magnitude $\frac{h}{2\pi}\sqrt{s(s+1)}$, where $s = \frac{1}{2}$. The effect of the spin is to introduce a fourth quantum number, m_s, which can assume the two values $\pm \frac{1}{2}$. The component of the magnetic moment along the z axis is $\frac{e}{mc} \cdot \frac{h}{2\pi} \cdot m_s$. The ratio of the magnetic moment of the spinning electron to the angular momentum of the spin is thus $\frac{e}{mc}$, and is double the corresponding ratio of the orbital angular momentum and its associated moment, which, as can be seen from p. 62, is $\frac{e}{2mc}$.

It is customary to relate the total angular momentum of the electron to a new quantum number j, such that $j = l \pm \frac{1}{2}$, except that when $l = 0$, only $j = +\frac{1}{2}$ is taken. The magnitude of the total angular momentum is then $\frac{h}{2\pi}\sqrt{j(j+1)}$, and the component in the direction of the z axis is $\frac{m_j h}{2\pi}$, where the quantum number m_j takes the values of $\pm \frac{1}{2}, \pm \frac{3}{2}, \pm \frac{5}{2}, \ldots \pm j$, so that m is always a half integer, whereas in the simple theory it is a whole number.

The total angular momentum, j, is thus compounded of the orbital angular momentum and the spin momentum, and the condition of affairs is often represented by a diagram of the type shown in Fig. II.3, where for abbreviation s^*, l^*, and j^* are written for $\sqrt{s(s+1)}$, $\sqrt{l(l+1)}$, and $\sqrt{j(j+1)}$, respectively, and the angle θ, made by j^* with the z axis, is given by $\cos \theta = m_j/\sqrt{j(j+1)}$. When the atom is placed in a weak magnetic field, j^* precesses round the z axis. At the same time s^* and l^* precess round their resultant j^*; this precession is due to the interaction between the spin and the "orbit," and takes place in the absence of an external field. In a weak magnetic field the rate of precession of j^* round the z axis is very much slower than the precessions of s^* and l^* round j^*. When the field becomes so strong that the frequencies of these precessions are of the same order the process becomes more complicated, but in still stronger fields it becomes simple again, and the so-called *Paschen-Back effect* is observed, but these effects lie outside the present work. The use of this kind of figure is open to the same objections as was Fig. II.2, and the Heisenberg Principle again prevents us from ever knowing the direction of j^* at a particular instant, although we may speak of a rotation of the angular momentum about the z axis. Since j^* is never completely determined, it is even more difficult to visualize the precession of s^* and l^* round j^*, and one can only say that the axes of the cones of s^* and l^* are rotating about the z axis, but must never be regarded as being at a particular place.

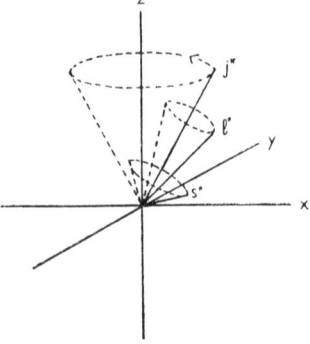

FIG. II.3.

The above remarks refer to the quantum numbers used to describe the state of a single electron in a hydrogen-like atom. Apart from these there are symbols describing the state of the atom as a whole and which are of especial importance in spectroscopy. Thus, the ground state of the boron atom may be written $1s^2 2s^2 2p^1\ ^2P_{\frac{1}{2}}$, or simply $^2P_{\frac{1}{2}}$, to indicate how the angular momenta of all the electrons are coupled together. The total spin quantum number S is given by a superscript of value $(2S+1)$ which is numerically equal to the number of possible values of M_S, the component in the z-direction, of the total spin. The total *orbital* angular momentum quantum number L is given by the symbols S for $L = 0$, P for $L = 1$, D for $L = 2$, &c.† A given L corresponds to possible values of M_L, the z-component of the total orbital angular momentum, equal to $-L, -(L-1) \ldots 0 \ldots +L$, and the allowed M_L values are given by Σm_l over all the electrons. For the $1s^2$ and $2s^2$ groups in boron Σm_s and Σm_l can only be zero, so the whole atom must have $M_s = \Sigma m_s = \pm \frac{1}{2}$ and $M_L = \Sigma m_l = -1$ or 0 or $+1$, only the one $2p$ electron contributing. Various possible combinations of L and S to give J, the *total* angular momentum quantum number, are still possible, however, and that given by $J = L - S = \frac{1}{2}$ has lowest energy for boron. The J value appears as the subscript in $^2P_{\frac{1}{2}}$.

In the carbon atom $1s^2 2s^2 2p^2$, a wider range of possibilities exists since each p electron has three possible values of m_l and two of m_s as follows:

n	l	m_l	m_s
1	0	0	$+\frac{1}{2}$
1	0	0	$-\frac{1}{2}$
2	0	0	$+\frac{1}{2}$
2	0	0	$-\frac{1}{2}$
2	1	$+1$ or 0 or -1	$+\frac{1}{2}$ or $-\frac{1}{2}$
2	1	$+1$ or 0 or -1	$+\frac{1}{2}$ or $-\frac{1}{2}$

† As for l, the orbital angular momentum of the atom is given quantum mechanically as $h/2\pi\sqrt{L(L+1)}$ but it can take as its component in a specified direction only the values $\frac{h}{2\pi}L$, $\frac{h}{2\pi}(L-1)$, &c.

The Structure of the Free Atom

provided only that all four quantum numbers are not the same for the last two electrons. The possible values of M_L are now all the 9 possible ways of adding m_{l_1} and m_{l_2} subject to all allowed combinations of m_{s_1} and m_{s_2}.

$m_l = +1$	↑↓			↑	↑		↑	↑	
$m_l = 0$		↑↓		↓		↑	↑		↑
$m_l = -1$			↑↓		↓	↓		↑	↑
$M_L = \Sigma m_l$	+2	0	−2	+1	0	−1	+1	0	−1
		$S = 0$						$S = 1$	

($+\tfrac{1}{2}$ is represented by ↑ in the table and $-\tfrac{1}{2}$ by ↓).

States of different spin are separated in energy and the ground state for carbon is that with $S = 1$, the M_L values for which ($+1, 0$, and -1) form the components of a P state with $L = 1$; this is denoted ^3P (read triplet P) meaning that $(2S + 1) = 3$ and $L = 1$. J can then be 2, 1, or 0 but these have different energy, the $J = 0$ state lying lowest.

Since all components M_L of a given L must have the same spin the group required for $L = 2$

$$M_L = +2, +1, 0, -1, -2$$

must make use of $M_L = +1, 0$, and -1 groupings of zero S and is denoted $^1D_2$. (One of the nine possible values of M_L remains and can be seen to be a $^1S_0$ configuration.)

2. Representation of Electron States.

In the preceding section we have seen how the electrons in free atoms are divided into groups and sub-groups, each electron being characterized by a definite energy. We have next to consider how the electrons are distributed in space round the nuclei of the different atoms. The problem of the hydrogen atom is naturally the most simple, and since the potential energy of an electron of charge $-e$, moving in the field of a nucleus of charge $+e$, is $-e^2/r$ (see p. 15), the problem of the motion of an electron of total energy E round the nucleus is equivalent to an optical problem involving the passage of light in a spherical bowl, in which the refractive index varies in such a way that, as the light travels outwards from the centre of the sphere, it tends to be bent back again, and does not escape, but remains travelling in a restricted volume. The mathematical treatment is then made more easy by transferring from rectangular to polar co-ordinates, in which the position of a point is described in terms of its distance (r) from the origin, and the two angles θ and ϕ, as shown in Fig. II.4. It is then found that the wave function can be split up into a product of three functions $R(r)$, $\bar{\Theta}(\theta)$, $\Phi(\phi)$, which involve only r, θ, and ϕ, respectively. That is to say, $R(r)$ is a function only of r, and does not involve θ or ϕ, and similarly for $\bar{\Theta}(\theta)$ and $\Phi(\phi)$. The probability of finding the electron in any element of volume ΔV is then proportional to the value of:

$$\Psi\Psi^*\Delta V = [R(r)R^*(r)] \times [\bar{\Theta}(\theta)\bar{\Theta}^*(\theta)] \times [\Phi(\phi)\Phi^*(\phi)]\Delta V$$

at the place considered, where $R^*(r)$, $\bar{\Theta}^*(\theta)$, and $\Phi^*(\phi)$ are the conjugate complexes of $R(r)$, $\bar{\Theta}(\theta)$, and $\Phi(\phi)$, respectively. This means that the

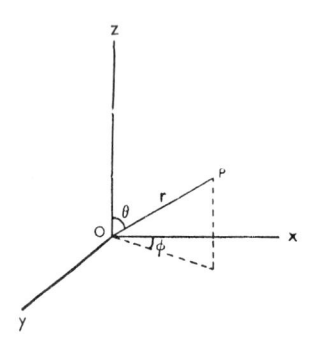

FIG. II.4. To illustrate polar co-ordinates. The position of the point P is given by the length (r) of OP, the angle θ made by OP with the z axis, and the angle ϕ made by the projection of OP on the xy plane with the x axis.

electron density at any point in the atom can be expressed as the product of three terms, one of which involves only the distance (r) from the nucleus, whilst the other two involve only the angles θ and ϕ, respectively. We have already explained (p. 37) how it is necessary to choose acceptable solutions of the wave equation, and the mathematical treatment then shows that the equation for $\Phi(\phi)$ yields acceptable solutions only if a certain parameter m is equal to 0, ± 1, ± 2, ... &c. When m has been given a particular value, it is found that the equation for $\bar{\Theta}(\theta)$ yields acceptable solutions only by the introduction of a whole number l, which enters in the form $l(l+1)$, and may have any value (numerically) greater than m, or equal to m. Finally, on introducing values for m and l, it is found that the equation for $R(r)$ yields acceptable solutions only by the introduction of a third whole number n, which must not be less than $(l+1)$. In this way the solution of the wave equation for the hydrogen atom leads to the introduction of the three whole numbers, n, l, and m, and it can be shown that these are equivalent to the quantum numbers n, l, and m_l described in the preceding chapter. This is an example of the fact mentioned in Section I.5 that in wave mechanics,

The Structure of the Free Atom

quantum numbers appear, not as arbitrary assumptions imposed on the system, but as conditions which must be satisfied if satisfactory solutions are to be obtained.

In the case of atoms containing more than one electron, the problem becomes extremely complicated, but approximate solutions have been obtained in many cases by the methods of the so-called " self-consistent field." We may illustrate the principle by considering the helium atom, which contains two electrons. For this we may assume that the nucleus of charge $+2e$, and one of the electrons (No. 1) give rise to a particular field, and we may investigate the behaviour of the second electron (No. 2) in this field. The resulting wave functions for electron No. 2 may then be used to calculate the field in which electron No. 1 would move, and then the field resulting from this motion of electron No. 1 may be compared with that originally assumed at the beginning of the calculation, and if there is not a good agreement the whole process must be repeated until the results are self-consistent. Atoms containing more than 50 electrons have been dealt with approximately by these methods, due to Hartree, and some of the results will be referred to later. The approximations involved are considerable, because the methods used do not properly take into account the way in which the electrons influence each other's motion. This point is discussed by Slater,† and is connected with the effects described on p. 83.

In the case of the hydrogen atom the value of $\Psi\Psi^*\Delta V$ gives the probability of finding the one electron in the volume ΔV at the place considered, and in the case of other atoms the total electron density is given by the sum of the values for the individual electrons. As explained on p. 43, the picture of the atom which we obtain is that of an electron cloud or probability pattern in which increasing density of the cloud implies a greater probability of finding an electron at the place concerned. It is customary to express the form of the probability patterns in two ways. The first of these involves the so-called radial electron density $U(r)$, which is defined so that $U(r)dr$ is the number of electrons contained in a spherical shell of radius r and thickness dr. Curves of this kind show the probability of finding an electron at different distances from the nucleus. Such curves are often very useful provided that the definition of $U(r)$ is borne in mind, but they are easily misleading, and must never be taken to indicate the density of the electron cloud of the atom. Since the surface of a sphere of radius r is equal to $4\pi r^2$, a given number of electrons per unit volume implies a value of $U(r)$ which becomes continually smaller as r decreases, and consequently

† J. C. Slater, " Quantum Theory of Matter," p. 133. **1951:** New York and London (McGraw-Hill).

68 Atomic Theory

$U(r)$ is very small near the centre of the atom, even though the electron density may be very great.

The second method of expressing the results is in terms of ρ, the charge per unit volume. In atoms whose electron clouds are spherically symmetrical, $U(r)$ and ρ are connected by the simple relation:

$$\rho = \frac{U(r)}{4\pi r^2}$$

and for some purposes it is therefore more convenient to express results in terms of $4\pi\rho$, instead of in terms of ρ itself. In these spherically symmetrical atoms the electron-cloud density can be expressed by a single curve showing the variation of ρ with r.

Thus a curve of the form shown in Fig. II.5 represents a spherically

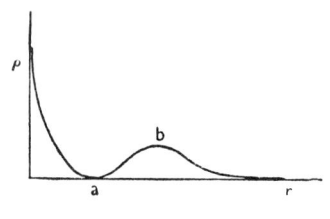

Fig. II.5.

symmetrical ball of negative electricity, the density of which diminishes to a value of zero at the distance $r = a$ from the nucleus, rises to a maximum at a distance $r = b$, and then gradually falls away to zero. The point a in Fig. II.5 thus corresponds to a spherical node, *i.e.*, to a point at which the amplitude is zero. The existence of a spherical node implies that no electrons are present at the places concerned. A spherical node thus implies that the atom contains a spherical shell in which there is no chance of finding an electron, although an electron may be found both inside and outside the spherical node.

In atoms whose electron clouds are not spherically symmetrical, the relations between ρ and r are different in different directions. It is therefore very difficult to obtain a satisfactory picture of the electron-cloud density, since the electron cloud itself occupies three dimensions, and a fourth dimension is required in order to show the electron density. To surmount this difficulty, use is made of the fact that the wave equation can be split up into the three components $R(r)$, $\bar{\theta}(\theta)$, and $(\Phi\phi)$, which involve only r, θ, and ϕ, respectively, the complete electron density ρ being proportional to the product

$$[R(r)R^*(r)][\bar{\theta}(\theta)\bar{\theta}^*(\theta)][\Phi(\phi)\Phi^*(\phi)].$$

The Structure of the Free Atom

The first term, which may be called the radial probability factor, or the radial density factor, depends on r only, and is therefore spherically symmetrical, and can be expressed in the form of a graph such as that shown in Fig. II.6(a). The values of the second term $\bar{\theta}(\theta)\bar{\theta}^*(\theta)$ can

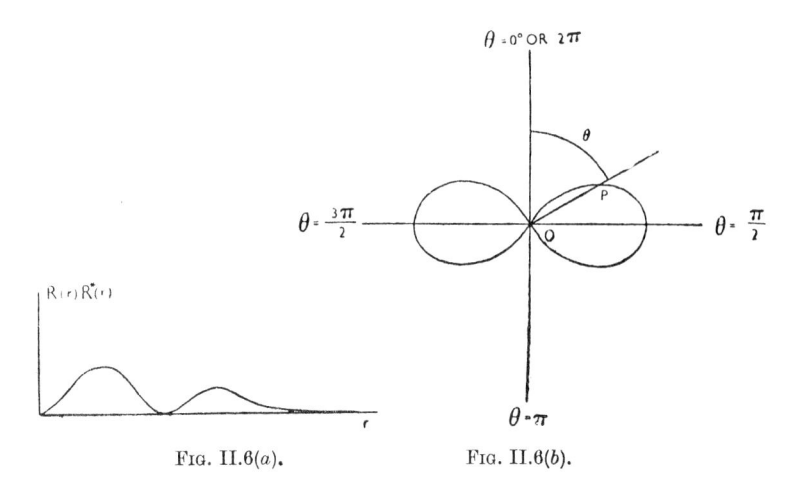

Fig. II.6(a).   Fig. II.6(b).

then be expressed in the form of so-called Polar Diagrams of such a nature that if a straight line is drawn from the origin at an angle θ, and cuts the curve of the diagram at the point P, then the length of OP is proportional to the value of $[\bar{\theta}(\theta)\bar{\theta}^*(\theta)]$ for the value of θ concerned. A diagram of this kind is shown in Fig. II.6(b), and for this particular example it will be seen that OP has maxima at $\theta = \pm \pi/2$, and is equal to zero for $\theta = 0$ or $\theta = \pi$. Since the total electron-cloud density is proportional to $OP \times R(r)R^*(r) \times \Phi(\phi)\Phi^*(\phi)$, the polar diagram of Fig. II.6(b) means that the electron-cloud density is zero along the z axis, because this axis corresponds with $\theta = 0$. A similar 2-dimensional diagram can be drawn to show the variation of $\Phi(\phi)\Phi^*(\phi)$ with ϕ, and the actual electron density at a point with co-ordinates (r, θ, ϕ) is then obtained by multiplying the value of the radial probability factor at the distance r by $OP_\theta \times OP_\phi$. The student has thus to learn to visualize the electron-cloud density in the space surrounding the nucleus as the product of three terms, involving r, θ, and ϕ, respectively. This visualization is difficult, and the process may be made easier by combining the two 2-dimensional diagrams referring to θ and ϕ into a single 3-dimensional polar diagram such that if a straight line is drawn from the origin at an angle (θ, ϕ) and cuts the surface of the diagram at the point P, the length of OP is proportional to the value of $[\bar{\theta}(\theta)\bar{\theta}^*(\theta) \times$

$\Phi(\phi)\Phi^*(\phi)$] for the direction concerned. Fig. II.7 shows a three-dimensional diagram of this kind for the case in which the variation with θ is similar to that of Fig. II.6(b), and in which the value of $\Phi(\phi)\Phi^*(\phi)$ is a constant independent of ϕ. Since in this particular example we have assumed $\Phi(\phi)\Phi^*(\phi)$ to be a constant, the electron density will be independent of ϕ and the figure must therefore be symmetrical about the z axis, and Fig. II.7 may be regarded as Fig. II.6(b)

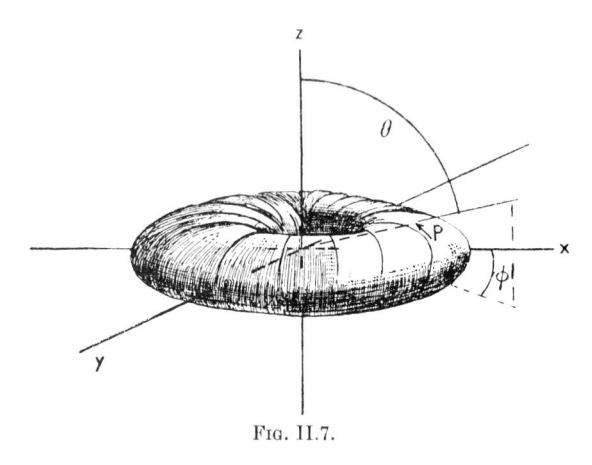

Fig. II.7.

multiplied by an appropriate factor and revolved about the z axis. Figures such as Fig. II.7 do not express the " shape of the atom," since the values of OP from the three-dimensional diagram of Fig. II.7 have to be multiplied by the radial probability factor in order to give the complete electron density. That is to say, the electron-cloud density at a given point is proportional to $OP \times R(r)R^*(r)$, at the point concerned, so that the values of OP for the different directions in Fig. II.7 have to be multiplied by the value of $R(r)R^*(r)$ from a diagram such as that of Fig. II.6(a) in order to indicate the total electron-cloud density. The electron-cloud pattern indicated by a combination of Figs. II.6(a) and II.7, would have a spherical node (i.e., a zero electron density) of radius $r = a$, because the radial curve of Fig. II.6(a) has a zero value at $r = a$. At other distances the electron-cloud density would be greatest in the xy plane, because the value of OP in Fig. II.7 is greatest in this plane, and the electron-cloud density at all distances would be zero along the z axis, because, in Fig. II.7, $OP = 0$ in this direction. The student must learn how to visualize the product of the radial factor and the polar (directional) factors in order to appreciate the " shape " of the electron cloud, which is naturally often very complicated. If the

The Structure of the Free Atom

radial factor indicates the presence of one or more spherical nodes, the inside of the atom will contain spherical shells on which the electron-cloud density is zero. Generally speaking, however, the nodes of the radial probability factor concern the interior rather than the exterior of the atom, and at the outside of the atom the radial factor is diminishing continuously in a way similar to that shown in Fig. II.6(a). When, therefore, we are dealing with the outside of the atom, diagrams such as those of Fig. II.7 do give a general indication of the shape of the atom in the state concerned. At a given distance from the centre of the atom, the electron density is proportional to the lengths of OP in the different directions of Fig. II.7, and hence the surfaces of constant electron density will project furthest in the directions for which OP in Fig. II.7 has its greatest values, and the outside of the atom may be regarded as having the shape of Fig. II.7 with the sharp bounding surface replaced by a diffuse electron cloud, the density of which diminishes as we proceed outwards. Alternatively, we could draw a series of surfaces or contours such that the volumes within successive contours contained, say, 50, 60, 70, 80, and 90% of the total electron density and, in this way, obtain a picture of both the inside and outside of the electron cloud.

It will be seen, therefore, that according to the purpose in hand, an electron state may be represented in different ways, of which a mathematical calculation for Ψ is one extreme. For spherically symmetrical atoms the state is for many purposes satisfactorily represented by a plot of Ψ or $\Psi\Psi^*$ against the distance from the nucleus, while for other atoms plots of RR^*, $\Theta\Theta^*$, and $\Phi\Phi^*$, are required. For some purposes it is sufficient to give a single radial probability function for either spherical or non-spherical atoms. Alternatively, we may represent the electron cloud by a contour diagram or, for some simple purposes, by a boundary surface containing, say, 90% of the electron cloud; this last method is sometimes used in attempts to describe the " shapes " of the electron clouds of molecules.

3. The s, p, and d States.

In the preceding section we have seen how the electron-cloud density in the space surrounding the nucleus of an atom is obtained as the product of the radial factor $R(r)R^*(r)$ at the distance concerned, and the length OP in the direction concerned in the 3-dimensional polar diagram, this diagram being a combination of the two 2-dimensional polar diagrams for the angles θ and ϕ. These polar diagrams indicate the variation of the electron-cloud density with direction, and were first calculated by means of the Schrödinger theory. This did not include the effects of electron spin, and the theory was later elaborated by Dirac to include the effects of spin. For the present purpose we shall concern ourselves almost exclusively with the Schrödinger diagrams, because in almost all cases of metallic and other combination, the perturbing forces are so much greater than the effects of spin that the Schrödinger theory is sufficient.

We may consider first the *s*-states, and for these the theories of both

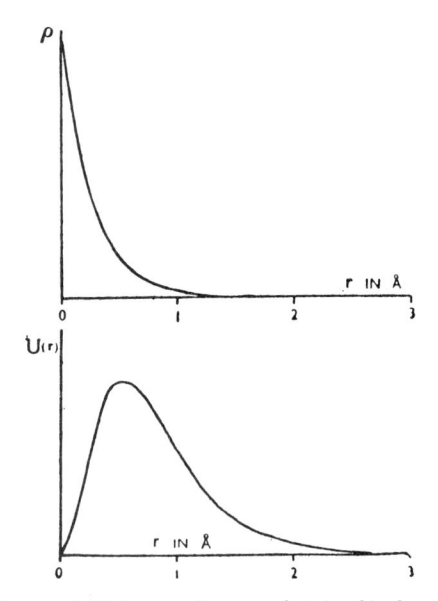

FIG. II.8. ρ and $U(r)$ curves for normal state of hydrogen atom.

Schrödinger and Dirac require the electron clouds to be spherically symmetrical, and in the Schrödinger theory they are of such a nature that if n is the principal quantum number, the electron cloud has $(n-1)$ spherical nodes. The values of $\bar{\Theta}(\theta)\bar{\Theta}^*(\theta)$ and $\Phi(\phi)\Phi^*(\phi)$ are thus both constants, and the electron-cloud density can be represented by a single curve showing the variation of ρ, or of $U(r)$, as a function of the distance from the nucleus.

Fig. II.8 shows the relation between ρ and r for the normal state, $(1s)^1$, of the hydrogen atom, and the corresponding relation between $U(r)$ and r. The picture of the atom represented by these curves is that of a nucleus surrounded by a spherically symmetrical ball of negative electricity, the density of which fades away rapidly at distances

The Structure of the Free Atom 73

greater than about 2 A. Fig. II.10(a) (Plate I) is an attempt to show this photographically by a method in which the brightness of illumination is proportional to the electron-cloud density.* This picture of an atom is clearly very different from that of the older theories of electrons in orbits, since these indicated an atom with the characteristics of a flat disc or ring. The older theories did, however, express part of the truth, for the radius of the first Bohr orbit of the hydrogen atom was 0·529 A, and as will be seen from Fig. II.8, the $U(r)$ curve shows a maximum at this point, so that the electron may be said to spend more of its time at a distance from the nucleus equal to that of the first Bohr orbit than at any other distance. In general, however, attempts to interpret electron-density figures in terms of orbits are dangerous, since an orbital model suitable for one purpose may not serve another.

Figs. II.9(a) and (b) show the ρ and $U(r)$ curves for the hydrogen atom in the $(3s)^1$ state. Since $n = 3$, the pattern of the $(3s)^1$ state is characterized by $3 - 1 = 2$ spherical nodes, and the picture presented by the curves of these two figures is that of a spherically symmetrical ball of negative electricity surrounded by two spherical shells, the outer of which is very diffuse. These diagrams show the danger of attempting to visualize the electron density from the $U(r)$ curve. In the $U(r)$ curve, the outermost maximum on the curve has the highest value of $U(r)$, although the value of the electron density ρ is comparatively small. This is merely the result of the fact that in a spherically symmetrical atom $U(r) = 4\pi r^2 \rho$, and consequently ρ is multiplied by a term $(4\pi r^2)$ which becomes continually larger as we proceed outwards. Fig. II.10(c) (Plate I) is an attempt to illustrate the electron-cloud pattern photographically by a method similar to that of Fig. II.10(a), but this is really more like a cross-section through the spherically symmetrical pattern, and the pattern is, of course, 3-dimensional with the nodes occupying spherical shells. The pattern for the $(3s)^1$ state occupies a larger region of space than that for the $(1s)^1$ state, showing that increasing quantum number produces a larger atom, just as in the older theories increasing quantum number resulted in a larger orbit. It must, however, be remembered that the whole pattern for the $(3s)^1$ state still represents only one single electron, and consequently the electron density must on the whole be much less than in the pattern for the $(1s)^1$ state. As the description indicates, the scale of Fig. II.10(c) is $\frac{1}{10}$th of that of Fig. II.10(a), whilst the intensity of illumination of Fig. II.10(c) has been multiplied by 500 in order to prevent the outermost parts of the pattern from being too faint to be seen. In the same way the vertical scales in

---

* The actual electron cloud is, of course, 3-dimensional.

Figs. II.9(a) and (b) have been greatly magnified as compared with the vertical scales in Fig. II.8. Fig. II.10(b) shows the corresponding pattern for the $(2s)^1$ state of the hydrogen atom, and this possesses

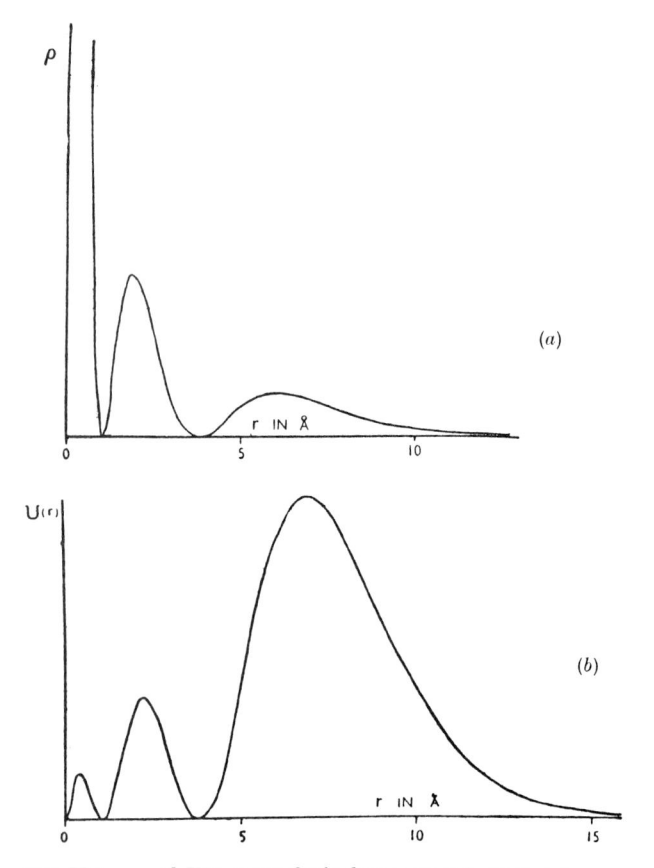

Fig. II.9. ρ and $U(r)$ curves for hydrogen atom in $(3s)^1$ state.

$2 - 1 = 1$ spherical node; the scale and intensity of reproduction of Fig. II.10(b) are intermediate between those of Figs. II.10(a) and (c).

Electron-cloud patterns of the type shown in Fig. II.9 have been objected to on the grounds that the existence of spherical nodes implies a definite probability of finding the electron in each region between the nodes, but no possibility of the electron travelling from one region to another. This criticism is, however, quite invalid, for as we have explained before (p. 47) the Indeterminancy Principle shows that it

The Structure of the Free Atom 75

is meaningless to speak of the electron travelling about within an electron-cloud pattern representing an exact value of the energy. Since the energy is specified exactly, the time is completely indeterminate, and all that can be stated is the relative probability of finding the electron in different places in the space surrounding the atom.

For the radial part of the wave functions of the s-states, the curves of the Schrödinger and Dirac Theories are almost indistinguishable, except that the absolute spherical nodes of the Schrödinger theory are replaced by spherical shells in which the electron-cloud density, although very small, retains a finite value.

In atoms other than hydrogen, the symmetry characteristics of s-states are still retained, although the details of the radial probability curves are profoundly influenced by the different charges on the nuclei, and the screening effects of the different shells of electrons. An electron in an s-state always has a spherically symmetrical electron cloud with $(n - 1)$ spherical nodes, where n is the principal quantum number. In the simple theory where electron correlation is ignored, quantum groups of 8, 18, or 32 electrons possess electron clouds with spherical symmetry. In these theories atoms of the alkali metals, and also of copper, silver, and gold in their normal states are spherically symmetrical,* since they are regarded as consisting of the spherically symmetrical electron cloud of the valency electron of an s state, superimposed on the spherically symmetrical atomic core with its group of 8, 18, or 32 electrons. This simple picture is not really true, because the different electrons interfere with each other's movements. This interference results firstly from the electrostatic or Coulomb repulsion between electrons, which prevents them from approaching very close to one another. Apart from this, the Pauli Exclusion Principle (p. 140) necessitates the use of wave functions which prevent two electrons of parallel spins from occupying the same region of space. As the result of this, the true " shape " of the electron cloud of a group of 8 or 18 electrons is not really known. The assumption of spherical symmetry for groups of 8, 18, or 32 electrons is frequently made in order to simplify calculation, and is often a reasonable assumption.

When we deal with electrons in metals, it frequently happens that an electron may be regarded as in a state derived from the s-state of the free atom. In such cases the electron-cloud pattern of an s-electron near to an ion has the same general characteristics as in the corresponding free atom, although the details are profoundly different at places

---

* As explained on p. 61, it is meaningless to speak of the shape of an electron cloud unless axes are defined, and this is done by reference to the direction of a very weak magnetic field.

where the electron is influenced to comparable degrees by two or more ions. Thus, in copper most of the valency electrons exist in states derived from the 4s state of the free atom, and the corresponding electron-cloud patterns *in the solid* have three nodes surrounding each ion. At distances very near to a nucleus, the ρ(r) curves for the 4s electrons *in the solid* closely resemble those for the free atom, since here the electron is essentially under the influence of the one nucleus concerned. But at a point mid-way between two ions in the solid, the electron is attracted equally in two directions, and the wave-function and ρ(r) curves are quite different from those of the free atom. Similar considerations apply to electrons in states derived from the *p*- or *d*-states of free atoms.

We may next consider the electron-cloud pattern of the hydrogen atom in *p*-states, *i.e.*, for states in which $l = 1$. For each value of the principal quantum number n there are three *p*-states, corresponding with $m = 0$, and $m = \pm 1$, respectively. In the Schrödinger theory, the radial probability factor $R(r)R^*(r)$ for a *p*-state is different for each value of the principal quantum number n, but is the same for each of the three states ($m = 0$ and ± 1) of a given n. These radial probability factors are characterized by the existence of $(n - 2)$ spherical nodes, and also possess a node at the centre of the atom. Fig. II.11, for

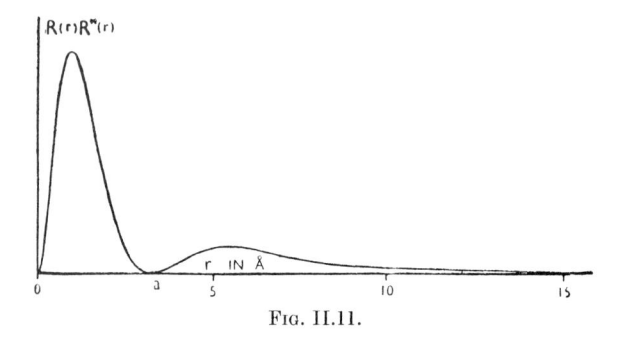

Fig. II.11.

example, shows the radial probability factor for the $(3p)^1$ state of the hydrogen atom, and the point a corresponds with a spherical node. The probability factor $\bar{\Theta}(\theta)\bar{\Theta}^*(\theta)$ involves the angle θ, and this involves the orientation of the atom. As explained before (p. 61), we cannot really speak of the orientation of a perfectly free atom, and the policy adopted is to imagine that a magnetic field acts along the z axis, and that when the field is reduced to zero this axis persists, and that the same mathematical treatment is applicable. The theory then shows that the φ

The Structure of the Free Atom 77

probability factor $[\Phi(\phi)\Phi^*(\phi)]$ is a constant, and that the electron clouds of the p-states are symmetrical about the z axis.

We may consider first the $2p$ state for which, since $n - 1 = 1$, there are no spherical nodes, and the radial probability factor is of the form

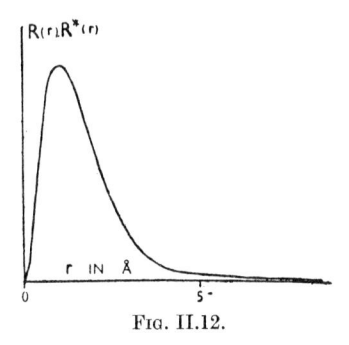

FIG. II.12.

shown in Fig. II.12. There are then the three cases $m = 0$ and $m = \pm 1$, and the complete 3-dimensional polar diagrams are as shown in Fig. II.13, and possess axial symmetry. The electron density at any point (r, θ, ϕ) is then obtained by multiplying the value of the radial probability factor at the distance r by the value of the polar factor for the direction concerned in Fig. II.13. As previously explained, the outermost parts

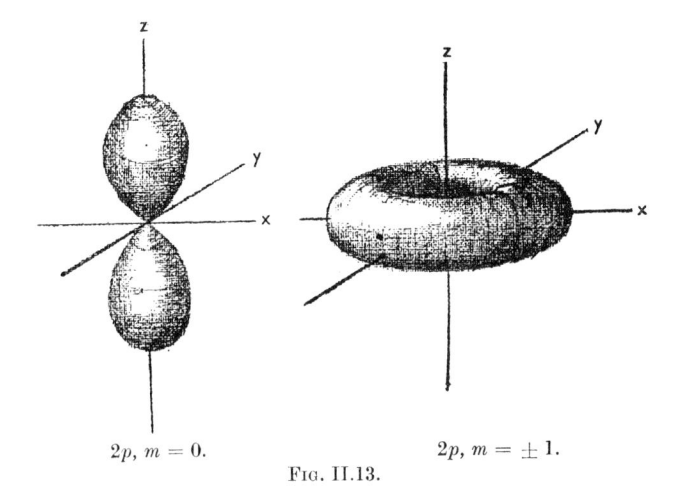

$2p, m = 0.$ $2p, m = \pm 1.$
FIG. II.13.

of the atom, where the radial probability curve is dying away gradually, may be regarded as having the " shape " of Fig. II.13, although the inner parts will be affected by the characteristics of the $(R(r)R^*(r))$ curve, and cannot be visualized simply. Fig. II.14 (Plate II) is an attempt to show the ($2p$, $m = 0$) state photographically, but this is more like a cross-section than a true 3-dimensional pattern.

When $n > 2$, the electron clouds of the p-states show $(n - 2)$ spherical nodes, and the radial probability factors analogous to Fig. II.12 show $(n - 2)$ nodal points as well as the node at $r = 0$. The θ and ϕ probability factors are still given by the same polar diagrams of Fig. II.13, and the electron cloud for, say, the $3p$ state may be looked on as more or less of the same nature as that of the $2p$ state, but with a spherical node inside the atom. Fig. II.15 (Plate II) is an attempt to show this for the ($3p$, $m = 0$) state, but the photograph is not very successful because the node at the centre is not clearly revealed. It must again be emphasized that the whole pattern represents only one electron.

For free atoms, the electron-cloud pattern for the p-states with $m = +1$ and $m = -1$, are identical, and differ only in that the Φ functions are $e^{+i\phi}$ and $e^{-i\phi}$, respectively. These differ in the sense of rotation, *i.e.*, the direction of the angular momentum, but each is perfectly symmetrical about the z axis, and the picture presented is that in a weak magnetic field the orbital angular momentum vector precesses round the z axis, this precession being the cause of the axial symmetry of Fig. II.13. Just as the z axis has to be defined with reference to something external to the atom (*e.g.*, the magnetic field), so the x and y axes can only be given a real physical meaning if the atom is submitted to the action of one or more additional fields, as, for example, by placing the atom in a crystal. Under these conditions, the electron clouds of the three p-states are mutually perpendicular, and the angular probability factors are of the form shown in Fig. II.16. In this figure the two right-hand patterns are the real and imaginary parts of progressive patterns similar to the one shown on the right hand of Fig. II.13. The fact that the electron clouds of the three p-states are mutually perpendicular is of great importance in understanding the structures of some molecules and crystals.

It will be noted that in Fig. II.13 the nodes of the one pattern are in the direction of the maximum value of the other, and the two combine to give an electron cloud of spherical symmetry. In the same way the three patterns of Fig. II.16 add together to form a cloud of spherical symmetry. In the simple theory, where electron correlation is ignored, this means that an atom such as that of nitrogen, with one electron in each of the three p-states, is spherically symmetrical, whilst an $(ns)^2(np)^6$ group of electrons is also spherically symmetrical. As explained on

The Structure of the Free Atom 79

p. 75, this is not really true, and the effects of electron correlation must be taken into account.

We shall not deal here with the Dirac theory of the p-states, since, as explained above, the differences between the Dirac and Schrödinger

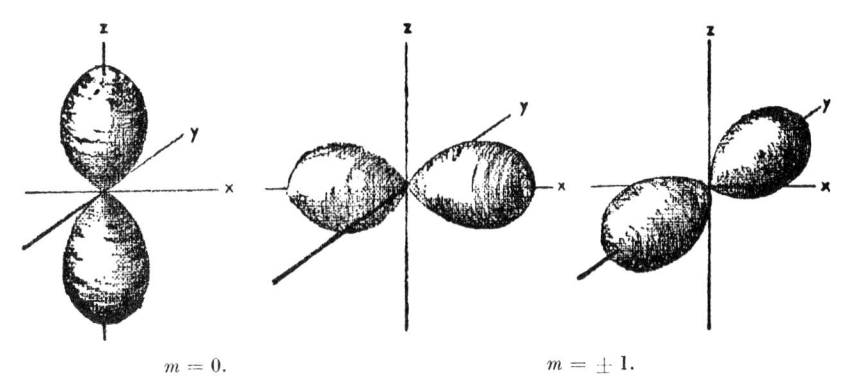

$m = 0.$ $m = \pm 1.$

Fig. II.16.

theories only concern the atom when the external fields are so weak that they are small compared with the effects of the electron spin. It should be noted, however, that the Dirac theory sometimes leads to a greater symmetry than the Schrödinger theory. Thus, in the Dirac theory one of the p-states is spherically symmetrical, and this applies not only to the hydrogen atom, but also to atoms of boron, aluminium, gallium, indium, and thallium, and atoms of these elements are spherically symmetrical in the free state when the orientation is defined by a field weak compared to the spin effects.† In slightly stronger fields the Schrödinger theory is sufficient.

The preceding description of the p-states has shown how the electron-cloud density is obtained from the product of the radial probability factor $R(r)R^*(r)$, and the angular factors $\bar{\theta}(\theta)\bar{\theta}^*(\theta)$, $\Phi(\phi)\Phi^*(\phi)$, and we shall therefore only give a very brief description of the d-states. In the Schrödinger theory of the hydrogen atom, the radial probability factor of the d-states is characterized by a node at the centre of the atom, and by $(n-3)$ spherical nodes.‡ The 3-dimensional polar diagrams of the three states of the d-electron ($m = 0, 1, 2$) are shown in Fig. II.17. For

† For further details see White's " Introduction to Atomic Spectra," p. 142.
‡ It will be noted that the principal quantum number n has the meaning that $(n - l - 1)$ is the number of nodes of the radial probability factor excluding that at the centre of the atom.

the free atom as defined by a weak magnetic field along the z axis, the electron clouds have axial symmetry.

The state ($l = 2$, $m = 0$) is characterized by nodes at $\theta = 52°$ and $\theta = 128°$, and these nodes form the surfaces of two cones of half

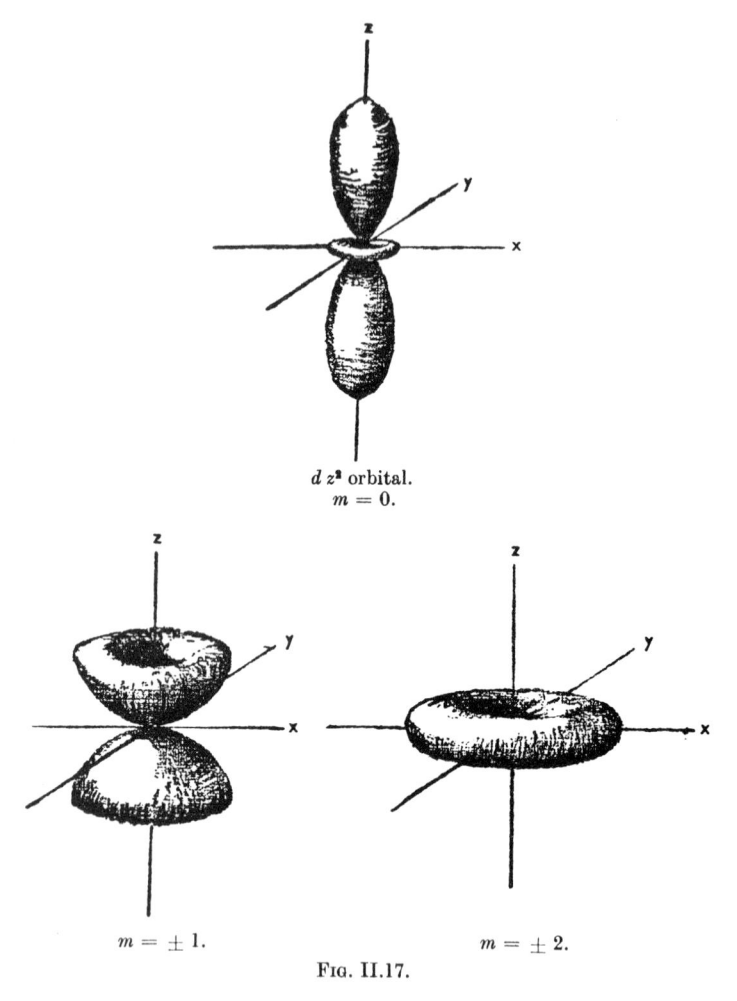

$d\, z^2$ orbital.
$m = 0$.

$m = \pm 1$.

$m = \pm 2$.

Fig. II.17.

angle 52° about the z axis. The pattern has a maximum value for $\theta = 0$, i.e., along the z axis, and there is a second, but much smaller maximum in the xy plane.

The state ($l = 2$, $m = 1$) in the free atom is characterized by

PLATE I.

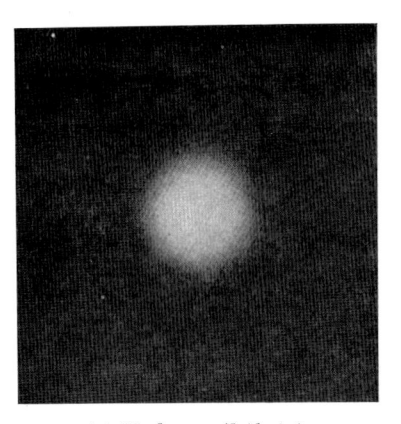

(a) Hydrogen $(1s)^1$ state.

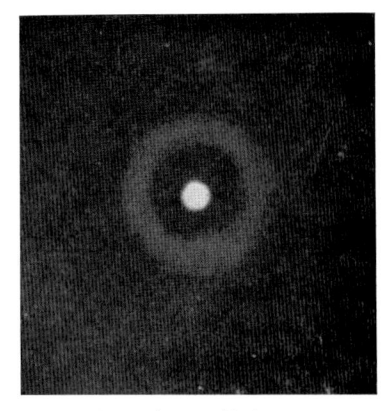

(b) Hydrogen $(2s)^1$ state.

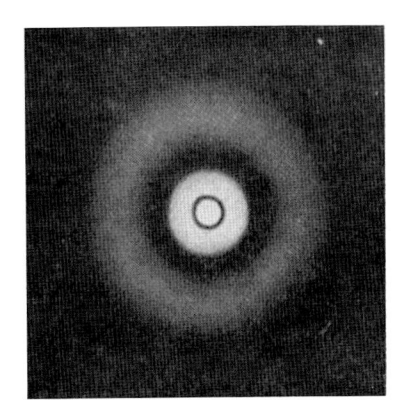

(c) Hydrogen $(3s)^1$ state.

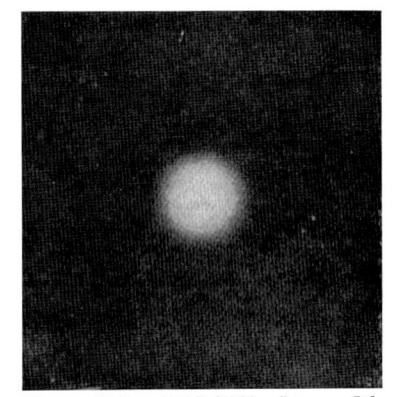

[*Courtesy McGraw-Hill Publishing Company, Ltd.*]

(d) Helium $(1s)^2$ state.

FIG. II.10. Distribution of the function $\psi\psi^*$ for certain states of hydrogen and helium.

Fig. II.10(a) is an attempt to show the electron-cloud density of the hydrogen atom in the normal $(1s)^1$ state by a method in which the intensity of illumination is proportional to the electron-cloud density. Figs. II.10(b) and (c) show the $2s$ and $3s$ states, and the scales on which these are shown are one-fifth and one-tenth of that of Fig. II.10(a); from this it will be realized that the increase of the quantum number results in an enormously increased extent of the electron cloud. In the higher-quantum states, the electron cloud extends over a much larger region of space, but since the probability refers to one electron, the density must be very much smaller, and the intensities of Figs. II.10(b) and (c) have therefore been increased by factors of 50 and 500 respectively, as compared with that of Fig. II.10(a). If this had not been done, the outer rings of Figs. II.10(b) and (c) would be too faint to be seen. Fig. II.10(d) is a similar attempt to show the electron-cloud density of the helium atom in the normal $(1s)^2$ state. The scale of this is about the same as that of Fig. II.10(a), but the intensity of the centre of Fig. II.10(d) should be about 100 times as great as that at the centre of Fig. II.10(a). The increased nuclear charge in helium has therefore drawn the electron cloud towards the nucleus, whilst the edge of the cloud is less diffuse.

PLATE II.

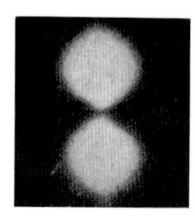

2p, m = 0.
FIG. II.14.

[Courtesy Physical Review.
3p, m = 0.
FIG. II.15.

FIGS. II.14 and II.15. Photographs of the electron cloud for various states of the hydrogen-like atoms.

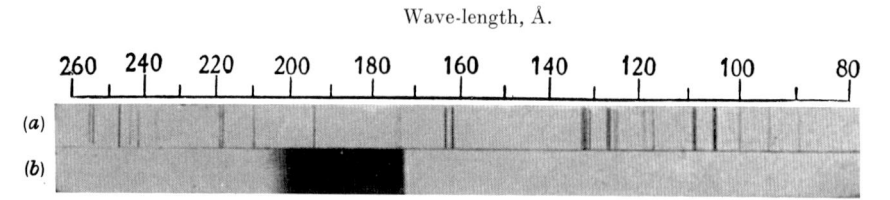

Wave-length, Å.

(a)
(b)

[Courtesy Physical Society.
FIG. III.2. Emission spectra of aluminium, (a) from vapour and (b) from solid.

The Structure of the Free Atom   81

maxima at $\theta = 45°$ and $\theta = 135°$, and there is a node along the z axis, whilst the xy plane is a nodal plane. When the atom is in a crystal, or under some other restraint which gives the x and y axes physical significance, the states $m = \pm 1$ split into two, and the resulting angular-probability factors take the form shown in Fig. II.18. The two

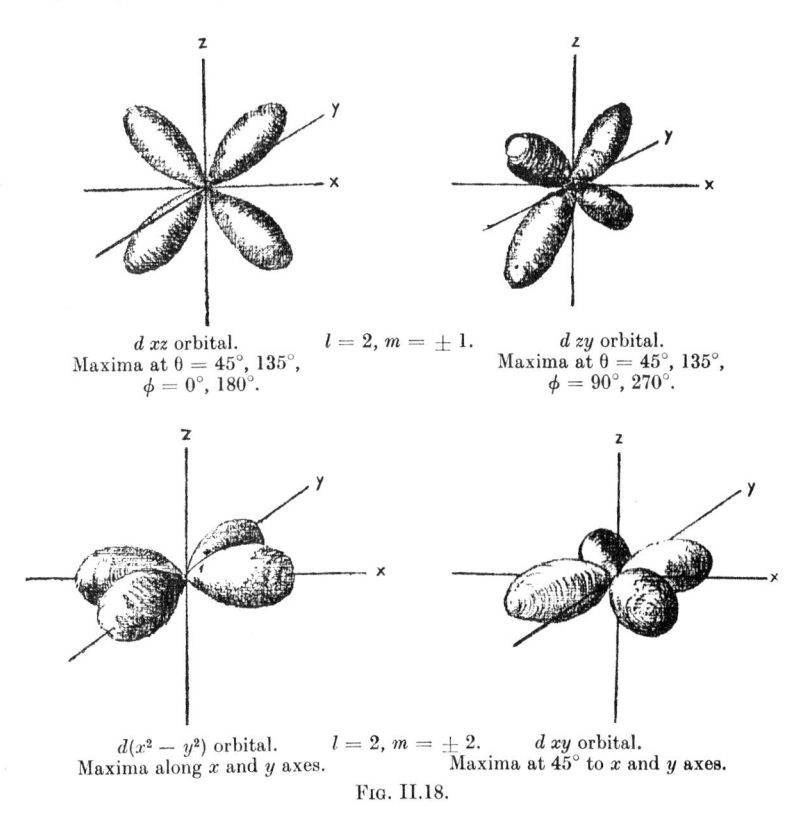

$d\ xz$ orbital.
Maxima at $\theta = 45°, 135°$,
$\phi = 0°, 180°$.

$l = 2, m = \pm 1.$

$d\ zy$ orbital.
Maxima at $\theta = 45°, 135°$,
$\phi = 90°, 270°$.

$d(x^2 - y^2)$ orbital.
Maxima along x and y axes.

$l = 2, m = \pm 2.$

$d\ xy$ orbital.
Maxima at $45°$ to x and y axes.

Fig. II.18.

patterns in the upper half of Fig. II.18 may be regarded as the real and imaginary parts of the $m = \pm 1$ pattern of Fig. II.17.

The state $(l = 2, m = \pm 2)$ of the free atom is characterized by a node along the z axis, and the electron density is greatest in the xy plane, as shown in Fig. II.17. When the atom is in a crystal or under some other restraint which gives the x and y axes a physical meaning, the electron clouds of the two states are mutually inclined at $45°$, as shown in Fig. II.18, and these two patterns may be regarded as the real and imaginary parts of the single diagram in Fig. II.17.

The complete set of d patterns when superimposed on one another gives a pattern of spherical symmetry, so that, in the simple theory where the effects of electron correlation are ignored, a group of 10 d-electrons is spherically symmetrical, and a group of 18 electrons is also spherically symmetrical, since it is composed of the superimposed patterns of the spherically symmetrical $(ns)^2$, $(np)^6$, and $(nd)^{10}$ sub-groups. As explained above (pp. 75 and 78), this is not really true, and the effects of electron correlation must be considered.

It will be noted that in the p- and the d-states, increasing value of the quantum number m_l results in the electron cloud becoming more and more concentrated in the xy plane, and this is a general principle for the higher $(f, g, \ldots)$ states.

Atoms of Higher Atomic Number.

The electron-cloud photographs and the radial-density curves referred to above are for the hydrogen atom. For atoms containing more than one electron, the problem of electronic motion becomes enormously more complicated, and no accurate calculations have been made for atoms of atomic number higher than 2 (helium).

In an atom of atomic number Z, a given electron moves in the field of the nucleus of charge $+ Ze$, and in the fluctuating fields resulting from the motions of the remaining $(Z - 1)$ electrons. The different electrons affect each other's motions, first on account of the Coulomb repulsive forces which prevent two electrons from approaching one another very closely. The electrons also affect each other's motions owing to the Pauli Exclusion Principle which implies that two electrons of parallel spin cannot be in the same place at the same time (see p. 140).

In the theory of the hydrogen atom described above (pp. 65-71) the wave equation is split into the three components $R(r)$, $\bar{\theta}(\theta)$, and $\Phi(\phi)$, which involve only r, θ, and ϕ, respectively, and the conditions for acceptable solutions lead to the introduction of the quantum numbers n, l, and m_l. If we consider a hypothetical atom with one electron only, moving round a nucleus of charge $+ Ze$, the same treatment can be adopted, and quantum numbers n, l, and m_l are again introduced. The increased nuclear charge means that the electron is held more firmly. The negative energy for a particular value of n is proportional to Z^2, and the whole scale of the atom is reduced by a factor proportional to Z, *i.e.*, the electron cloud is smaller owing to the increased attraction of the electron as Z is increased.

In the approximate theories of poly-electronic atoms, each electron is assumed to move in the field of the nucleus, and of all the other

The Structure of the Free Atom

electrons, whose motions are regarded as smeared out so as to produce a spherically symmetrical field. In this approximation, the motion of the one electron is again in a central field, and quantum numbers n, l, and m_l still appear, and the s, p, d ... states give rise to electron clouds with the same shapes as those for the hydrogen atom. The K, or $1s$, electrons are so close to the nucleus that the interference or *screening effect* of the remaining electrons is very small, and for the K electrons the energies and sizes of the electron clouds are nearly those expected for the motion of a single electron round a nucleus of charge $+ Ze$. For the outer electrons part of the nuclear attraction is screened off by the remaining electrons, and this effect is greater the further the average distance of the electron from the nucleus; it is, thus, far greater for an M electron than for an L electron,* and greater for a $3d$ than for a $3s$ state. In this approximation the wave function of an electron can be written as that which obtains for a central field due to a nucleus of charge $(Z - S)$, where S is the *screening constant*. This is not strictly true, and detailed examination shows that if a hypothetical charge $(Z - S)$ is assumed, different values of S may be required to give the best agreement for different physical properties.

To a higher degree of accuracy, the position is less satisfactory. We may first ignore the effect of electron spin, and consider the motion of any one electron in a group of 8. The remaining 7 electrons will not give rise to a spherically symmetrical electron cloud, and the assumption that the one electron considered is moving in a symmetrical field is not strictly true. This assumption is, however, made in the Hartree method (see p. 67). It is in this way that solutions involving the quantum numbers n, l, and m_l are obtained, and there is much evidence to show that the approximation is reasonable.

If electron spin is considered, the quantum number l is replaced (p. 64) by j, the quantum number of the total (orbital and spin) angular momentum. For light atoms, the spin–orbital coupling† is not important, and the quantum number l can still be used. For heavy atoms at the end of the Periodic Table, this is no longer true. The spectroscopic evidence indicates that the electrons are still divided into groups and sub-groups. These may still be interpreted as involving s, p, d ... electrons, but if this is done it corresponds to a much rougher level of approximation than for light atoms.

* It is customary to use the symbols K, L, M, ... to describe the groups of electrons with principal quantum numbers $n = 1, 2, 3, \ldots$
† This coupling gives to states of different j for given l and s somewhat different energies; its size increases with atomic number.

PART III. ASSEMBLIES OF ATOMS.

INTRODUCTION.

IN Part II we have described the structures of free atoms, and how these are interpreted in terms of the new theories. We have now to consider the nature of the interatomic forces involved in the formation of solids, liquids, and compound molecules. It will be appreciated that the problem of the cohesive forces in solids and liquids is only part of the general problem of interatomic cohesion, and, as we shall see later, there are some solids in which the forces between the atoms are of exactly the same nature as those involved in the formation of compound molecules. In many senses a crystal of a metal is to be looked on as an immense molecule of the atoms concerned, and in order to understand the structure we need first of all to determine the type of crystal structure and then the energy characteristics of the electrons in the solid. The determination of the crystal structure is carried out by means of X-ray diffraction methods, and in the present book we shall accept the results of this work and not concern ourselves with the experimental methods. Just as the energy characteristics of molecules are deduced from the analysis of molecular spectra, so the energies of electrons in solids can be studied by using the X-ray emission spectra of solids. Historically, this is a comparatively new branch of science, but it is of such fundamental importance that we shall consider it in the next section. We shall then review, in Section III.2, some of the characteristics of the elements, and show that the latter fall into four main classes, in which distinct types of interatomic forces can be recognized as responsible for the formation of either molecules or crystals; these can be considered as assemblies of like atoms. In Section III.4 we shall consider what happens in mixtures or compounds of unlike atoms, and shall see that in many cases the interatomic forces are clearly of the same nature as those found in the molecules or crystals of the elements, but that a new type of force, the *ionic bond*, is also met with, and we shall show how this kind of force exerts an influence in many alloy structures. In Sections III.5–III.8 we shall consider in greater detail the electronic processes involved in three of the other main types of interatomic attraction, namely the van der Waals or Polarization Forces, the Exchange Forces, and the Wave-Mechanical Resonance Forces. The metallurgical student should appreciate that, although much of the description refers to non-metallic substances, the processes described

Assemblies of Atoms

are often present in solid metals and alloys, even though they are confused by the simultaneous presence of other types of interatomic attraction.

1. The Soft X-Ray Spectra of Solids.

In Part II we have seen how the electrons in a free atom occupy a series of energy states, each characterized by the appropriate values of the four quantum numbers n, l, m_l, and m_s. In the normal state of a free atom, the electrons are divided among the different quantum states in such a way as to give the lowest possible energy, subject to the Pauli restriction principle that not more than one electron occupies a state as defined by a given combination of all four quantum numbers. It is these electronic configurations or groupings which are given in Table II (p. 52), and these groupings give the electronic structures of the free atoms in their normal states. Under appropriate conditions, such as the high temperatures of flames or electric discharges, a free atom can undergo electronic excitation into a state of higher energy. Thus the normal state of the sodium atom is $(1s)^2(2s)^2(2p)^6(3s)^1$, but under appropriate conditions the valency electron can be excited into a state of higher energy, so that an excited atom with a configuration such as $(1s)^2(2s)^2(2p)^6(3p)^1$ is formed, and this has a higher energy than the normal state, because the $(3p)$ state is higher than the $(3s)$ state. If the electron then falls back from the excited state to the normal state, the atom emits radiation of frequency ν, given by the relation:

$$E_1 - E_2 = h\nu \quad \ldots \ldots \quad \text{III (1)}$$

where E_1 and E_2 are the energies of the excited and normal states respectively. In the case described above the excitation affects the valency electron only, but this is not necessarily so, and an excited state such as $(1s)^2(2s)^2(2p)^5(3s)^2$ might be formed, and this would then revert to the normal state by one of the $(3s)$ electrons dropping back into the $(2p)$ sub-group. Speaking generally, the ordinary optical spectra correspond to transitions in which an excited electron falls into one of the outer quantum shells of a free atom. Since visible light has wave-lengths of the order of 5000 A. and the velocity of light is of the order of 3×10^{10} cm/sec, the frequencies of the optical spectra are of the order 6×10^{14}, and since $h = 6 \cdot 6 \times 10^{-27}$ erg/sec, the transitions involved concern energy differences of the order of 10^{-12} erg/atom.

Under appropriate conditions it is possible to excite the electrons in the lower energy states of a free atom. Thus the free atom of sodium might be excited into a state such as $(1s)^1(2s)^2(2p)^6(3s)^1(3p)^1$ in which

86 *Atomic Theory*

one of the (1s) electrons has been excited into a (3p) state. In this case if the (3p) electron drops back into the (1s) level, the energy change involved is of the order of 10^{-9} erg/atom, and is thus very much greater than those taking place in the optical spectra referred to above. This kind of transition involving the inner levels of an atom gives rise to X-ray spectra, in which the wave-lengths are of the order of 1 Å unit. It is customary to use the symbols K, L, M ... to denote the groups of electrons with principal quantum numbers $n = 1, 2, 3$ ... and the K X-ray emission spectra are those in which the excitation expels an electron from the K shell, and emission occurs when an electron falls back from one of the outer shells. Thus, in the case of aluminium, the electronic structure of the normal state of the free atom is:

$$\underbrace{(1s)^2}_{K} \underbrace{(2s)^2(2p)^6}_{L} \underbrace{(3s)^2(3p)^1}_{M}$$

and if an electron is expelled from the (1s) or K-shell, the $K\alpha$ emission spectra are the result of an electron falling back from the L shell to the K shell, whilst the $K\beta$ emission spectra result from an electron falling from the M shell to the K shell. Since the energy difference between the K and M shells is greater than that between the K and L shells, it follows from equation (1) that the $K\beta$ emission line has a higher frequency, *i.e.*, is harder, than the $K\alpha$ emission line.

A free atom can thus exist in a number of energy states, each characterized by a definite energy and by a definite distribution of the electrons among the possible quantum states. The lines of the optical or X-ray spectra of free atoms are due to transitions of the atom from one state to another, and the sharpness of these lines is the result of a definite energy being associated with each state. It is not, however, possible for transitions to occur between any two electron states of an atom, and the possible transitions are controlled by what are known as *Selection Rules*. Of these the most important is that the probability of a transition occurring is very small unless the quantum number l changes by 1. Thus transitions can occur from a p-state to an s-state, or from a d-state to a p-state, but not between two s-states. There is no restriction on the change of the principal quantum number n. In the case of K X-ray emission spectra, for example, the electron jumps into the (1s) state from a state in an outer quantum shell. Here the final state after the transition is (1s), for which $l = 0$, and so we do not obtain an X-ray line corresponding to a transition from a (2s) state to a (1s) state, but we do obtain transitions from a (2p) state to a (1s) state, since here l changes by 1; the resulting X-ray line is called the $K\alpha$ emission line, and is a doublet (*i.e.*, two lines of nearly the same wave-length) because the (2p)

Assemblies of Atoms

sub-group of the free atom is split by spin-orbit coupling into levels that have very slightly different energies.

In the L X-ray spectra, an electron is excited from the L shell, and emission results from one of the outer electrons jumping into the unoccupied state. In this case the possibilities are more numerous, since the L or 2-quantum shell contains both $2s$ and $2p$ electrons. If one of the $2s$ electrons is excited, the selection rule indicates that X-ray spectra may result from an electron falling back from $3p$, $4p$, &c., subgroups. If, on the other hand, one of the $2p$ electrons is excited, X-ray lines are not found for transitions from a $(3p)$ to a $(2p)$ state, since for this l would change by 0, but X-ray lines are found for transitions from $(3s)$, $(3d)$, $(4s)$, or $(4d)$ states to the $(2p)$ state, since the transitions $s \rightarrow p$ or $d \rightarrow p$ mean that l changes by 1. There is thus a whole group of L-emission lines corresponding to the different possible transitions.

The above remarks refer to free atoms, and it is by the study of the frequencies and intensities of the emission lines that the spectroscopist has been able to deduce the energies of the different electronic states of atoms and the probabilities of transitions occurring from one state to another. When we deal with crystals, we are concerned with immense assemblies of atoms at distances of the order of 10^{-7} to 10^{-8} cm from one another. The conditions are thus very different from those of free atoms, but speaking generally, the wave functions of an electron in the periodic field of a crystal may be regarded as derived from the wave functions in the free atom, and retain some of the symmetry characteristics of the s, p, ... states. In the case of sodium, for example, the valency electrons of the free atoms are in $(3s)$ states, and in crystals of sodium there are electrons in states derived from the $(3s)$ states of the free atom, and possessing $n - 1 = 3 - 1 = 2$ nodes round the centre of each atom. In this case the innermost two nodes will be approximately spherical, but the outermost node will be distorted, although the total number of nodes will remain at three per atom. In the same way the $(1s)$, $(2s)$, and $(2p)$ states of the free atom give rise to corresponding electronic states in the solid. Just as the energy characteristics of the electrons in a free atom can be determined from a study of the emission spectra of free atoms which have been excited in an electric discharge, so the energies of electrons in a crystal can be deduced by investigating the spectra emitted by solids in which excitation has been produced by suitable means, the most usual method being by electron bombardment. The experimental methods are naturally complicated, but the underlying principles are very simple. When a metal such as aluminium is bombarded by electrons, the atoms receive an intense stimulus, as a result of which an electron may be expelled from one of the lower electron shells, say the K or (1s) shell. An electron may then fall back from the L shell to the K shell, with the production of a quantum of radiation of frequency given by equation III (1). The frequencies of this emission, the $K\alpha$ emission, then indicate the differences in energies of the K and L electrons *in the solid crystal*, as distinct from their differences in the free atom. Alternatively, an electron might fall back from the ($3p$) state to the ($1s$) state, and the resulting $K\beta$ emission would indicate the differences between the energies of the ($3p$) and ($1s$) electrons in the solid. The selection principles referred to above still apply, and the K emission spectra of the solid could not be used to deduce the characteristics of the ($3s$) electrons in aluminium, since a ($3s$) $\rightarrow$ ($1s$) transition is prohibited. For this purpose the spectral region corresponding to the L_3 emission would be studied, since this emission corresponds to the transition ($3s$) $\rightarrow$ ($2p$). In this way, by choosing the appropriate X-ray emission spectra, the physicist attempts to explore the energy characteristics of electrons in solid crystals, just as the earlier spectroscopy investigated the energy characteristics of free atoms. After preliminary work by several investigators, the first great advance in soft X-ray spectroscopy of solid metals was made by Skinner,* who obtained results for several metals and, by using targets coated with liquid air, was able to examine the effect of temperature on the energies of electrons in a metal. As will be appreciated from Section V.4, the interpretation of the results leads to difficulties, but the importance of the work is so great that we shall consider here some of the general conclusions and underlying principles.

By a free atom, we mean in practice an atom in a very dilute gas, and if we imagine an immense assembly of free atoms to be gradually compressed to form a dense gas, or a liquid, or a solid, the general effect is for each sharp energy level of the free atom to be broadened into a band or range of energy levels. The details of the process depend on the electron state and the atom concerned, and on the structure of the solid or liquid which is formed, but the broadening of the sharp levels of a free atom is a quite general process.

For the lowest electron states of all but the lightest elements, the broadening of the energy levels is negligible at the interatomic distances with which we are concerned in solids and liquids. It is for this reason that we still obtain sharp $K\alpha$ emission *lines* from solid elements such as copper, since these correspond with transitions between the K and L shells, which in the heavier elements are too deep down in the atom to be affected appreciably by the influence of other atoms at distances of

---

* H. W. B. Skinner, *Phil. Trans. Roy. Soc.*, 1940, [A], **239**, 95.

the order of 10^{-8} cm. For the same reason these emission lines are practically independent of the state of combination of the copper atom, because although the chemical combination affects the outer electrons profoundly, the inner electrons are so predominantly under the influence of the one nucleus that they are practically undisturbed by changes in the electron distribution at the outside of the atom. It is for this reason that in X-ray crystal analysis the wave-length of the copper $K\alpha$ radiation is unaffected by superficial oxidation or contamination of the copper target.

When the assembly of free atoms is gradually compressed, the broadening of the energy levels of the outermost electrons eventually becomes appreciable, and at interatomic distances of the order of 10^{-7} to 10^{-8} cm each energy level of the valency electrons has broadened into a wide band of allowed energies. In its simplest form this may be represented by a diagram of the type shown in Fig. III.1, in which the

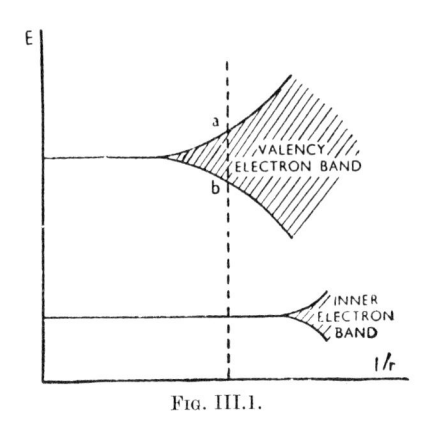

FIG. III.1.

widths of the energy levels are plotted against $1/r$, where r is the distance between two atoms in the assembly. The dotted line indicates an interatomic distance of the order 3×10^{-8} cm at which the broadening of the valency electron level is appreciable, whilst the inner electron level is practically unaffected.

From a figure of the above type it will be appreciated that if we consider radiation caused by the transition of an electron from the valency electron level to the K shell, then at interatomic distances of the order of 10^{-8} cm there is no longer one constant difference in energy between the states before and after the transition. When the interatomic distances are of the order of 10^{-8} cm, the energies of the valency electrons of the assembly extend over a range, and consequently, according to equation III (1), there will be a range of possible frequencies, and the emission will consist of an emission *band*, instead of a single sharp line. In actual practice this is what occurs, and Fig. III.2 (Plate II), which is due to Skinner, shows the sharp X-ray *line* spectrum of the free atoms of aluminium compared with the X-ray band spectrum obtained from solid aluminium. Since it is known that the innermost electron levels are unaffected by the compression of atoms to interatomic distances of the order of 10^{-8} cm, the above type of X-ray emission spectrum shows conclusively that the sharp energy levels of the valency electrons have broadened into a band or range of possible energy states, and for solid metals the width of the valency electron band is of the order 1–10 eV. The study of the energy distribution in these emission bands enables us to deduce the energy characteristics of the valency electrons in the solid metal.

Fig. III.1 shows clearly that as the assembly of atoms is compressed more and more closely, the valency electron band becomes wider, and speaking generally it may be said that in an assembly of atoms, the width of the band depends essentially on the mean interatomic distance, whilst the number of energy states within the band increases with the number of atoms in the assembly, and hence with the size of the crystal if we are dealing with a solid metal. We may suppose, for example, that in Fig. III.1 *ab* represents the width of the band at the actual interatomic distance of a particular metal. If we then compare a large crystal with a small crystal, the width of the band *ab* will be the same in both cases, but the number of individual energy states in the range *ab* will be the greater for the larger crystal.

The ideas expressed in Fig. III.1 lead at once to a further and most important conclusion. We may suppose that Fig. III.3(*a*) shows diagrammatically the energy levels of the free atom of sodium. Then in the free atom of sodium the electrons will occupy the $(1s)^2(2s)^2(2p)^6(3s)^1$ levels, and there will be one electron in the 3*s* state and none in the 3*p* state (see Table II). Fig. III.3(*b*) shows diagrammatically how the energy levels broaden when an assembly of sodium atoms are brought together, and the line *xy* indicates the interatomic distance in the crystal of sodium. At this distance the broadening of the (1*s*), (2*s*), and (2*p*) levels is negligible, but the (3*s*) and (3*p*) levels have broadened to such an extent that they overlap. When this stage is reached, a process known as "hybridization" occurs, and the valency electrons exist, not in (3*s*) or (3*p*) states, but in a condition which may be described by the superposition of the two wave functions. Such electrons are said to exist in *hybrid orbitals*, and in sodium, for example, the valency electrons exist in (3*s*)(3*p*) hybrid orbitals. In this case the work of Skinner shows that the valency electrons of lowest energy are in a condition closely resembling that of (3s) electrons as modified by the crystal lattice, and that with increasing energy the (3p) characteristics become more and more pronounced. Strictly speaking, we should not describe such electrons as existing in s or p states, but should always think of the hybridized condition. It is, however, common practice to refer to electrons in metals as belonging to s-bands, d-bands, &c., and for many

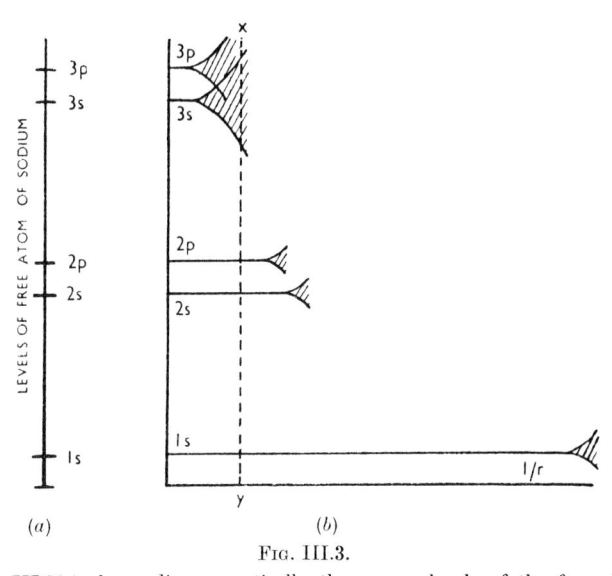

FIG. III.3.

Fig. III.3(a) shows diagrammatically the energy levels of the free atom of sodium. The vertical scale is purely diagrammatic and does not represent accurately the relative distances between the different energy levels. An accurate drawing would require the 1s level to be very much lower, and the difference between the 3s and 3p levels to be very much less than that shown.

Fig. III.3(b) shows diagrammatically how the levels broaden as the atoms are brought together. The dotted line corresponds to the interatomic distance in the crystal of the metal.

purposes this is permissible provided that the true state of affairs is borne in mind. Thus the work of Skinner shows that solid sodium gives rise to a $K\beta$-emission spectrum, and this is commonly referred to as corresponding with an electronic transition from a (3p) state to a (1s) state. Strictly speaking, we should say that the $K\beta$ spectrum reveals the 3p fraction of the (3s)(3p) hybrid, but as an approximation we may say that a fraction of the valency electrons is in (3p) states, and similarly in other cases.

The above considerations will indicate how important it is to realize that the electronic configurations of Table II are the electronic structures of *free atoms*, and that the details of the distribution of the outer electrons are often quite different in a solid or liquid metal. Even in a simple univalent metal, such as sodium, the solid metal contains its valency electrons in $(3s)(3p)$ hybrid orbitals, and some $(3p)$ characteristics are developed, whilst in the divalent element magnesium the valency electrons are also in $(3s)(3p)$ hybrids, and for some of the electrons of higher energy the $(3p)$ characteristics are predominant, although in free atoms of magnesium the two valency electrons occupy the complete $(3s)^2$ sub-group, and there are no electrons in $(3p)$ states. Here we may say that a considerable proportion of the valency electrons in solid magnesium exist in $(3p)$ states, although the true condition is one of hybridization. In the same way, in the transition elements the relative numbers of electrons in the $(n-1)d$ and ns states may be quite different in the free atoms and the solid metals. Thus the free atom of nickel has the configuration $(1s)^2(2s)^2(2p)^6(3s)^2(3p)^6(3d)^8(4s)^2$, but in solid nickel there is on the average only 0·6 electron per atom in $(4s)$ states, and 9·4 electrons per atom in $(3d)$ states. In some metallurgical papers there has been an unfortunate tendency to ignore these differences, and quite misleading conclusions have been drawn by assuming that the details of Table II apply to the outer electrons of solid or liquid metals. It will be appreciated that when we deal with the inner electrons of an atom, the electronic distributions in the solid metal and the free atoms are identical. In the case of nickel, for example, the free atoms and the solid metal have the same electron distribution as regards the $(1s)^2(2s)^2(2p)^6(3s)^2(3p)^6$ groups.

It will be seen, therefore, that the experimental work on the soft X-ray spectroscopy of solids leads to the direct conclusion that the energies of the valency electrons of a metal are distributed over a band or range of energies, in contrast to the sharp energy levels in free atoms. By measurements of the variation of intensity across an emission band, it is possible to deduce the distribution of the electrons among the various possible energy states of the valency band, *i.e.*, across the range of energies ab in Fig. III.1. To describe these results, it is necessary to have some method of expressing the energy distribution, and this is done by means of what is called an $N(E)$ curve. For this purpose we may first define $n(E)dE$ as the number of electrons per unit volume of metal with energies between E and $E+dE$, and it is customary to plot $n(E)$ against E. In Fig. III.4(a), for example, the curve indicates that the electrons have energies extending over the range ab, and that the number of electrons with energies in the neighbourhood of a particular value E increases continuously with increasing E, until the energy reaches the value b, beyond which $n(E) = 0$, indicating that no electrons have energies greater than $E = b$. In this case the high-energy end of the band would be sharp. Fig. III.4(b) would indicate that the electronic energy distribution for the valency electrons of low energy was the same as that of Fig. III.4(a). The steep fall in the $n(E)$ curve at x indicates that the number of electrons with energies beyond this point becomes

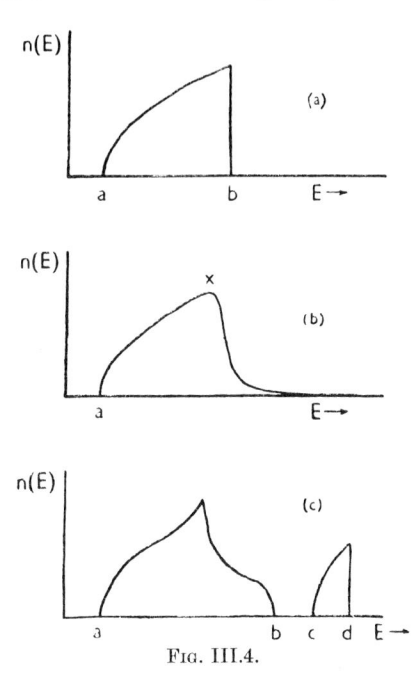

Fig. III.4.

very small and gradually vanishes, but there is no sharp upper limit to the energies of the electrons, and the head of the emission band would be diffuse. The curves in Fig. III.4(c) would indicate that the valency electrons had energies extending over the ranges ab, and cd, but that no electrons had energies in the range bc. These $n(E)$ curves are therefore nothing more than ways of showing how the electrons are distributed over the ranges of energy concerned.

The probability of a transition occurring from one electron state to another is proportional, among other things, to the number of electrons present in the first state, and hence it will be readily understood that the $n(E)$ curves are of roughly the same form as the curves which show the variation of intensity over the emission bands of the soft X-ray spectra concerned. Actually the interpretation is not quite so simple, and some of these complications* are dealt with later (p. 255); for the present we shall assume that the intensity curves are a direct indication of the $n(E)$ curves. Fig. III.5 is due to Skinner, and shows the L_3 emission band for sodium; since the L_3 emission corresponds to a transition from the $(3s)$ band to the $(2p)$ state,† this curve gives the form of the $n(E)$ curve for the valency electrons of solid sodium existing in $(3s)$ states in the solid.‡ Fig. III.5 also shows the K emission band for lithium, and since this corresponds to transitions from the $(2p)$ to the $(1s)$ state, the curve gives the form of the $n(E)$ curve for the $(2p)$ electrons in solid lithium; a curve of more or less the same shape would give the $n(E)$ curve for the $(3p)$ electrons in solid sodium.

The form of these curves at the low-energy end is uncertain, partly owing to experimental difficulties. As will be shown later (p. 172), some of the simple theories of electrons in metals require the $n(E)$ curve at the bottom of a band to be parabolic in form, so that $n(E) \propto (E - E_0)^{\frac{1}{2}}$, where E_0 is the energy at the bottom of the band. In much of the early work it was customary to fit the middle part of the emission curve to the most probable parabola, and then extrapolate this back to zero intensity and so to obtain the bottom of the band. It is now known that some $n(E)$ curves may not be parabolas, and in such cases the position of the bottom of the band, and hence the width of the band, are sometimes uncertain.

As can be seen from Fig. III.5, the K and L_3 curves for magnesium and aluminium are more complicated in form at the high-energy end than those for sodium. Fig. III.6 shows the effect of temperature on the high-energy end of the emission band, and from this it will be seen that the curve falls more steeply as the temperature is lowered.

It must be emphasized that the above curves are purely experimental. They lead directly to the conclusion that the energies of the valency electrons in metals extend over a range, and that the high-energy end of the range terminates more and more abruptly as the temperature becomes lower. At the absolute zero the $n(E)$ curve would terminate sharply with a vertical line at the high-energy end. At other temperatures, as shown in Fig. III.6, the end of the band becomes diffuse, and the width of this temperature broadening is of the order of $k\theta$, where θ is the absolute temperature; so that at room temperatures the diffuse head of the band extends over about $\frac{1}{10}$th eV, as compared with a total band width of the order of 5–10 eV. The curves also make it clear that with

* The reader may consult H. W. B. Skinner, *Phil. Trans. Roy. Soc.*, 1940, [A], 239, (801), 95.
† Strictly speaking the $(2p)$ state has, of course, broadened into a band, but the broadening is so slight that we may talk of a single $(2p)$ state.
‡ The actual $n(E)$ curve does not tail off at the low-energy end of the band.

FIG. III.5. K and L_3 bands of metals.

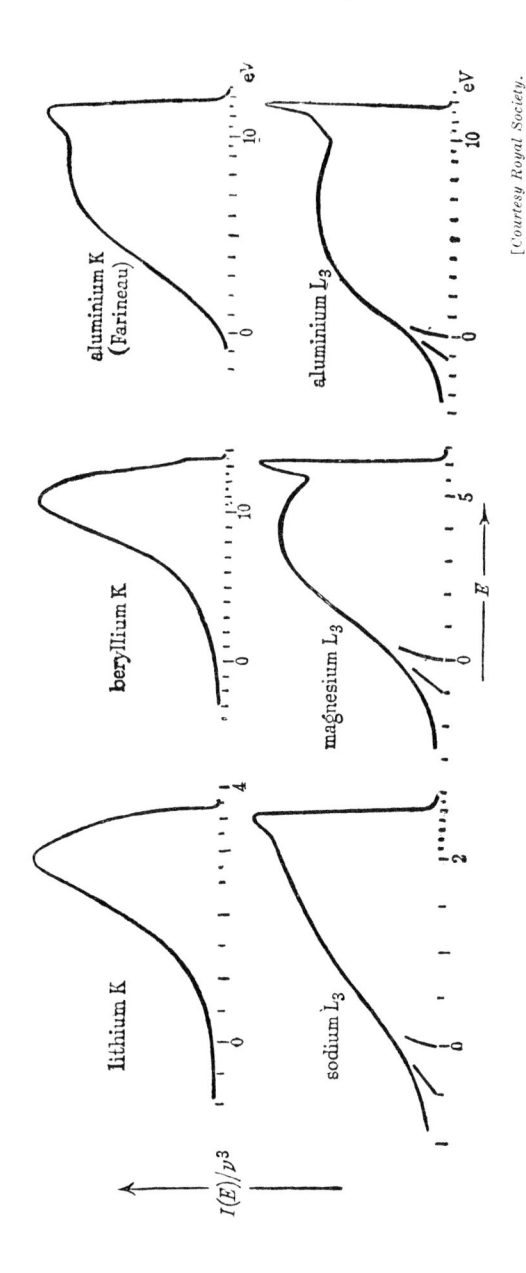

In this diagram $I(E)$ is the intensity of the radiation in the emitted band in the range of quantum energies E to $(E+dE)$ and ν is the frequency. The quantity $I(E)/\nu^3$ is a measure of the $n(E)$ referred to in the text; for details see the paper by Skinner mentioned on p. 88.

increasing valency, the characteristics of the energy distribution become more complicated. All this is now definitely established fact, and has to be explained by any satisfactory theory of metals. We may here anticipate some of the later theory and explain that a metal crystal is to be regarded as containing an immense number of possible electron

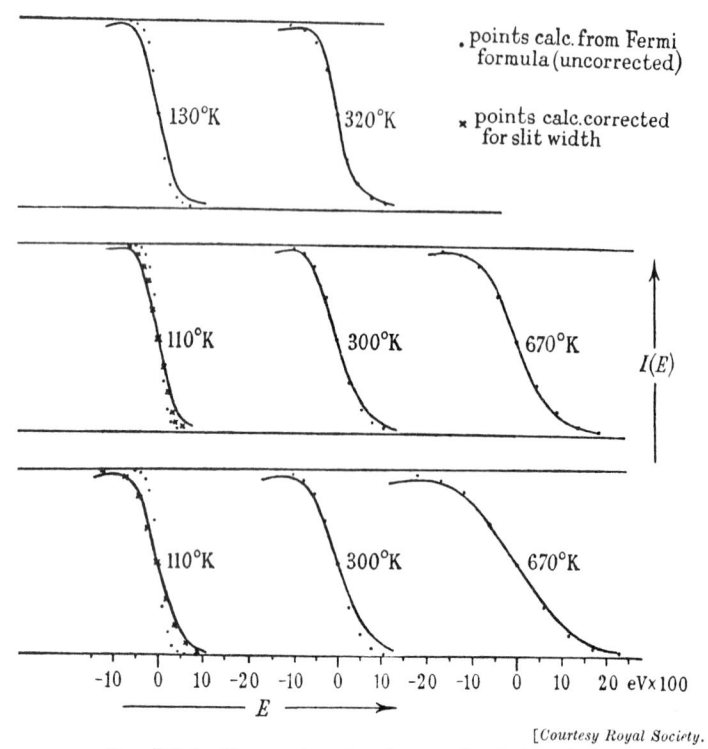

FIG. III.6. Temperature-broadening of emission edges.

This diagram shows the effect of temperature on the high-energy end of some of the emission bands of Fig. III.5. At the absolute zero the end of the band would be a vertical line.

states, and that at the absolute zero of temperature the N valency electrons of a crystal occupy the $N/2$ lowest energy states of the valency electron band, each state containing two electrons, one of each spin. At the absolute zero of temperature all the energy states up to a particular limit are occupied, and all those above this limit are unoccupied. The electronic-energy distribution curve thus shows a sharp limit, and it is this which corresponds with the vertical fall of the curve in Fig. III.4(a). At higher temperatures some electrons are excited into higher

Assemblies of Atoms 97

energy states, with the result that the $n(E)$ curve no longer falls so steeply.

The $n(E)$ curves referred to above show the way in which the actual *electrons* in the metal are distributed over the range of energies of the band, but for many purposes it is of interest to know how many *possible energy states* exist in the different regions of energy, and for this purpose $N(E)dE$ may be defined as the number of energy states per unit

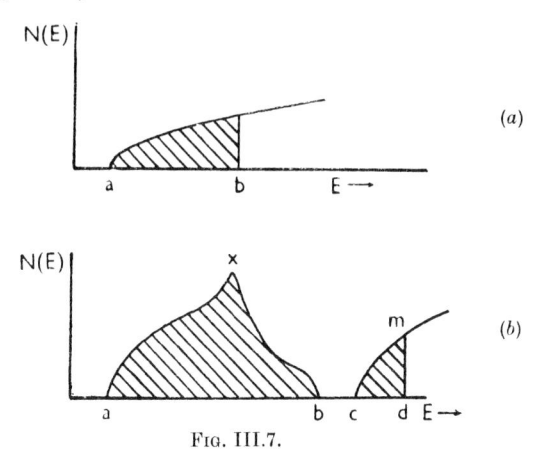

FIG. III.7.

volume of metal having energies between the limits E and $E + dE$. In this case the $N(E)$ curves show how the possible energy states are distributed over the ranges of energy concerned, in just the same way that the $n(E)$ curves described above show how the electrons are distributed over the range of energy. As will be seen later, much of the theory has been developed first for the absolute zero of temperature, where the N valency electrons in a crystal occupy the $N/2$ lowest energy states of the valency band. In this case it is customary to draw the $N(E)$ curve and to shade the area of the occupied states. Thus in Fig. III.7(a) the $N(E)$ curve shows that the number of states per unit energy range* increases continually, and that the states are fully occupied up to the limit $E = b$, beyond which no states are occupied. In this case the curve in the $N(E)$ diagram is the same as the $n(E)$ curve of Fig. III.4(a), except that $n(E) = 2N(E)$, since each state contains two electrons. Similarly, the curve of Fig. III.7(b) shows that there are possible electron states in the ranges $a \to b$ and from c onwards, but that there are no states in the range $b \to c$. Since the $N(E)$ curve rises to a maximum in the region of x, there are far more energy states with ener-

* This is often called the " density of states."

gies in the region $E = x$ than at any other values of the energy, and the shaded area which terminates abruptly at $E = md$ indicates that all the electron states up to this point are occupied, and all those beyond are unoccupied. This method of drawing a vertical line at some point of the $N(E)$ curve is, of course, justifiable only for the absolute zero of temperature where there is a sharp boundary between the occupied and unoccupied states. At higher temperatures the $N(E)$ curve continues to show the distribution of the possible energy states, but the distribution of the electrons is shown by the $n(E)$ curve. In many books the symbol $N(E)$ is used both for what we have called $n(E)$ and $N(E)$, and the reader must be careful to distinguish between cases where the curves refer to the distribution of electron states and those referring to the distribution of the states occupied by electrons.

The methods of soft X-ray spectroscopy indicate the energy distribution of electrons in solids, and it should be noted that this method by itself begins to give a rough classification of the different types of substance. In Fig. III.7(b) we have shown an $N(E)$ curve such that there are possible electron states from a to b and from c onwards, but none in the range b to c, whilst the shaded areas indicate that all the possible states are occupied up to the limit md. In such a case we may say that the electrons with energies between a and b form a closed group which completely fills the first *band* or *zone* of possible energy states, whilst the second zone is only partly filled. In general it may be said that *metals* are characterized by having $N(E)$ curves in which the outermost of the occupied zones is only partly filled, whilst insulators are substances in which the number of electrons is sufficient completely to fill one or more zones, the outermost of which is separated by an energy gap from the next highest zone. If, for example, in Fig. III.7(b) the number of electrons present were just sufficient to fill the first zone, *i.e.*, if the region axb alone were shaded, the substance would be an insulator (see Part V, p. 219). The technique of soft X-ray spectroscopy is not yet sufficiently developed to give a very sensitive classification of substances, and in the next two sections we shall therefore indicate briefly the main types of chemical elements as classified empirically and then some of the classes of substance which are of interest in the study of alloys.

SUGGESTIONS FOR FURTHER READING
Elementary.

E. C. Ellwood, D. J. Fabian, and L. M. Watson, *Metals and Materials*, 1967, **1**, 333.

More Advanced.

D. H. Tomboulian, *Handbuch der Physik*, 1957, **30**, 246.
L. G. Parratt, *Rev. Mod. Physics*, 1959, **31**, 616.

Assemblies of Atoms 99

2. THE ELEMENTS.

In the preceding pages we have described the energy characteristics of solid metals as these are revealed by the methods of soft X-ray spectroscopy, and it remains to consider the nature of the forces which give rise to cohesion between atoms. The problem of interatomic cohesion is naturally immensely complicated, and includes the whole range of processes from the combination of atoms to form molecules to the formation of liquids or solids from atoms or molecules. In the general case, different processes are superimposed, but fortunately it is possible to find examples where one kind of interatomic force is so predominant that other processes may be ignored; the substance concerned may then be regarded as typical of one kind of interatomic force, and its properties used to study the characteristics of this type of interatomic binding. Proceeding in this way we may recognize the following main types of element as throwing light on the interatomic cohesion with which we are concerned.

The first and most characteristic group of elements is that of the inert gases, helium, neon, argon, krypton, xenon, and radon. These are monatomic in the gaseous state, and have very low boiling points: He $4°$ K, Ne $27°$ K, A $87°$ K, Kr $121°$ K, Xe $174°$ K, Rn $212°$ K. They form no stable chemical compound except for those of the clathrate type in which an inert-gas atom is held in a hole in a framework of other atoms. Of all the elements the inert gases possess the least power of interatomic cohesion. In Part II we have seen that complete octets of electrons in free atoms form very stable groups, and it has long been recognized that the properties of the inert gases are due to their atoms having outer groups of eight electrons. These groups of eight electrons (or two in the case of helium) are so stable that they are only very slightly affected by the presence of other atoms, and the motion of their constituent electrons cannot be changed sufficiently to give rise to strong binding forces. The forces which bind the atoms of inert gases together to form the liquid or the solid state are the so-called van der Waals Forces, and are described later (p. 131).

The next group consists of the elements immediately preceding the inert gases, and these fall into two main classes. The elements hydrogen, nitrogen, oxygen, and fluorine form very stable diatomic molecules and have low boiling and melting points. G. N. Lewis (1916) first suggested that in molecules of this type the interatomic forces were due to the sharing of electrons, so that each atom acquired a full complement of eight electrons (or two in the case of hydrogen). In fluorine, for example, each atom possesses seven electrons, and in the diatomic

100 Atomic Theory

molecule it is regarded as sharing one electron with its neighbour and so obtaining a stable octet. Similarly, in the O_2 molecule each oxygen atom shares two electrons, and in the N_2 molecule each nitrogen atom shares three electrons, and in this way stable octets are built up. The low boiling points of these gases are the result of the fact that when once the atom has acquired its octet, the group of electrons is so stable that its electrons cannot give rise to further bonding, and the forces between the molecules in liquid and solid hydrogen, nitrogen, oxygen, and fluorine are again of the van der Waals type. The kind of atomic linkage due to electron sharing is called *homopolar* or co-valent bonding, and is found in many other substances, although the boiling points may be very much

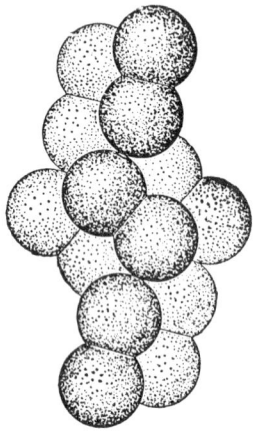

[*Courtesy Reinhold Publishing Corp.*

FIG. III.8. The crystal structure of iodine, showing the presence of pairs of atoms formed by diatomic I_2 molecules.

higher, and their other properties very different from those of hydrogen, nitrogen, oxygen, or fluorine.

Fluorine is the first member of the halogen family, and the remaining halogens, chlorine, bromine, and iodine all form stable diatomic molecules, whilst the melting points and boiling points rise with increasing atomic number. Here the diatomic molecules are again formed by simple homopolar bonds, and the forces which hold the molecules to one another in the liquid or solid are of the van der Waals type, which clearly becomes much more pronounced as the atomic number increases—iodine is a solid at room temperature. Fig. III.8 shows the crystal structure of iodine, and here it is clearly possible to recognize the diatomic molecules, the distances between two atoms in a molecule being

Assemblies of Atoms 101

considerably smaller than the remaining interatomic distances. The crystal structures of the remaining halogens are different, but so far as they have been determined they all contain diatomic molecules, and they may all be regarded as structures where homopolar bonding produces diatomic molecules which are themselves held together by van der Waals forces.

When we turn to the remaining elements of Groups VB and VIB, we find that arsenic, antimony, bismuth, sulphur, selenium, and tellurium in the gaseous state can all form stable diatomic molecules, but

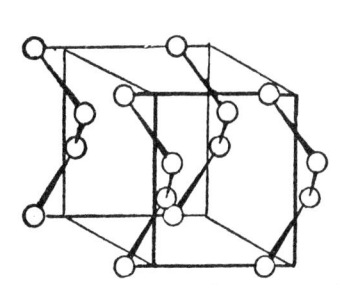

FIG. III.9. The crystal structure of tellurium or selenium, showing the spiral chains of atoms in which each atom has two close neighbours.

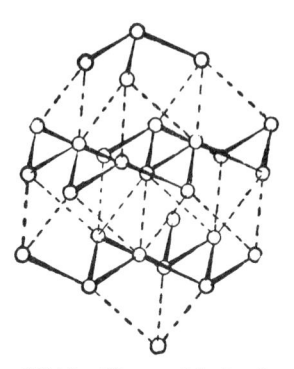

FIG. III.10. The crystal structure of antimony, showing the double layers of atoms, in which each atom has three close neighbours, and three at a greater distance in the next layer.

more complicated molecules (*e.g.*, S_8, As_4) are also found, whilst the solid state is much more stable; antimony, for example, has a melting point of 630° C. The crystal structures of arsenic, antimony, bismuth, selenium, and tellurium* are of great interest, because they illustrate what has been called the $(8 - N)$ rule. In the halogens we have seen that the crystal structures show diatomic molecules, in which each atom shares an electron with its one close neighbour and thereby builds up a complete octet. The crystal structure of the Group VI elements selenium and tellurium is shown in Fig. III.9 and it will be seen that the atoms form spiral chains in which each atom has two close neighbours,

---

* The structures of sulphur and phosphorus are more complicated, but we omit these because they are non-metallic. The structure of black phosphorus is different from that of arsenic, but is such that each atom has three close neighbours. The structure of rhombic sulphur contains S_8 molecules in the form of closed puckered rings in which each atom has two close neighbours. These structures are thus further examples of the $(8 - N)$ rule.

the remaining interatomic distances being considerably greater. This clearly suggests that in these crystals each atom is building up an octet of electrons by sharing one electron with each of its two close neighbours to form normal co-valent bonds. Other forms of selenium exist in which the atoms are arranged in closed rings, and here again each has two close neighbours. The crystal structure of arsenic, antimony, and bismuth in Group V is shown in Fig. III.10, and here it will be seen that the structure consists of double layers in which each atom has three close neighbours. Since the atoms of the elements of Group VB contain 5 electrons, they require 3 more electrons to complete an octet, and the crystal structure suggests clearly that this is accomplished by each atom forming three normal co-valent bonds and sharing one electron with each of its three close neighbours. It will be seen that if N is the number of the group in the Periodic Table, these crystal structures, and those of the halogens, are characterized by the fact that each atom has $(8 - N)$ close neighbours. In these structures we may regard the chains of tellurium atoms and the double layers of antimony atoms as constituting immense molecules in which the atoms are held together by simple co-valent linkages. The nature of the forces which bind the chains or double layers to one another is not so well understood. They were first thought to be of a van der Waals type, but in selenium the closest distance between atoms in adjacent chains (3·49 Å) is less than would be expected for normal van der Waals binding, and the bonding between adjacent chains probably involves some metallic character,* owing to resonance (see p. 158) between molecular and ionic configurations. Whatever view is taken, it is generally recognized that in crystals such as those of tellurium and antimony, two kinds of cohesive force exist. It should be noted that, under high pressure, bismuth, antimony, and tellurium form new modifications which are metallic and exhibit super-conductivity.

According to the $(8 - N)$ rule, the elements of Group IVB should crystallize so that each atom has $8 - 4 = 4$ close neighbours. Here we find that carbon (diamond), silicon, germanium, and grey tin crystallize in the well-known diamond type of structure in which each atom has four close neighbours arranged tetrahedrally, as shown in Fig. III.11. In this case we are to regard the whole crystal as constituting an immense co-valent molecule, in which each atom builds up an octet of electrons by sharing one electron with each of its four neighbours. It will be noted that the stability of these co-valent bonds decreases with increasing atomic number, as can be seen from the following list of melting points:

* A. R. von Hippel, *J. Chem. Physics*, 1948, **16**, 372; F. de Boer, *ibid.*, p. 1173.

Assemblies of Atoms 103

Diamond . . . . . . 3570° C (approx.)
Silicon . . . . . . 1421° C
Germanium . . . . . 937° C
Grey tin . . . . . . Stable only at low temperatures.

The same principle applies to the co-valent bonds in the crystals of the antimony, tellurium, and iodine types. Thus the co-valent linkages between the atoms in the iodine I_2 molecules are weaker than those in

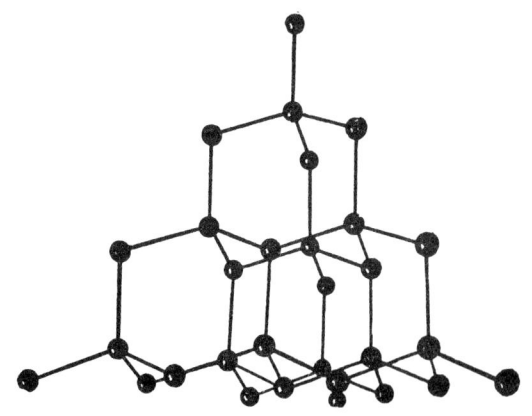

[Courtesy George Bell and Sons.]

Fig. III.11. The crystal structure of the diamond, showing how each atom has four equidistant neighbours.

bromine Br_2 molecules, and the co-valent linkages between adjacent atoms in the spiral chains of tellurium (Fig. III.9) and the double layers of antimony (Fig. III.10) are weaker than the corresponding bonds in selenium and arsenic. The van der Waals or other forces between the molecules do not, however, behave so simply; thus iodine is a solid, and bromine a liquid at room temperatures, but bismuth has a lower melting point than antimony.

It is to be noted that crystals which possess structures of the $(8 - N)$-rule type are either insulators (p. 219) or "semi-conductors" (p. 227), or semi-metals (p. 399). Thus, diamond is an insulator, whilst silicon shows a metallic lustre and possesses a slight electrical conductivity, but the conductivity increases with rise of temperature in contrast to the behaviour of normal metals.* On moving back to Groups IIIB,

* In a semi-conductor the resistance decreases when small amounts of impurity enter into solid solution, whereas in a normal metal the resistance diminishes with increasing purity.

H

IIB, and IB of the Periodic Table and then to the transition elements and elements of the A sub-groups, we find that with the exception of boron all these elements are true metals which conduct electricity and have normal temperature coefficients of resistance, *i.e.*, the resistance increases as the temperature rises. Nearly all these elements in one or other of their allotropes crystallize in one of the three typical metallic structures, the face-centred cubic, the body-centred cubic, and the close-packed hexagonal structures, which are illustrated in Fig. III.12. The

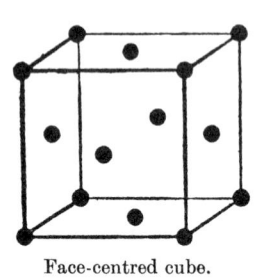

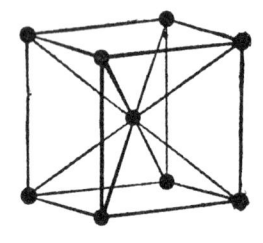

Face-centred cube.   Body-centred cube.

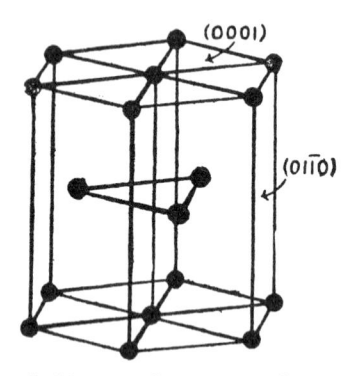

(0001)

(01$\bar{1}$0)

The close-packed hexagonal structure with axial ratio 1·633.

FIG. III.12. The three typical metallic structures.

face-centred cubic structure is one which would result from the stacking of close-packed layers of spheres so that the fourth layer is vertically over the first—the stacking sequence may be denoted $ABCABC$ ... If close-packed layers of spheres are stacked so that the third layer is vertically over the first, the stacking sequence may be denoted

Assemblies of Atoms

ABAB ..., and the structure obtained is the close-packed hexagonal structure of Fig. III.12, with axial ratio* equal to 1·633. In most metals, the axial ratio is slightly less than 1·633, and the structure corresponds to the close-packing of spheroids.

In the elements of the A sub-groups, the α and β forms of manganese crystallize in complex structures, but δ-manganese, which is the stable form at the melting point, is body-centred cubic like δ-iron. γ-manganese is face-centred cubic, although on quenching this is changed to a face-centred tetragonal structure. Tungsten has the normal body-centred cubic structure, which is often called α-tungsten because for many years it was thought that a second form, β-tungsten, also existed, but it is now known that this is in reality a tungsten-oxide phase. In uranium the α and β modifications are complex, but γ-uranium is again body-centred cubic.

The rare-earth elements after europium crystallize in one or more of the typical metallic structures, but in some of the earlier rare earths the structure corresponds to the stacking of close-packed layers with stacking sequences *ABACABAC* ... (Ce, Nd) or *ABABCBCAC* ... (Sm). This suggests that, in the earlier rare earths, the bonding hybrids involve a proportion of $4f$ wave functions, but that in the later rare earths the $4f$ states are not involved. Many of the actinides crystallize in complicated or unusual structures which are not understood, but it is probable that the bonding hybrids involve a greater proportion of $5f$ functions than the proportion of $4f$ functions involved in the rare earths.

The three structures of Fig. III.12 are fundamental and it is remarkable that they cover such a large number of the metallic elements.

In the B sub-groups zinc and cadmium crystallize in a curious modification of the close-packed hexagonal structure which is shown in Fig. III.13. This is similar to the normal close-packed hexagonal structure, except that it is extended in the direction of the c-axis, so that the axial ratio (p. 319) is much greater than that for close-packed spheres (1·633). Each atom has six close neighbours in the basal plane and six others at a greater distance, three in the layer above and three below. Mercury does not crystallize in this structure, but forms instead a simple rhombohedral structure as shown in Fig. III.14. Here again each atom has six close neighbours, so that, considered empirically, the $(8 - N)$ rule applies to the elements of Group IIB as well as to those of Groups VB, VIB, and VIIB and the earlier members of Group IVB. There can,

---

* If c is the height of the unit cell in Figs. III.12 and III.13, and a is the side of the hexagon in the basal plane, the axial ratio is equal to c/a. For close-packed spheres the axial ratio equals $\sqrt{(8/3)} = 1 \cdot 633$.

however, be no question of the formation of simple co-valent linkages in the crystals of zinc, cadmium, and mercury, because, since each atom has only two valency electrons, there are insufficient electrons to form

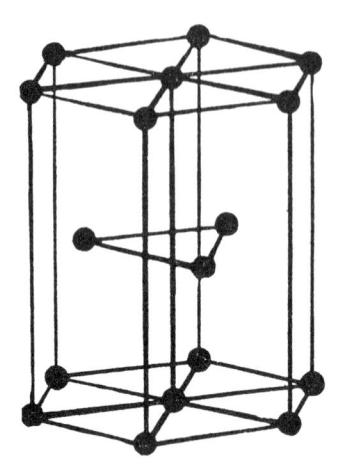

FIG. III.13. The crystal structure of zinc with axial ratio 1·856.

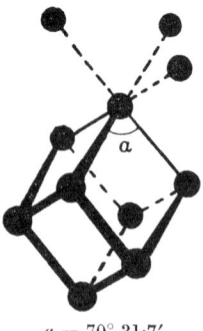

$a = 70° 31·7'$.

FIG. III.14. The crystal structure of mercury.

One cell of the simple rhombohedral structure is shown. There is one atom at each corner of this cell, and three atoms from adjacent cells have been included in order to show that each atom has six equidistant neighbours.

Assemblies of Atoms 107

six co-valent bonds per atom. The metallic properties of these three elements are also normal.

In Group IIIB, gallium forms an abnormal crystal structure. Indium crystallizes in a face-centred tetragonal structure, but as can

Face-centred cube.   Indium structure.

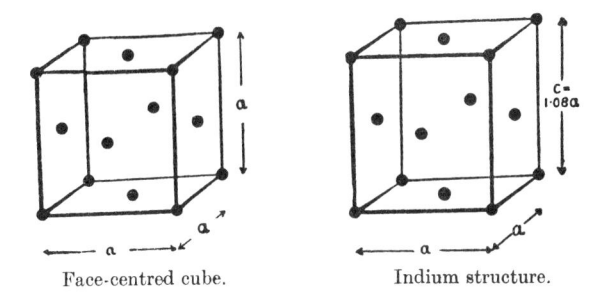

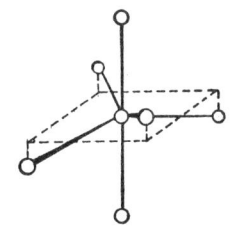

Fig. III.15. Structure of white tin.

be seen from Fig. III.15, the axial ratio, c/a, is very nearly unity, so that the structure is really only a very slightly distorted form of the typical face-centred cube. The next member of the sub-group, thallium, crystallizes in typical metallic structures, and the same applies to lead in Group IVB. White tin has an abnormal structure in which the atoms are arranged as shown in Fig. III.15. This is a greatly distorted tetrahedral arrangement, and is not understood, but it appears in some way to be a fundamental structure because, under high pressure, it is the stable structure for silicon, germanium, and tin, as well as for the compounds InSb, GaSb, and AlSb. It is to be noted that although the closest interatomic distance in grey tin (2·8 Å) is less than those in white tin (3·01–3·02 Å), the atomic volume of grey tin is greater than that of white; high pressure thus favours the structure with the smaller volume, in accordance with Le Chatelier's Principle, and the transition temperature (grey $\rightleftharpoons$ white tin) is lowered by pressure.

It will be seen, therefore, that when the elements are surveyed as a whole, they fall into four main classes:

(1) The inert gases.

(2) The elements hydrogen, nitrogen, and oxygen, which form very stable diatomic molecules and exist as gases down to very low temperatures.

(3) The elements preceding the inert gases. These form diatomic, and often more complex, molecules in the gaseous state, and crystallize according to the $(8 - N)$ rule, so that their valencies are satisfied in the solid crystal. These are insulators, semiconductors, or semi-metals.

(4) The metals, which nearly all crystallize in one or more of the three typical metallic structures. At the present time important developments, especially by Heine and his co-workers, in theoretical investigations of the electronic structures of metals by the pseudo potential method (p. 238) are yielding convincing explanations in fundamental terms of the causes of some of the distorted metal structures, such as those in In, Ga, Hg, and even Zn and Cd.

It will be noted that in the First Long Period of the Table of the Elements, germanium crystallizes in the diamond type of structure. In the next Period, the modification of tin (grey tin) which crystallizes in this structure is stable only at low temperatures, whilst in the Third Long Period, lead does not crystallize in this structure at all, but forms a face-centred cubic structure. The reason for this difference is to be found in the relative stabilities of the $(4s)^2$, $(5s)^2$, and $(6s)^2$ sub-groups of electrons which have already been discussed in Section II.1 (p. 56). The results of X-ray crystal structure analysis show that the closest distances of approach of the atoms in thallium and lead (*ca.* 3·5 Å) are considerably greater than in gold (2·88 Å), whereas the closest interatomic distances in gallium and germanium are slightly less than that in copper. This has been taken to indicate that in thallium and lead, the $(6s)^2$ sub-group of electrons is sufficiently stable to persist in the solid crystals of these elements, so that thallium and lead in the crystalline state are analogous to univalent and divalent metals, respectively, rather than to trivalent and tetravalent elements; this explains why lead does not crystallize in the diamond structure which involves four valency electrons per atom. Later work suggests that this view is too extreme and that, although some " memory " of the $(6s)^2$ sub-shell will persist in the metallic state, it is incorrect to regard these two electrons as forming part of an undistorted ion core. Interatomic distances suggest that

Assemblies of Atoms 109

similar, although weaker, effects are present in indium and white tin. It must, however, be emphasized that this is a characteristic of the crystals of the elements, and that in many of their alloys these metals may act according to their normal valencies of three and four; e.g. when thallium dissolves in solid magnesium it does so as a typical trivalent metal. In the liquid state, also, these elements are essentially three- and four-valent.

3. Bonding Energies.

In the previous section we have dealt with the crystal structures resulting from the different kinds of bonding force, and we may now discuss the magnitudes of the bonding energies involved. In the case of a metallic crystal the heat of sublimation is the same as the heat of atomization, i.e., the energy required to convert the crystal to a monatomic gas. It is convenient to deal with the heat of atomization, expressed in kcal/g-atom, and this may be denoted by ΔH_{AT}. If a metallic crystal is regarded as an array of charged ions held together by v electrons per atom, the bonding energy per electron $\Delta H_E = \Delta H_{AT}/v$. In simple metals such as the alkalis and alkaline earths, v is the same as the chemical valency of the metal, but in other metals the predominant valency of the chemist may be different from v.

The crystal structures of hydrogen, oxygen, nitrogen, and of the halogens involve diatomic molecules that are held together by van der Waals forces. The process of atomization of the crystals involves first the overcoming of the relatively weak van der Waals forces to give a diatomic gas, followed by the much greater energy change needed to dissociate the diatomic molecules. We may therefore write $\Delta H_{AT} = \Delta H_{SUB} + \Delta D$, where ΔH_{SUB} is the heat of sublimation and ΔD the heat of dissociation, each referring to one gram atom.* All values given refer to sublimation at 298° C or at the melting point, whichever is the lower.

Van der Waals Forces.

Table IV shows the values of ΔH_{AT} for the crystals of the inert gases, in which the atoms are bound together by van der Waals forces. The values are small, and the increase with atomic number is due to the greater deformability of the electron clouds as they become larger or, to put it in another way, to the decreased isolation of the s^2p^6 shell from higher energy levels.

* The dissociation of a diatomic molecule produces 2 atoms, and consequently ΔD equals one half the energy of dissociation of the X–X bond. Similarly, ΔH_{SUB} is one-half the heat of sublimation per gram molecule.

Table V gives the values of ΔH_{AT}, ΔH_{SUB}, and ΔD for some crystal structures involving diatomic molecules, and also the values in Å of the intermolecular and interatomic distances d_m and d_a. For crystals of hydrogen, nitrogen, oxygen, fluorine, and chlorine, the values of ΔH_{SUB} are small and of the same order as those of ΔH_{AT} for the inert gases; in

TABLE IV.

Element	He	Ne	A	Kr	Xe	Rn
ΔH_{AT} .	0·05	0·5	1·8	2·6	3·7	—

both cases van der Waals forces are involved. For these diatomic molecular crystals, the values of d_m/d_a are large, and the ΔH_{AT} values are determined predominantly by those of ΔD. (In passing through the halogen group, d_m/d_a decreases with increasing atomic number, until for iodine, $d_m/d_a = 1 \cdot 3$, and ΔH_{SUB} (9) is as much as one half of ΔD (18).

TABLE V.

Element	H_2	N_2	O_2	F_2	Cl_2	Br_2	I_2	At_2
ΔH_{SUB} .	0·1	0·8	0·9	1	3	5	9	—
ΔD .	52	113	59·5	19	29	23	18	—
ΔH_{AT} .	52	113	60	20	32	28	27	—
d_m(Å) .	3·5	3·4	3·5	3	3·3	3·30	3·54	—
d_a(Å) .	0·7	1·1	1·21	1·4	1·99	2·28	2·67	—
d_m/d_a .	5	3	3	2·1	1·5	1·45	1·3	—

Here, therefore, the bonding energy resulting from the intermolecular forces is of the same order as that from the intramolecular bonds, and some authors would regard the two kinds of force as merging into one another and no longer sharply distinguished.

Inspection of the Periodic Table reveals a trend towards metallic character on descending any of the later columns, whether the resultant bonding energy increases by loss of simple van der Waals character (as in fluorine → iodine) or decreases by loss of covalent character (as in carbon → lead). This can be understood, because an increasing size of charge cloud on the individual atoms means an increasing tendency for the localized directional character of co-valent bonds to be replaced by a more generalized sharing of electrons.

The above ΔD energy values result from co-valent bonding, and the increase on passing from F(19) → O(59·5) → N(113) is due to the change from single (F–F) to double (O=O) to triple (N≡N) bonds in which 2, 4, and 6 electrons, respectively, are concerned in the bond. In organic

Assemblies of Atoms

compounds, the bond energies (per atom of carbon) of the single, double, and triple carbon–carbon bonds in H_5C-CH_3, $H_2C=CH_2$, and $HC\equiv CH$ are 40, 60, and 115 kcal/g-atom of carbon. The F_2 molecule has an abnormally low value* of ΔD but, apart from this, Table V shows that the strengths of the covalent bonds in the diatomic halogen molecules become weaker as the atomic numbers increase. The same applies to the double and triple bonds in the elements of Groups VIB and VB, as can be seen from Table VI. From these Tables it will be seen that, for

TABLE VI.

Molecule	O_2	S_2	Se_2	Te_2	N_2	P_2	As_2	Sb_2	Bi_2
ΔD^*	59·5	50	33	26·5	113	58	45	35·8	24

* To preserve uniformity with the previous and later discussion, the above values are the heats of dissociations per gram atom, i.e., ½ the heat of dissociation of the diatomic molecule.

the elements immediately preceding the rare gases, in any one Period the bond energies are in the order single-bond < double-bond < triple-bond. Taken as a whole, and omitting hydrogen, the energies of single co-valent bonds are of the order of 20–30, those of double bonds from 25 to 60, and those of triple bonds from 24 to 115 kcal/g atom.

Turning to the metallic elements, we find that abnormally high values of ΔH_{AT} are shown by the elements of the First Short Period. Table VII shows the values of ΔH_{AT} and ΔH_E, for the alkali and alkaline-earth elements, and for those immediately following them in the Short Periods. It will be seen that for Na, K, and Rb the values of ΔH_{AT} ($=\Delta H_E$) are roughly the same as those of the single co-valent bond energies of the halogen element at the end of the same Period. Further, with the single exception of Mg, the bond energy per electron (ΔH_E) is almost the same for an alkali metal and its following alkaline earth. In the Short Periods, again with Mg as an exception, the ΔH_E values are roughly the same for the elements of all the first Groups. These figures strongly support the view of Pauling (p. 162) that the bonding forces in metals are closely related to those in normal co-valent bonds.

Table VIII shows the values of ΔH_{AT} and ΔH_E for the early transition elements. For the elements of Groups IA–VA, there is general agreement that all the outer electrons are concerned in the cohesion, so that v is equal to the Group Number. The figures show that for Groups IA and IIA the value of ΔH_E is ~ 20 kcal/g electron, and there is then an in-

* This abnormally small value is due to the small value of the interatomic distance in the F_2 molecule which leads to repulsion between the inner electrons.

crease of 50–70% on passing to Group IIIA, after which ΔH_E remains roughly constant on passing through Groups IIIA → IVA → VA, and then falls on passing to Group VIA.

TABLE VII.

Element	Li	Be	B	C	F
ΔH_{AT}	38	78	135	171	
ΔH_E	38	39	45	43	$\Delta D = 19$
	Na	Mg	Al	Si	Cl
ΔH_{AT}	26	35·6	77·5	108	
ΔH_E	26	18	26	27	$\Delta D = 29$
	K	Ca			Br
ΔH_{AT}	21·5	42·2			
ΔH_E	21·5	21			$\Delta D = 22·7$
	Rb	Sr			I
ΔH_{AT}	19·5	39·1			
ΔH_E	19·5	19·6			$\Delta D = 18$
	Cs	Ba			
ΔH_{AT}	19	42·5			
ΔH_E	19	21·2			

There is no general agreement as to the explanation of the variations in ΔH_E along the Long Periods, this being due to the fact that, when electrons of widely different character contribute to bonding, there is no *a priori* reason for expecting the bond energy to be simply related to the number of electrons involved in the bonding. Thus, in all three Long Periods, ΔH_E increases between Groups IIA and IIIA, and this is

TABLE VIII.

Element	K	Ca	Sc	Ti	V	Cr	Ni	Cu
ΔH_{AT}	21·5	42·2	91	113	123	95	103	81
ΔH_E	21·5	21	30	28	25	16	—	—
	Rb	Sr	Y	Zr	Nb	Mo	Pd	Ag
ΔH_{AT}	19·5	39·1	101·4	146	173	158	91	68
ΔH_E	19·5	19·6	34	37	25	26	—	—
	Cs	Ba	La	Hf	Ta	W	Pt	Au
ΔH_{AT}	19	42·5	103·5	150	187	200	135	87
ΔH_E	19	21	34	38	37	33	—	—

Assemblies of Atoms

probably associated with the increased d-character of the electrons in the metal. Some authors postulate specific types of hybrid bonds, while others do not regard this as meaningful. Beyond Groups VA, correlation of ΔH_{AT} with Group Number breaks down, and there is again no general agreement as to the interpretation. Some authors regard the wave functions of all the electrons outside the inert-gas shells as being significantly modified in the solid state, so that the Group Number is equal to the number of electrons per atom involved in the bonding process. Others (e.g. Pauling, Engel, Brewer) regard only some of the electrons as having wave functions modified in this way, and distinguish between bonding and non-bonding d-electrons, the latter having wave functions resembling those in free atoms.

It is to be noted that the values of ΔH_{AT} for Cu, Ag, and Au, are of the order of 70–90 kcal/g-atom, and are very much higher than would be expected for univalent metals. There is general agreement that, in these metals, the cohesion due to the single valency electron per atom is being reinforced by interaction between the outermost electrons of the ions.

The above figures show that, in kcal/g-atom the values of ΔH_{AT} vary by a factor of 10, from ~ 20 for the alkali metals to 200 for W. On the other hand, where the values of v are certain, the values of ΔH_E lie within the range 20–45 kcal/g atom, and the range of variation is comparatively small.

According to Trouton's rule, the entropy of evaporation of all liquids is approximately constant and equals 22 cal/deg g mole, so that if L_e is the latent heat of evaporation of a metal of boiling point T_e (°abs.), we may write in kcal:

$$L_e = T_e \times 22/1000$$

This rule breaks down for substances of very low boiling point (e.g. He, N_2), or when there is association or dissociation of the substance in the gaseous phase.

With metals of large co-ordination number (8–12) in the crystalline state, the entropies of fusion are roughly constant and $\simeq 2\cdot 2$ cal/deg g atom, so that we may write in kcal

$$L_f \sim T_f \times 2\cdot 2/1000$$

where L_f is the latent heat of fusion, and T_f the melting point (°abs.). Since $T_f < T_e$, the latent heat of fusion, is less than one tenth of the latent heat of evaporation.

4. Assemblies of Unlike Atoms.

The Ionic Bond.

The preceding section has dealt with the characteristics of the elements, and we have now to consider the classes of substance which are found when two or more elements unite to form a system. In the case of the elements we have seen that the interatomic forces can be divided into the three main classes: (a) van der Waals Forces, (b) co-valent bonds, and (c) metallic linkages. When we consider the forces which exist between unlike atoms in solid or liquid solutions, compound molecules, &c., examples of all the above types are found, but apart from these a new type of interatomic force is met with, namely the *ionic* or polar bond. We have already seen that the inert gases correspond with very stable groupings of electrons. An atom of sodium has one valency electron outside a stable group of eight, whilst an atom of chlorine has seven valency electrons, and thus needs one electron to complete a group of eight. In this case chemical combination occurs by the sodium atom giving up its valency electron to the chlorine atom so that the latter acquires a complete group of eight electrons. The resulting compound is said to be polar or ionic, since the transference of the electron from one atom to another produces a positive and a negative ion, and these are held together by their mutual electrostatic attraction which pulls the ions together until the attractive force is balanced by the repulsion produced when the electron clouds of the ions overlap. Since the resulting molecules each consist of a positive and a negative ion, they possess a relatively large dipole moment, in contrast to co-valent molecules such as H_2, where the electron cloud of the valency electrons is much more uniformly distributed between the two atoms.

To understand the nature of a dipole moment we may imagine two equal and opposite charges, $+e$ and $-e$, separated by a distance l. These form an electrical dipole whose moment, μ, is defined by the relation $\mu = el$. In a molecule we can only obtain the probable electron distribution or electron density in the space surrounding the nuclei, and if the molecule is asymmetric, the charge distribution will in general possess a permanent electric moment, or dipole moment, which can be measured experimentally. Interatomic distances in molecules are of the order of 10^{-8} cm, and since the electronic charge $e = 4 \cdot 80 \times 10^{-10}$ e.s.u., molecular dipole moments are of the order of 10^{-18} e.s.u., and this quantity is called a *debye*. The dipole moment of the HCl molecule may be written $\overset{+\rightarrow}{\text{HCl}} = 1 \cdot 04$ debyes, where the arrow indicates that the negative charge is at the chlorine end of the molecule. It should be

Assemblies of Atoms 115

appreciated that when a co-valent bond is formed by electron sharing, the shared electrons may spend a greater proportion of their time near to the one atom (A) than to the other (B), and in this case the atom A will be associated with a negative charge; the concept of shared electrons does not imply that the electrons are associated equally with the two atoms concerned.

The crystal structure of sodium chloride is shown in Fig. III.16.

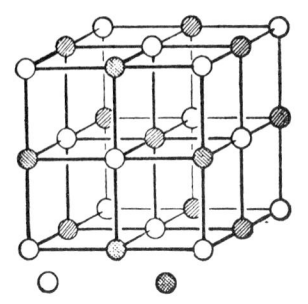

FIG. III.16. Sodium chloride.

Here it will be seen that all trace of diatomic Na^+Cl^- molecules has disappeared, and the structure consists of an array of positive and negative ions which are held in contact by their mutual electrostatic attraction, and held apart by the repulsion which results from the overlapping of the electron clouds of the ions. Magnesium oxide, $Mg^{++}O^{--}$, also possesses the NaCl type of structure; here the divalent magnesium atoms have each given up two electrons to the oxygen atoms, which thereby acquire complete groups of 8 electrons. Since the ions in MgO are doubly charged, we can understand why the cohesion in magnesium oxide is much stronger than in sodium chloride, and the high melting point of magnesia is the result of this increased attraction. A very satisfactory theory of the properties of simple polar structures of this kind has been developed by Born and others. In this work the potential energy of the crystal is written in the form:

$$\Phi = -\frac{a}{r} + \frac{b}{r^n}.$$

Here the first term is the potential energy resulting from an electrostatic attraction varying as the inverse square of the distance (r) between the ions, and the constant a can be calculated for the different types of crystal structure. The second term is purely empirical and represents a repulsion. Comparison with experimental values of various physical properties requires the constant n to be of the order of 10 for salts, so that the repulsion increases very rapidly at small distances. Later developments of the theory indicate that a repulsion term of the form $e^{-r/a}$ is more probable.

From the point of view of the theory of metals, it is important to realize that there is an almost continuous transition from typical ionic compounds such as those described above, to the purely metallic linkage of normal metals. There is, for example, a whole series of compounds such as MgO, MgS, MgSe, MgTe, MnSe, PbSe, all of which crystallize in the NaCl type of structure. For these it is a general principle that the compound is more stable, and its melting point higher, the more electropositive the metal and the more electronegative the element from Group VI. In applying this principle the sizes of the ions must also be

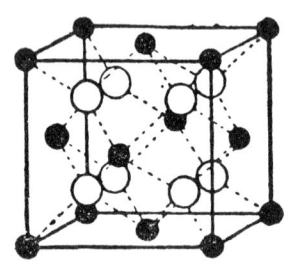

FIG. III.17. Calcium fluoride.

taken into consideration. Thus, in the compound PbSe, both ions are larger than those in the compound MgO, and the attractive forces are thus weaker in PbSe. In this case the size-effect reinforces that of the electrochemical factor, but this is not always so.

Fig. III.17 shows the crystal structure of calcium fluoride, CaF_2, which is a typical ionic compound of the type AB_2. This structure is anti-isomorphous* with the crystal structures formed by the compounds Mg_2Si, Mg_2Ge, Mg_2Sn, and Mg_2Pb. Here again the melting points fall as the metalloid element becomes less electronegative, so that Mg_2Pb has the lowest and Mg_2Si the highest melting point of the series. These compounds exhibit a metallic lustre and possess a slight electrical conductivity, and according to Robertson and Uhlig† Mg_2Pb is like a normal metal of high resistance with a normal temperature coefficient. On the other hand, Mg_2Sn is a semi-conductor with a negative temperature coefficient of resistance, and is probably an intrinsic semi-conductor, i.e., a semi-conductor whose conductivity is a genuine characteristic of

---

* i.e., the metal ions in CaF_2 occupy the position of the negative ions in Mg_2Si.
† W. D. Robertson and H. H. Uhlig, *Trans. Amer. Inst. Min. Met. Eng.*, 1949, **180**, 345.

Assemblies of Atoms

the substance itself and is not due to impurities. These compounds can readily be understood as a continuation of the typical ionic compounds into the region of weakly electronegative elements, and it will be noted that they have the formulae to be expected from the normal valencies of the elements concerned. As will be discussed later (p. 221) this may only be part of the truth, and the new electron theories suggest that these compounds may be looked on from a quite different point of view. Whatever may be the truth about the particular series of compounds Mg_2Si, Mg_2Ge ... it is undoubtedly a general principle that stable compounds tend to be formed when one element is very electropositive compared with the other, and these compounds may be regarded as the result of an electrochemical factor. Even in cases where definite compounds do not exist, the electrochemical factor may exert an important influence on the type of alloy structure formed and some of these effects are described below.

Co-Valency Compounds.

Turning from the effects of polar or ionic bonding to that of co-valent bonding, it is, of course, well known that the idea of electron sharing introduced by G. N. Lewis and developed in terms of quantum mechanics, has explained the structures of many chemical compounds. This work has been discussed in many books, and we shall, therefore, not deal with it here.* We may note that just as the structure of the diamond is one immense molecule in which each atom exerts its normal valency of four, so crystals with this, or a similar structure, are formed by combinations of atoms, provided that the average number of valency electrons per atom is four. Thus, the compound AlSb crystallizes in the diamond type of structure and, since aluminium has three valency electrons per atom, whilst antimony has five, the average number is four per atom. Fig. III.18 shows the zinc blende and wurtzite structures, which are characterized by each atom having four neighbours arranged tetrahedrally. Here we find the zinc-blende structure is formed by a whole series of compounds, such as GaAs, CdSe, and CuBr, in which the characteristic is that there are eight valency electrons to two atoms, so that the ratio of four valency electrons to one atom is preserved. Some of these compounds show metallic properties, and, just as in the case of the ionic and metallic bonds, we cannot draw any hard and fast line between the metallic and co-valent linkage, and it is very probable that in many intermetallic phases forces of a partly

---

* The reader may consult N. V. Sidgwick's " Co-valent Link in Chemistry " and L. Pauling's " The Nature of the Chemical Bond."

covalent nature are present. It must also be emphasized that although for convenience it is usual to describe purely ionic and purely co-valent linkages, there are innumerable cases in which the linkage is of an intermediate form in the sense that, although an electron is shared between two atoms, it spends a greater proportion of its time nearer to one atom than to the other, and so adds a negative charge to the one atom.

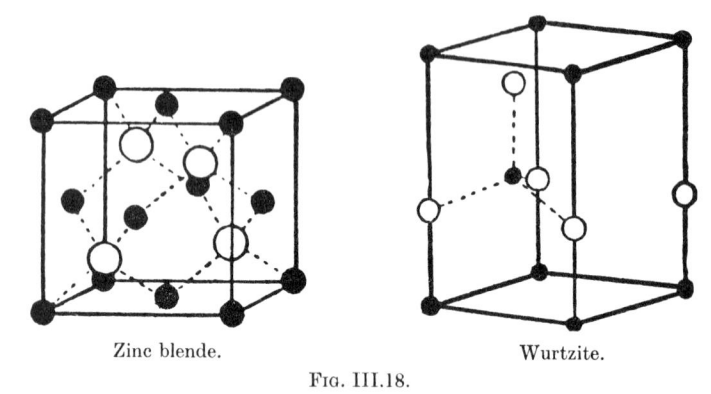

Zinc blende.   Wurtzite.
Fig. III.18.

Mooser and Pearson[*] have given an interesting explanation of the factors that determine whether a given AB compound forms a structure of the NaCl (ionic) or tetrahedral (wurtzite) type. They introduce an average quantum number, $\bar{n}$, which is the mean of those of the outermost electrons of the two atoms, e.g. for ZnO, $\bar{n} = (4 + 2)/2 = 3$, and argue that, as $\bar{n}$ increases, co-valency and directional (tetrahedral) bonding will decrease. If x_A and x_B are the electronegativities of the two atoms, an increase in $(x_A - x_B)$ means an increase in the ionic nature, and hence a decrease in directional bonding. The matter is slightly complicated by the effect of the radius ratio r_a/r_c of anion and cation but on plotting $\bar{n}$ against $\Delta x \cdot r_a/r_c$ it is found that a smooth curve divides the points for the NaCl-type compounds from those for the tetrahedral types. When the same policy is adopted for the Mg_2Si-type structures, metallic conduction is confined to large $\bar{n}$ and small Δx.

Alloy Structures: Primary Solid Solutions.

Before proceeding further it is convenient to review briefly the types of phase usually encountered in alloys, and for this purpose we have

---

[*] E. Mooser and W. B. Pearson, *Acta Cryst.*, 1959, 12, 1015; W. B. Pearson, *J. Physics Chem. Solids*, 1962, 23, 103.

Assemblies of Atoms 119

reproduced the equilibrium diagram of the system copper–zinc in Fig. III.19. We shall assume that the reader is familiar with diagrams of this kind,* showing the phases which exist in alloys of different compositions when they are in equilibrium at different temperatures. It will be seen that the copper-rich alloys of this system consist of the α

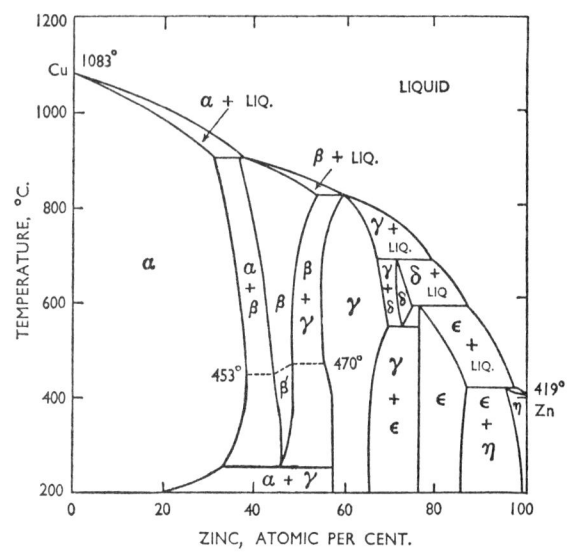

Fig. III.19. Equilibrium diagram of the system copper–zinc.

solid solution of zinc in copper, whilst at the zinc-rich end of the system there is a narrow range of solid solution of copper in zinc. Solid solutions of this kind may be called *Primary Solid Solutions* or *Terminal Solid Solutions*, and they retain the crystal structure of the parent metal. These primary solid solutions are of two types, the so-called *substitutional* and *interstitial* solid solutions. In the interstitial solid solutions, the solute atoms fit into the interstices or spaces between the atoms of the solvent. This kind of solid solution is naturally confined to systems where one atom is very much smaller than the other. The most important example is the solid solution of carbon in γ-iron, which forms the

---

* Reference may be made to C. H. Desch, "Metallography," W. Rosenhain and J. L. Haughton, "An Introduction to the Study of Physical Metallurgy," W. Hume-Rothery, J. W. Christian, and W. B. Pearson, "Metallurgical Equilibrium Diagrams and Experimental Methods for Their Determination," **1952**: London (Institute of Physics), or A. Prince, "Alloy Phase Equilibria." **1966**: (Amsterdam Elsevier).

I

basis of the steels, whilst boron and nitrogen can also form interstitial solid solutions with some transition elements.

In the more general case of substitutional solid solutions the solute atoms replace those of the solvent, so that the two kinds of atom are situated on a common lattice. It is only natural that the formation of an appreciable range of substitutional solid solution is confined to cases where the two kinds of atom do not differ too greatly in size, and this is confirmed experimentally. It might at first be thought that the modern picture of an atom would make it impossible to allot any definite size to the atom of a particular element, since the electron cloud has no sharp boundary. Actually, however, considerable progress has been made by assuming that the " atomic diameter " of an element is given by the closest distance of approach between two atoms in the crystal of the element. It was first shown by Hume-Rothery, Mabbott, and Channel-Evans (1934) that if the atomic diameters are defined in this way,* the solid solubility of one element in another becomes very restricted if the atomic diameters of the two elements concerned differ by more than about 15%, whilst within this range solid solutions may be formed provided that other factors are favourable. This principle requires amplification in the case of elements such as lead which may change their valency state on entering into solid solution (p. 108), and also with some elements such as gallium where an abnormal crystal structure gives one very close distance of approach to the atoms; but in general the principle of what may be called favourable and unfavourable size-factors is well established. In view of frequent misunderstanding, it must be emphasized that considerations of atomic diameters alone do not enable one to predict that one metal will dissolve in another in the solid state. This will readily be appreciated from Fig. III.19, where there is a wide solid solution of zinc in copper, but only a very restricted solid solution of copper in zinc. The concept of a favourable size-factor is thus most useful as a preliminary test which must be satisfied if there is to be any chance for one metal to dissolve appreciably in another, but whether a solid solution is actually formed is often controlled by other factors. Of these, the one most readily understood is that if the two metals differ greatly in the electrochemical series, there is a very general tendency for them to form stable intermetallic compounds at the expense of the primary solid solutions.

The extent of primary solid solutions varies from complete miscibility in the solid state, as in the systems silver–gold and copper–nickel, to a solubility which is too small to be measured, and whose presence can

* See Appendix II, p. 422.

Assemblies of Atoms

only be deduced indirectly. Primary solid solutions at high temperatures are usually described as having a random arrangement of the different kinds of atom on the common lattice, and this is true in the sense that there is no regular arrangement of the two kinds of atom which extends over distances large compared with the atomic diameter. It was long suspected that in most primary solid solutions there is short-range order, in the sense that a given atom tends to have more neighbours of the opposite kind than would result from a purely random distribution. This has now been proved experimentally by analysis of the variations in the intensity of general background scattering on X-ray diffraction films. In some cases (*e.g.*, the solid solution of zinc in aluminium) the reverse effect is found, and there is a tendency for like atoms to cluster together to an extent greater than would result from a random distribution.

Alloy Structures: Superlattices.

In cases where the primary solid solution is of wide extent, there is a very general tendency for alloys in the regions of 25 and 50 at.-% of solute, on slow cooling or on annealing at a low temperature, to

Fig. III.20. The Cu_3Au superlattice structure.

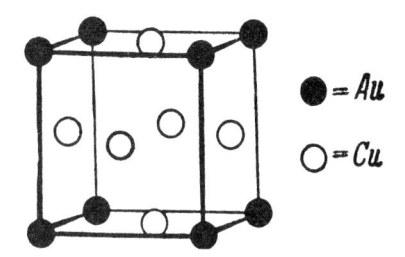

● = Au

○ = Cu

undergo an atomic rearrangement in which, although the structure as a whole remains that of the solvent, the two kinds of atom occupy definite positions relative to one another, so that long-range order is established. In the system copper–gold, for example, the alloy of composition Cu_3Au consists at high temperatures of a disordered face-centred cubic structure, but on slow cooling, the two kinds of atom arrange themselves as shown in Fig. III.20. This kind of structure is called a *Superlattice*, and is usually characterized by a tendency for the solute atoms to keep as far away from one another as possible.* The

---

* This principle has been discussed by W. Hume-Rothery and H. M. Powell, *Z. Krist.*, 1935, **91**, 23. See also H. Lipson, "Progress in Metal Physics," Vol. II, p. 1. 1950: London (Butterworths Scientific Publications).

Atomic Theory

formation of a superlattice may be regarded as a process in which the lattice strain produced by the solute element in the random solid solution is reduced by the formation of an ordered structure. It is therefore natural that, provided the solvent and solute atoms are of sufficiently similar size to permit the formation of a wide solid solution, the tendency to form superlattices increases with increasing difference in atomic diameters, since the greater this difference, the greater the strain to be relieved. Thus, superlattices are found in the system copper–gold, where the atomic diameters are Cu = 2·54 Å, Au = 2·88 Å, but not in the system silver–gold, where the sizes of the two atoms are nearly the same. But apart from this simple effect of atomic diameter, the formation of a superlattice is favoured by increasing electrochemical factor. It is probable, therefore, that there is a continuous range from superlattices involving only the metallic bond, and formed merely to relieve the lattice strain resulting from atoms of unequal size, to superlattices with increasing degrees of ionic bonding, and the presence of bonding of a partly ionic nature in solid solutions must always be borne in mind. Up to the present, this phenomenon has not been developed mathematically. In the alloys of the transition elements with one another, superlattices are often found for reasons which are not yet understood. Thus, superlattices occur in the systems iron–cobalt and iron–nickel, although in these the atomic diameters and general chemical characteristics of the elements are very similar (see also p. 358).

Alloy Structures: Intermediate Phases.

Reference to Fig. III.19 will show that in the system copper–zinc increasing zinc content results in the primary α-solid solution being succeeded by a two-phase region. There is then a homogeneous region, denoted β, the composition of which is in the region of 50 at.-% of zinc (CuZn). This has a disordered body-centred cubic structure at high temperatures, and at low temperatures an ordered body-centred cubic structure of the caesium chloride type (Fig. III.22). The change from order to disorder takes place in the neighbourhood of 460° C, and is a simple superlattice transformation. At higher percentages of zinc there is a phase denoted γ with a characteristic type of structure, and then the ε-phase, which has a close-packed hexagonal structure. The mean composition of the ε-phase is about 82 at.-% of zinc, whilst its boundary on the copper-rich side is at about 79 at.-% of zinc. In the system copper–zinc the intermediate phases are all of variable composition, but in many alloy systems intermediate phases are found, the composition of which is fixed, or at any rate varies over such a narrow range that no variation can be detected experimentally. Such phases are

Assemblies of Atoms 123

usually called intermetallic compounds, and their composition generally corresponds with a reasonably simple whole-number ratio of the constituent atoms. Considerable discussion has taken place as to whether the intermediate phases of variable composition are to be regarded as intermetallic compounds, and to some extent this is a matter of definition. The usual text-book distinction between mixtures and compounds does not take into account the possibility of a compound forming a solid solution with its constituent elements. If we regard a metal as an assembly of positive ions held together by the attraction of the shared valency electrons, a crystal of a metallic element may itself be considered as a giant molecule, in which the atoms or ions are held together by the shared electrons. From this point of view all the phases in alloy systems, including the primary solid solutions, are to be regarded as involving chemical combination of one kind or another, although some of them may not correspond with the formation of polyatomic molecules in the liquid or vapour states. If, on the other hand, we prefer to regard the metallic linkage as something distinct, then there is no reason why some intermediate phases, although differing in crystal structure from the parent metals, should not have their atoms held together by purely metallic linkages, and we may expect a continuous transition from purely metallic intermediate phases to those involving ionic or co-valent bonds.

In the great majority of cases, the formulæ of intermetallic compounds do not agree with those to be expected from the normal principles of valency, and this is readily understood when it is remembered that these phases are often normal conductors of electricity. We have already seen that in the ionic and co-valent compounds the electrons tend to be bound into stable groups, and the ordinary principles of valency are the result of the fact that definite numbers of electrons form stable groupings. Consequently, if free electrons are to be left to give the substance metallic properties, the combining ratios must in general be different from those in the normal compounds of chemistry, where all the valency electrons form stable groups, so that the metallic properties are lost.

The development of the science of alloy structures makes it possible to understand some of the principles concerned. In the first place, as we have explained above, when two metals differ markedly in electrochemical characteristics there is a general tendency to form stable compounds, many of which have the formulæ to be expected from the normal valencies, and which may be regarded as an extension of the simple ionic or co-valent compounds into the region of the weakly electronegative elements, e.g., compounds such as Mg_2Sn and Mg_3Sb_2. In the second place, some intermediate phases have been shown to have structures which are clearly determined to a large degree by the relative sizes of the atoms concerned. For example, the so-called Cu_2Mg type of structure is formed by Cu_2Mg, W_2Zr, KBi_3, and Au_2Bi; here an approximately constant ratio of atomic diameters is shown, and it is particularly interesting to note that the bismuth atom can play the part of the smaller atom in KBi_2, and of the larger atom in Au_2Bi. Phases of this kind, in which the structure is determined primarily by the relative sizes of the atoms, are sometimes called *Laves Phases*, because they were established by the brilliant X-ray investigations of F. Laves; an interesting review of some of this work has been given by G. V. Raynor.* In the same way the ternary aluminium–copper–magnesium alloys give rise to a phase of composition approximately $Mg_3Cu_7Al_{10}$ and this is isomorphous with the phase Mg_2Zn_{11} in the binary system magnesium–zinc. The characteristic here is an approximately constant ratio of large magnesium atoms to small atoms of copper, aluminium, or zinc, since Mg_2Zn_{11} has 33 small atoms to 6 large magnesium atoms, whilst in $Mg_3Cu_7Al_{10}$ there are 34 small atoms to 6 large magnesium atoms. The determination of crystal structure by X-ray diffraction methods has enabled several groups of intermetallic phases to be classified in this way, but up to the present it has not been possible to understand the other factors which are concerned; so that one cannot yet explain why the above types of structure, for example, are formed in some alloy systems but not in others where the atomic diameters have the ratio which is clearly the main characteristic of the structure.

If hard spheres of a given size are packed together, a given atom cannot be in contact with more than 12 neighbours, *i.e.*, the maximum co-ordination number is 12. If, however, spheres of two or three different sizes are packed together, it is possible to have co-ordination numbers greater than 12 (*e.g.* 14, 15, or 16). The significance of this was first recognized by Frank and Kasper,† who showed that not only the Laves phases but also some of the complex transition-metal intermetallic compounds (*e.g.* σ phases) correspond with structures in which the atoms were packed together as compactly as possible. For further details of these the reader may consult the book edited by J. H. Westbrook.‡

---

* G. V. Raynor, " Progress in Metal Physics," Vol. I, p. 1. **1949**: London (Butterworths). R. L. Berry and G. V. Raynor, *Acta Cryst.*, 1953, **6**, 178. F. Laves, " Theory of Alloy Phases," p. 124. **1956**: Cleveland, O. (American Society for Metals).
† F. C. Frank and J. S. Kasper, *Acta Cryst.*, 1958, **11**, 184; 1959, **12**, 483.
‡ " Intermetallic Compounds " (edited by J. H. Westbrook). **1967**: New York and London (John Wiley).

Assemblies of Atoms

Alloy Structures: Electron Compounds.

We have discussed above the main characteristics of the system copper–zinc, whose equilibrium diagram is shown in Fig. III.19, and examination has shown that alloys with somewhat similar characteristics are formed by many alloys of copper, silver, and gold with the elements of the earlier B Sub-Groups of the Periodic Table, as well as with beryllium, magnesium, aluminium, and silicon. Thus, in the system copper–gallium there is a body-centred cubic β-phase stable only at high temperatures, whilst a phase with the "γ-brass" structure appears at higher concentrations of gallium. The β-phase lies in the region of 25 at.-% of gallium and at low temperatures there is a close-packed hexagonal phase at roughly the same composition. Thus, on passing from zinc to gallium the proportion of the solute metal contained in the β-phase becomes less. The first step towards an understanding of this class of alloy was made in 1926 when Hume-Rothery* pointed out that the compositions of the β-phases in some of these alloys had the common characteristic that they corresponded to a ratio of 3 valency electrons to 2 atoms. In the systems copper–zinc, copper–aluminium, and copper–tin, the compositions of the β-phase are near to those given by the formulæ $CuZn$, Cu_3Al, and Cu_5Sn, and it will be seen that if copper, zinc, aluminium, and tin are given their normal valencies of 1, 2, 3, and 4 respectively, these compositions all correspond with a ratio of 3 valency electrons to 2 atoms. This principle was then confirmed and extended by the remarkable X-ray work of Westgren and his collaborators which showed that in alloys of copper, silver, and gold, body-centred cubic, γ-brass, and close-packed hexagonal structures tended to occur at electron:atom ratios of 3:2, 21:13, and 7:4 respectively. The electron concentration 3:2 also tends to give rise to phases with structures similar to that of β-manganese, or of the close-packed hexagonal type, and in some systems the body-centred cubic phase is stable at high temperatures, and at low temperatures a β-manganese (*e.g.*, Ag_3Al) or close-packed hexagonal (*e.g.*, Cu_3Ga) phase is formed. Later work by Ekman† showed that similar structures were formed in some alloys of the transition elements, provided that these were regarded as possessing a zero valency. Thus in the systems nickel–zinc and iron–zinc, phases with γ-brass structures occur at the compositions Ni_5Zn_{21} and Fe_5Zn_{21}, so that if the transition metal is allotted a zero valency the ratio of valency electrons to atoms, which may be called the *electron concentration*, is 42:26 or 21:13. Actually, these phases extend over a considerable range of composition, and it is probably better to say that the

---

* *J. Inst. Metals*, 1926, **35**, 309.
† *Z. physikal. Chem.*, 1931, [B], **12**, 57.

transition metal acts as though it possessed a low valency, rather than to emphasize a valency of exactly zero. The apparently low valency of the transition metals in the γ-phases may be taken to indicate that if they contribute electrons to the crystal lattice, they also absorb electrons into non-bonding groups (Section VI.4). It was shown by Bradley and Gregory that in ternary copper–zinc–aluminium alloys, the typical γ-brass structure could be obtained if the proportions of the elements were adjusted so as to maintain the electron concentration at the required value 21:13. Later, Hume-Rothery, Mabbott, and Channel-Evans showed that where the size-factors were favourable, the $\alpha/\alpha + \beta$ phase boundaries in many alloys of copper and silver could be roughly superimposed if the diagrams were drawn in terms of electron concentration instead of the more usual weight or atomic percentages of the elements. All this work led to the conclusion that in some alloys the structures of the phases and the forms of the equilibrium diagrams were determined primarily by the electron concentration, and as suggested by J. D. Bernal, phases of this kind may conveniently be called *electron compounds*. In general, this class of compound is formed only when the size-factor is reasonably favourable, but the degree of toleration is greater than for primary solid solutions. Thus in the system copper–cadmium, the cadmium atom is so large that the primary solid solution in copper is very restricted, but a γ-phase of slightly varying composition is formed at a composition near to Cu_5Cd_8. The very beautiful X-ray work of Bradley shows that although the general γ-type of structure is preserved, the detailed atomic arrangement in Cu_5Cd_8 is different from that in Cu_5Zn_8, where the size-factors are favourable. This clearly suggests that there may be a continuous transition from structures determined mainly by the electron concentration to those where the atomic diameters are the controlling factors. There are, however, many exceptions to these simple electron-concentration rules. Thus, in the system silver–lithium, a typical γ-brass structure is formed by an alloy whose composition is approximately Ag_3Li_{10}. No combination of univalent elements can give an electron concentration of 21:13, but some recent developments of electron theory (p. 300) offer a possible explanation of the abnormal behaviour of lithium.

We shall describe in Part VI the theoretical explanation of these electron compounds, and for the moment we may merely note that, in some alloys, the electron concentration appears to be the essential factor in determining the structure of the alloy. More detailed work, both experimental and theoretical, has shown that, in these alloys, the real characteristic is that phases of definite crystal structure occur at compositions corresponding to definite *numbers of valency electrons per*

Assemblies of Atoms 127

unit cell of the structure. In the general case, where all lattice points are occupied, the electron concentration (number of electrons per atom) is a simple fraction of the number per unit cell. There are, however, some alloys, such as the γ_3 phase of Cu–Ga, in which, on increasing the proportion of gallium, defect structures are formed in which atoms drop out of the structure in such a way as to maintain a constant number of valency electrons per unit cell, although the number of valency electrons per atom continues to increase.

Electron compounds may exist with ordered structures or superlattices, as in the low-temperature modification of β-brass, and also with what are usually called disordered or random structures, as in the high-temperature form of β-brass. As in the case of primary solid solutions, there is almost certainly some kind of short-range order in what are usually classed as disordered electron compounds. The tendency for an electron compound to develop order (both long range and short range) increases with increasing electrochemical factor, and in this connection it is instructive to examine the equilibrium diagram of the system silver–magnesium (Fig. III.21). In this case there is again a body-centred cubic phase in the equi-atomic region, but this now has a maximum freezing point at the equi-atomic composition, and possesses an ordered body-centred cubic structure in which the structure as a whole is body-centred cubic, but like atoms tend to avoid being closest neighbours, so that at the composition AgMg the alloy has the caesium chloride structure shown in Fig. III.22, where one kind of atom is situated at the centre and the other kind at the corners of the unit cube. Alloys in the β-phase area of the silver–magnesium system with compositions other than 50 at.-% of magnesium cannot have a perfectly ordered structure, but they do have partly ordered structures in which like atoms tend to avoid being closest neighbours. The work of Clarebrough and Nicholas has shown that at low temperatures a superlattice also exists in the α solid solution at the composition Ag_3Mg. When the systems copper–zinc and silver–magnesium are compared, we see that silver is more electronegative than copper, and magnesium is more electropositive than zinc, so that the electrochemical factor is much greater in the system silver–magnesium than in the system copper–zinc, and it is this which accounts both for the greater stability of the ordered structure and the maximum freezing point of the β-magnesium–silver phase. It is uncertain whether definite MgAg molecules exist in the liquid alloy, but it is clear that increasing electrochemical factor has made the β-magnesium–silver phase acquire some of the characteristics (maximum freezing point and ordered structure) which are usually associated with definite chemical compounds.

128   *Atomic Theory*

Even in systems where a phase with an ordered structure is not formed, the effect of the electrochemical factor can sometimes be seen. Fig. III.23 shows a comparison of the equilibrium diagrams of the

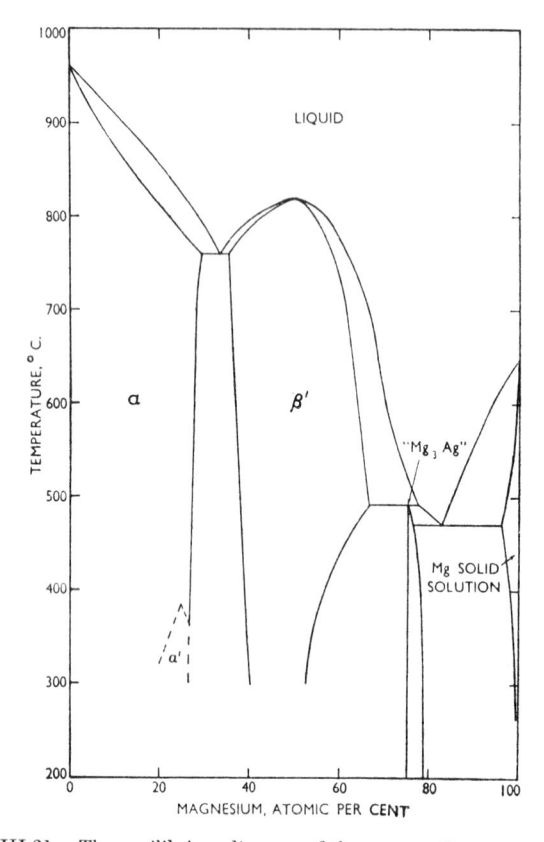

Fig. III.21. The equilibrium diagram of the system silver–magnesium.

The diagram is known accurately in the ranges 0–40 and 80–100 at.-% of magnesium. In the range 40–80 at.-% of magnesium the general form of the diagram is established conclusively, but the phase boundaries are not so accurate. The work of L. M. Clarebrough and J. F. Nicholas (*Australian J. Sci. Research*, 1950, [A], **3**, 284) shows that a superlattice exists in the α-solid solution at the composition Ag_3Mg. It is not yet known how far the ordered structures extend into the solid solution.

systems silver–cadmium and silver–zinc. The high-temperature regions of these are quite analogous to that of the system copper–zinc, and in each case increasing percentage of the solute results in the face-centred cubic α-solid solution being followed by a two-phase region and a dis-

Assemblies of Atoms

ordered body-centred-cubic β-phase. It will be noted that in the system silver–zinc, the liquidus curve shows a much greater change in direction at the temperature of the (α + liquid ⇌ β) peritectic horizontal than

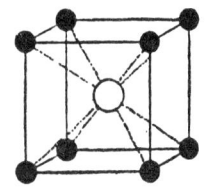

FIG. III.22. The CsCl structure.

in the system silver–cadmium. Zinc is more electropositive than cadmium, and consequently the electrochemical factor is greater in the system silver–zinc, and the equilibrium diagram near the β-phase in the

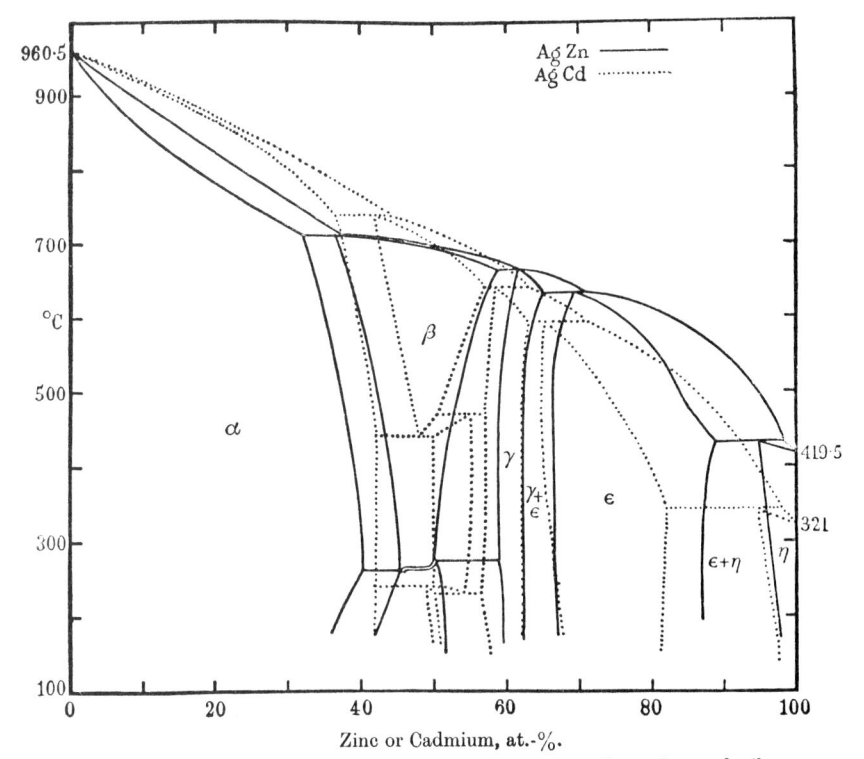

FIG. III.23. The equilibrium diagrams of the systems silver–zinc and silver–cadmium.

latter system is clearly beginning to acquire the characteristics of the silver–magnesium diagram, although the process has not been carried far enough to result in the production of a maximum in the liquidus curve. Other examples of this kind have been discovered* and it is important to realize that when the two constituents of an alloy differ considerably in electrochemical properties, the interatomic forces in the solid alloys are often partly of an ionic nature. In the elements copper, silver, and gold, the electronegative nature increases in the order:

$$Cu < Ag < Au.$$

This means that in alloys with electropositive elements such as magnesium or zinc, the electrochemical factor increases in the order $Cu < Ag < Au$, whereas in alloys with electronegative elements such as arsenic, the electrochemical factor increases in the order $Au < Ag < Cu$. It follows also that in alloys of copper, silver, or gold, with a series of elements such as:

$$Zn \to Ga \to Ge \to As \to Se$$

the element of Group IB begins by being the electronegative constituent of the alloy and finishes by being the electropositive constituent, and there is a point somewhere in the region of Group IVB where the electrochemical factor is very small.

Just as the electrochemical factor may result in an electron compound acquiring some of the characteristics of a polar compound, so there is no sharp boundary between electron compounds and compounds involving co-valent bonds, and there are undoubtedly cases where an electron compound involves atomic bonding of a partly co-valent nature. It will be realized, therefore, that the actual state of affairs in many alloy phases is very confused, since different types of interatomic cohesion may be superimposed, and may be modified by the difference in size of the atoms. In the case of the polar or ionic bond, we can obtain a reasonably simple picture of the process as the result of the attractions of opposite charges. The understanding of the co-valent bond is much more difficult, and we shall approach this subject by considering the general nature of the forces to be expected when two atoms are brought into contact, and then the nature of the cohesive forces in some simple molecules. The metallurgical reader will appreciate that although these molecules are non-metallic, it is only by considering such comparatively simple structures that we can begin to understand the nature of the different kinds of force which are superimposed on one another in the crystals of metals.

* For a review of this work, see the paper by Hume-Rothery, Reynolds, and Raynor, *J. Inst. Metals*, 1940, **66**, 191.

Assemblies of Atoms

SUGGESTIONS FOR FURTHER READING.

G. V. Raynor, " Progress in Metal Physics," Vol. I, p. 1. **1949**: London (Butterworths Scientific Publications).
H. Lipson, *ibid.*, Vol. II, p. 1. **1950**.
" Theory of Alloy Phases," **1956**: Cleveland, O. (American Society for Metals).
" Intermetallic Compounds " (edited by J. H. Westbrook). **1967**: New York and London (John Wiley).
" Phase Stability in Metals and Alloys " (edited by P. S. Rudman, J. Stringer, and R. I. Jaffee. **1968**: New York and London (McGraw-Hill).

5. Atomic Attraction and the Nature of van der Waals Forces.

In the preceding sections we have dealt with the characteristics of some of the main classes of substance, and we have now to consider the various processes which give rise to attraction between atoms, and hence to the formation of molecules, liquids, and solids. For this purpose we may consider first what happens when two atoms approach

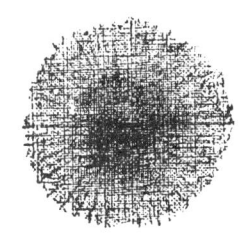

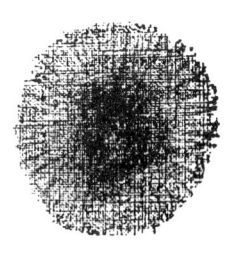

Fig. III.24.

one another. In Part II we have seen that an atom is to be considered as consisting of a positively charged nucleus surrounded by a cloud of negative electricity, and for a first approximation we may assume that the electron clouds of two similar atoms are spherical. Although there is no sharp boundary to the electron cloud of an atom, the cloud density diminishes so rapidly beyond a certain distance that we are justified in saying that some points are well outside the electron cloud of one atom, whilst others are inside.

We may now suppose that we have two atoms at a distance from one another which is greater than the extent of their electron clouds, as in Fig. III.24. If the spherical electron clouds were static and undeformable, there would be no force between the two atoms. Actually, however, the electron clouds result from the motion of the various electrons, and although we can no longer follow the trajectories of these in detail, we must still regard the individual electrons as being in

132 Atomic Theory

motion round their nuclei. We can see, therefore, that whilst an atom with a spherically symmetrical electron cloud will have no average electrical moment, or " dipole " moment, it will have a rapidly fluctuating dipole moment. Thus, in the sodium atom, the electron cloud of the 3s valency electron is spherically symmetrical, but at the instant when the electron is to the left of the nucleus, there will be a preponderance of negative electricity on this side of the atom, and hence a temporary dipole will be formed with a $+$ ve charge on the right, and a $-$ ve charge on the left. When, therefore, two atoms approach one another, the rapidly fluctuating dipole moment in each atom will affect the motion of the electron in the other atom, and a lower energy (*i.e.*, an

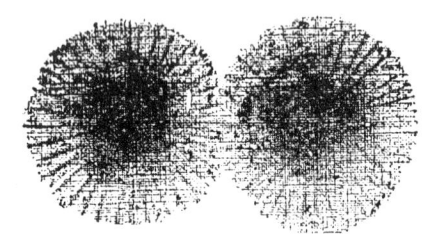

Fig. III.25.

attraction) can be produced if the electrons in the two atoms move in sympathy with one another. This phenomenon, in which the electronic motions in two atoms harmonize or sympathize with one another, gives rise to the " van der Waals forces "; this name is used because it is these forces which are responsible for the attractive term in the well-known gas equation of van der Waals. These forces are also sometimes called " polarization forces," and they produce a potential energy proportional to $-1/r^6$, if the zero of potential energy is taken to be that of the atoms at an infinite distance apart.

When two atoms approach one another from a large distance, forces of the van der Waals type are always produced, and we may regard them as drawing the atoms together until the electron clouds overlap slightly as shown in Fig. III.25. At this stage additional forces are introduced which, according to the nature of the atoms concerned, may produce either attraction or repulsion, quite apart from the van der Waals attraction. On the one hand, the fact that the electron clouds are overlapping slightly means that the outermost electrons of each atom are attracted by the nucleus of the other atom, and this produces an attraction between the two atoms. At the same time, the electron clouds repel one another, and as we shall see later (p. 150), wave

Assemblies of Atoms 133

mechanics indicates that this repulsion is particularly strong if the overlapping electron cloud is that of a complete group of 8 or 18 electrons.* As the atoms are drawn together the repulsion increases, until a stage is reached at which the attraction and repulsion balance one another, and equilibrium results.

The van der Waals forces are the only long-range forces which exist between neutral atoms. In many substances as the atoms approach one another, these forces are rapidly outweighed by those of other kinds, but the van der Waals forces are always present, and in the case of the inert gases with complete groups of electrons in the outermost shell, they are the only attractive forces which can be produced. It is thus the van der Waals forces which serve to hold the atoms together in the solid and liquid states of the inert gases, and in these substances we are to regard the atoms as drawn together by the van der Waals forces until these are balanced by the repulsive forces caused by the overlapping electron clouds of the atoms. In inert gases van der Waals forces are weak because the octets of electrons are so stable that the electronic motion in any one atom can be only slightly disturbed by the motion of the electrons in adjacent atoms. It will be noted that the boiling points of the inert gases rise, *i.e.*, the interatomic bonding becomes stronger, with increasing atomic number; this is because the electron clouds become larger and are more easily deformed or polarized. In quantum-mechanical treatments of van der Waals bonding, the factor of importance is the energy required to excite the atom to a higher configuration (*e.g.*, $ns^2np^6 \to ns^2np^5(n+1)s$), and spectroscopic data for the inert gases are in accordance with the dependence of boiling point on atomic number.

Forces of the van der Waals type are also responsible for the cohesion in liquid and solid hydrogen, nitrogen, and oxygen. Here the diatomic molecules are formed by simple co-valent bonds (electron sharing), and then the electrons in adjacent molecules move in sympathy or harmony with one another, and so give rise to van der Waals forces which draw the molecules together until the attraction is balanced by the repulsion due to the overlapping of the electron clouds of adjacent molecules. In the same way the cohesion between the iodine molecules in solid iodine (Fig. III.8) is regarded as due to van der Waals forces. As pointed out in Section III.2, the intermolecular forces are here clearly more pronounced, and comparable with the intramolecular ones, so that iodine is a solid at room temperatures.

Even when the main cohesion is due to forces of another kind, the van der Waals type of cohesion may still exist, and may exert an im-

* Or a group of 2 electrons in the first quantum shell.

portant influence. In the case of copper, for example, each atom has one valency electron outside a closed group of 18 electrons, and the general picture of the cohesion in solid copper is that the valency electrons are shared between adjacent atoms, and so serve to draw them together, until this attraction is balanced by the repulsion which results from the overlapping of the electron clouds of the ions. In this case, when we consider the motion of the valency electrons, we cannot regard these as completely independent of one another, because the electrostatic repulsion between two electrons prevents them from being in the same place at the same time. In this way the motion of the whole swarm of electrons is subject to what is called a *correlation effect*, in the sense that electrons tend to avoid each other. This correlation of motions affects the energy of the electrons, and these *correlation forces* are really an extension of the van der Waals forces to the case where the electrons are moving in the same region of space. According to one theory (see p. 305), the electrons in the outermost parts of an *ion* in solid copper are affected by the motion of the electrons in an adjacent ion, and if the two move in harmony there is a lowering of energy, and hence an attraction which is essentially of the van der Waals type.

6. The Interpretation of the Co-Valent Bond and the Nature of Exchange Forces.

In the above description we have seen how the van der Waals forces are to be understood as the result of a dynamic polarization of the electron cloud of one atom under the influence of adjacent atoms, and we have seen how this kind of attraction may exist in many types of substance, even where the main interatomic attraction is due to some other cause. We have now to consider how the simple co-valent bond is to be understood in terms of electron theory, and in particular why elements such as hydrogen, nitrogen, oxygen, and fluorine form stable diatomic molecules, whilst others such as helium or magnesium do not. The Lewis theory of the sharing of electrons so as to build up groups of 2, 8, or 18 electrons, is clearly satisfactory as a qualitative picture of the nature of valency, but it is entirely lacking in detail, and does not permit any quantitative estimation of properties such as heats of formation, interatomic distances, &c.

The Hydrogen Molecule Ion H_2^+.

The normal co-valent bond is formed by the sharing of two electrons between two atoms, as in the H_2 molecule. In order to illustrate some

Assemblies of Atoms

of the principles, we may consider first the much more simple case of the singly ionized H_2^+ molecule, in which the two protons are held together by one electron. In this case if we denote the two nuclei by 1 and 2, we have a condition of affairs in which at some instant the electron might be at the position shown in Fig. III.26, where R is the

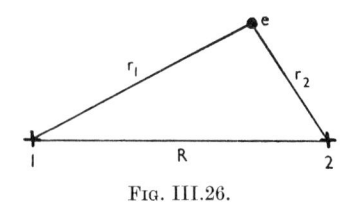

Fig. III.26.

distance between the nuclei, and r_1 and r_2 are the distances between the electron and the two nuclei. The two nuclei repel each other with a force equal to $\dfrac{e^2}{R^2}$, whilst the electron is attracted to the nuclei by forces equal to $\dfrac{e^2}{r_1^2}$ and $\dfrac{e^2}{r_2^2}$, respectively. The problem of the motion of the electron under these conditions is one that can be solved exactly by wave mechanics. If we consider a plane containing the two nuclei, the variation of potential will be of the form shown in Fig. III.27. When

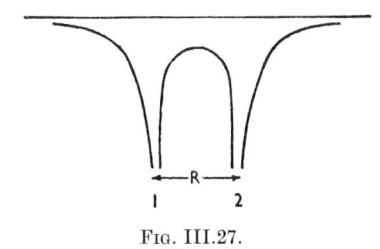

Fig. III.27.

R is large, and the two nuclei are far from each other, the potential round each will resemble that round the nucleus of a single isolated hydrogen atom, and will become identical with this in the limit $R \to \infty$. At the other extreme when $R = 0$, the two nuclear charges of $+e$ unite to form a charge of $+2e$, and the motion of the electron round this is the same as that of the one electron moving round the singly charged He^+ ion of helium.

The application of wave mechanics to the above problem shows that the two lowest energy states correspond to solutions in which the wave function is spatially symmetric or spatially anti-symmetric, as shown in Fig. III.28. As explained later (p. 139) it is customary to use the terms symmetric and anti-symmetric in a quite different sense connected with + or − signs in the combination of functions. As this easily leads to confusion we shall use the term *gerade* to denote the wave function which is symmetric in space, and *ungerade* for that which is spatially anti-symmetric, and shall use the terms symmetric and anti-symmetric in the sense defined on p. 139.

The *gerade* function of Fig. III.28 has no nodes, and gives a large probability of finding the electron in the region between the nuclei. When

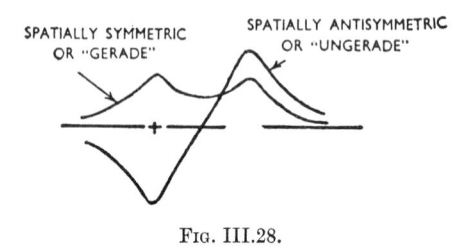

SPATIALLY SYMMETRIC OR "GERADE"   SPATIALLY ANTISYMMETRIC OR "UNGERADE"

Fig. III.28.

the nuclei become very far apart, the minimum in the curve drops to zero, and the function becomes the same as that of the (1s) states of the two separated atoms. In the other extreme, when the nuclei coalesce, the *gerade* function becomes the same as that of the (1s) state of the united He^+ ion.

In contrast to this, the *ungerade* curve of Fig. III.28 has a nodal plane through the point mid-way between the nuclei, and when the two nuclei coalesce the function becomes that of the (2p) state of the He^+ ion, with the electron cloud stretched out in the direction of the line joining the two nuclei.

It is important to note that the *gerade* or *ungerade* nature of the function refers to the wave function, and not to the symmetry of the electron cloud. The charge distribution for a stationary state is proportional to ψ^2, and so the *ungerade* function of Fig. III.28 gives a charge distribution which is *gerade* about the mid-point between the nuclei.

At all distances between the nuclei, the energy of the *gerade* function is less than that of the *ungerade* function. If we imagine the two nuclei to approach each other from a large distance apart, their mutual repulsion results in an increase in the energy, but for the *gerade* function this is counterbalanced by the accumulation of negative charge in the region between the two atoms. There is, thus, at first a decrease in the energy which sinks to a minimum, and then rises as the nuclei become

Assemblies of Atoms 137

so close together that the electrostatic repulsion is the predominant term. The curve connecting the total energy of the H_2^+ molecule with the internuclear distance is thus of the form of Fig. III.29, curve (a) when the electron has the *gerade* function. For the *ungerade* function, the electron cloud is not concentrated in the region between the nuclei, and as the latter approach one another there is nothing to compensate for the nuclear repulsion, and the energy increases continually as in curve (b) of Fig. III.29.

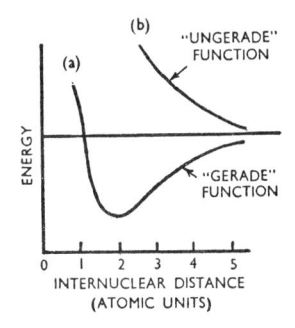

Fig. III.29.

The shape of the curve (a) of Fig. III.29 means that the *gerade* function gives rise to a stable molecule because, if the distance between the nuclei is that of the minimum point, work will be required either to move them further apart or closer together. It is customary to call this electron state an *attractive orbital* or a *bonding orbital*, although as in the case of free atoms we can no longer follow the electron in its path round the nuclei. The *ungerade* function is then said to give rise to a *repulsive* or *non-bonding* orbital.

It is a characteristic of all bonding processes that the curve connecting the total energy with the distances between the atoms passes through a minimum. In the case of a diatomic molecule, it is then possible for the molecule to have vibrational energy corresponding to vibrations of the atoms along the line joining their centres, and these vibrational states are stable, provided that the energy is not sufficient to make the system jump out of the potential valley. In this way the single energy value of the electronic state may be accompanied by a number of vibrational energies, and corresponding vibrational spectra are obtained. This illustrates a general principle that any system where the curve connecting energy with interatomic distance shows a minimum may possess vibrational energy up to a certain limit.

Atomic Theory

The Indistinguishability of Electrons and the Pauli Principle.

The problem of the singly ionized H_2^+ molecule is capable of solution because there is only one electron, and so the mutual influence of electrons on each other's motions does not have to be considered. All problems involving more than one electron are enormously more complicated, and accurate solutions have been obtained only for the helium atom and the H_2 molecule. The complications arise partly from the electrostatic repulsion between two electrons, and partly because empirical facts require the use of certain types of wave functions when more than one electron is present.

In Section II.1 we described the electronic structures of the atoms of the elements, and explained how these satisfied the Pauli Exclusion Principle, according to which no two electrons in an atom can be in exactly the same state as defined by all the four quantum numbers. In order to understand the nature of the co-valent bond, it is necessary to discuss the meaning of this Principle more fully. In the problem of the free hydrogen atom, the electron state is described by the three orbital quantum numbers n, l, and m_l, together with the spin quantum number m_s. For atoms of higher atomic number, no accurate solution exists, but for elements of lower atomic number it is a useful approximation to assume that any one electron can be regarded as moving in the field of the nucleus, together with a spherically symmetrical field resulting from the smoothing out of all the electrons except the one under consideration. In this case, the electron considered moves in a central field, and the solution of the wave equation leads to the introduction of orbital quantum numbers n, l, and m_l, similar to those for the hydrogen atom.

The quantum numbers of a molecule are complicated, and for the present purpose it is sufficient to say that, just as in the case of a free atom the quantum number m_l is a measure (in units of $h/2\pi$) of the component of the angular momentum in the direction of an applied field, so in the case of two atoms a quantum number λ is introduced, such that the component of angular momentum along the line joining the centres of the two atoms is equal to $h\lambda/2\pi$. It is then customary to denote states for which $\lambda = 0, 1, 2, 3 \ldots$ by the symbols σ, π, δ, ϕ. . . . An electron in a molecule has also a spin quantum number m_s which equals $\pm \frac{1}{2}$, and other quantum numbers which are more difficult to define.

In both free atoms and in molecules, the state of the electron is defined by the orbital quantum numbers on the one hand, and the spin quantum number on the other. In all cases the Exclusion Principle

Assemblies of Atoms

applies, and only one electron can be in a state as defined by all the quantum numbers. The Pauli Principle is thus one of the great principles underlying the motions of electrons, and wave functions must be chosen so as to satisfy the Principle.

The second fundamental principle is that which states that electrons are indistinguishable. This implies that, in a two-electron problem, it is meaningless to assert that electron No. 1 is in state a, and electron No. 2 in state b, although it is meaningful to assert that states a and b each contain one electron.

We may consider first a simplified problem of two electrons whose motions are statistically independent, so that the probability of finding one electron in state a and the other in state b is the product of the individual probabilities. In this case, if $\Psi_a(1)$ were the wave function of electron No. 1 and $\Psi_b(2)$ that of electron No. 2, a wave function of the form $\Psi_a(1)\Psi_b(2)$ would be a wave function for the complete system of two electrons. An alternative function $\Psi_a(2)\Psi_b(1)$ would imply that electron No. 2 was in state a, and electron No. 1 in state b. These functions, by themselves, are unsatisfactory because they distinguish between the two electrons.

The form of the Schrödinger equation is such that any linear combination of the above solutions is also a solution, and we can thus have the following two combinations:

$$\Psi_a(1)\Psi_b(2) + \Psi_a(2)\Psi_b(1) \quad \ldots \ldots \quad (S)$$
$$\Psi_a(1)\Psi_b(2) - \Psi_a(2)\Psi_b(1) \quad \ldots \ldots \quad (A)$$

The first of these is called the *symmetric function* * because it is symmetrical with respect to electron exchange. The second is called the *anti-symmetric function*. They have the interesting characteristic that they satisfy the principle of indistinguishability of electrons. For the S function the value of $\Psi\Psi^*$ for the system of two electrons will be given by:

$$[\Psi_a(1)\Psi_b(2) + \Psi_a(2)\Psi_b(1)][\Psi_a{}^*(1)\Psi_b{}^*(2) + \Psi_a{}^*(2)\Psi_b{}^*(1)]$$

and the product involves all combinations of 1 and 2 with a and b. The anti-symmetric function has the same characteristic.

Examination of the above combinations shows that the Pauli Principle requires the *anti-symmetric form*. In this case if $\Psi_a = \Psi_b = \Psi$, the function reduces to

$$\Psi(1)\Psi(2) - \Psi(2)\Psi(1) = 0$$

and thus vanishes. In contrast to this the symmetric form would not

---
* It is for this reason that we use the terms *gerade* and *ungerade* to describe the *spatially* symmetric and anti-symmetric functions.

vanish, and would therefore permit the two electrons to be in the same state.

As explained above, the total wave function involves the orbital and spin quantum numbers, and may be written:

$$\Psi_{Total} = \Psi_{Orbital}\Psi_{Spin}.$$

It follows therefore that if the spins are parallel the orbital wave functions must be anti-symmetric, and it can also be shown that if the spins are anti-parallel the orbital wave functions are symmetric.

Electron Correlation and the Exclusion Principle.

We have now to consider how the above principles affect the motions of electrons when more than one is present. We may imagine first a two-electron problem in which the electrons have the same spin, and the orbital wave functions must be anti-symmetric. In this case, if we ignore the electrostatic repulsion, the wave function will vanish if both electrons are at the same place (xyz), because the wave function for the one electron will then be $\Psi_a(xyz\sigma)$, and that for the other electron $\Psi_b(xyz\sigma)$, where σ refers to the spin, and the anti-symmetric combination reduces to:

$$\Psi_a(xyz\sigma)\Psi_b(xyz\sigma) - \Psi_a(xyz\sigma)\Psi_b(xyz\sigma) = 0.$$

The function will remain very small if the two electrons are very near to one another so that the co-ordinates of the one $(x_1y_1z_1)$ are very nearly the same as those of the other $(x_2y_2z_2)$. In this way the use of anti-symmetric combinations of orbital wave functions tends to keep the electrons of parallel spin away from one another. This effect is quite distinct from that due to the Coulomb or electrostatic repulsion between two particles with the same charge, and is a new principle resulting from the restrictions imposed on wave functions if the latter are to be reconciled with the Pauli Exclusion Principle and the Indistinguishability of Electrons.

It is also easy to show that the use of symmetric combinations of orbital wave functions tends to make the electrons crowd together. In the case of the free atoms, this crowding together of electronegatively charged electrons increases the electrostatic energy, and the symmetric combination of orbital wave functions gives a larger energy. It is for this reason that, in the filling up of the sub-sub-groups of electrons, the process occurs (p. 51) in such a way that the electrons have parallel spins as long as this is possible, because parallel spins imply the use of anti-symmetric orbital functions.

This effect of the Exclusion Principle on the motion of electrons with parallel spins is quite general, and is of importance in the electron

Assemblies of Atoms

theory of metals. As shown later the crystal of a metal can be considered as a gigantic molecule, in which the electrons occupy an immense number of energy states, and the Exclusion Principle means that electrons of parallel spin keep farther apart than they would do if the Principle did not operate.

The Hydrogen Molecule.

We may now consider the problem of the diatomic hydrogen molecule, H_2. Here there are two electrons moving round two nuclei, and at some instant the position might be as in Fig. III.30 where the two nuclei

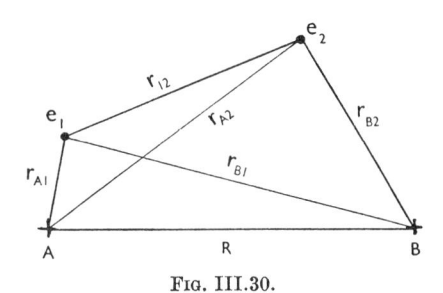

Fig. III.30.

A and B are separated by a distance R, and the electrons are at e_1 and e_2 with the distances as shown. There are then attractive forces between each electron and each nucleus, and repulsive forces between the two nuclei, and between the two electrons. The potential energy is thus of the form

$$\frac{e^2}{R} + \frac{e^2}{r_{12}} - \frac{e^2}{r_{A1}} - \frac{e^2}{r_{B1}} - \frac{e^2}{r_{A2}} - \frac{e^2}{r_{B2}}.$$

When this is substituted into the Schrödinger equation, it is not possible to obtain an accurate solution, and we may now consider the two chief methods which have been devised in order to obtain an approximation.

The Method of Molecular Orbitals.

In this method, each electron is imagined to move in the fields of the two nuclei, with the other electron smeared out so as to give a charge distribution which is *gerade* about the mid-point between the nuclei. Under these conditions each electron behaves in much the same way as the single electron in the problem of the H_2^+ molecule described above, except that the attraction between one electron and a nucleus is weakened owing to the screening effect of the other electron. For the H_2 molecule there are *gerade* and *ungerade* functions, analogous to those for the H_2^+ molecule, and it is again the *gerade* function which shows a minimum in the curve connecting the total energy with the internuclear distance. As the atoms are separated, the wave functions must resemble those for the isolated $(1s)^1$ atoms, and when the nuclei are so close together that they coalesce, the wave functions must become those of the $(1s)^2$ helium atom. As there are two electrons it is clear that these must have opposite spins if the functions are to be derived from $(1s)$ atomic states. If the spins are opposed, the orbital wave functions must be symmetric (p. 139), and as explained above, this leads to a concentration of electrons which is found in the region between the two atoms.

It is to be noted that the above splitting of the $(1s)$ level as two atoms of hydrogen approach one another is quite analogous to the broadening of the energy levels of valency electrons in solid metals which is revealed by soft X-ray spectroscopy (p. 88). In the hydrogen molecule there are only two atoms, and the $(1s)$ level splits into the two sub-levels—*gerade* $(\sigma_g\ 1s)$ and *ungerade* $(\sigma_u\ 1s)$. In a metal crystal, the number of atoms is of the order of $10^{23}/cm^3$ and the splitting of the level is correspondingly fine.

The splitting up of the atomic levels in this way does not alter the number of electron states per atom. The free hydrogen atom has its one electron in the $(1s)$ state, and the orbital is thus half filled, since this state can contain two electrons. In the H_2 molecule this single level of the free atom has split into the $(\sigma_g\ 1s)$ and $(\sigma_u\ 1s)$ states, each of which could contain two electrons. If a crystal of sodium contains N atoms, the $(3s)$ level of the free atom will split into N states, each of which could hold two electrons.

The simple method of molecular orbitals gives a minimum in the curve connecting the total energy of the molecule with the internuclear distance, but the energy values at large separations of the atoms are inaccurate. The reason for this is that, at large separations, the wave functions should approach those for single free hydrogen atoms, whereas the simplifying assumption means that in considering the behaviour of an electron near a particular nucleus, part of the nuclear charge is screened or neutralized by the second electron. This assumption is justified when the two nuclei are close together, but not when they are far apart, and it is a defect of the simple method of molecular orbitals that it breaks down at large internuclear distances.

This defect can be corrected by more detailed calculations. It is a general principle that if a system can exist in alternative configurations, the lowest energy of the system is not that of either configuration alone, but is formed by a superposition of functions. In the above description

Assemblies of Atoms 143

we have referred to the state of the H_2 molecule with both electrons in the $(\sigma_g \, 1s)$ state, and this has a lower energy than any other state. A still lower energy can be obtained if the function is regarded as slightly perturbed by the superposition of a small proportion of the state in which both electrons occupy the $(\sigma_u \, 1s)$ state. The details of this calculation lie outside the present book but, when the correction is made, the results are in good agreement with fact. The corrections are unimportant at the distances between the nuclei in the actual H_2 molecule, and it is a general principle that the method of molecular orbitals is satisfactory at small, but not at large, internuclear distances. The reason for this is that the simple molecular-orbital method gives excessive weight to the chances of finding two electrons near the same nucleus at the same time, so that it overestimates the contribution of ionic or polar structures $H_A{}^+H_B{}^-$. It can be shown that mixing of the $(\sigma_g \, 1s)$ state with a little of the $(\sigma_u \, 1s)$ state reduces this weight to the correct value.

In principle, the method of molecular orbitals described above for the H_2 molecule is analogous to the collective electron theories of metals, based on the use of Bloch functions (p. 235).

The Method of Heitler and London.

In the method of Heitler and London, the two atoms of the H_2 molecule are first assumed to be so far apart that the electrons can be described by wave functions similar to those of free atoms. We might write the two wave functions in the form $\Psi_A(1)$ and $\Psi_B(2)$ to indicate that electron No. 1 is near nucleus A, and electron No. 2 near nucleus B. As the atoms gradually approach each other it can be seen from Fig. III.31 that each electron comes under the influence of both nuclei, but as an approximation we may assume that the wave functions retain the forms $\Psi_A(1)$ and $\Psi_B(2)$.

The assumption that the function of the system as a whole is of the form $\Psi_A(1)\Psi_B(2)$ is in contradiction to the Exclusion Principle (p. 140), and does not lead to the correct energy of the molecule. Calculation shows that a lower energy is given by the symmetric *gerade* function:

$$\Psi_A(1)\Psi_B(2) + \Psi_A(2)\Psi_B(1)$$

than by the corresponding anti-symmetric *ungerade* function, and the symmetric function is therefore regarded as the correct form of the wave function, and gives a better agreement with the observed energy of the molecule. The use of a combination of this type implies that each electron may be found on either atom, so that although the Heitler–London method begins by using atomic wave functions it presents a picture in which, as in the molecular-orbital model, each electron belongs to the system as a whole.

This is sometimes expressed by saying that there is an oscillation or exchange of the electron between the two atoms. When the atoms are comparatively far apart, the interchange is slight, and we may regard a given electron as moving round one particular nucleus for a comparatively long time, and then passing over to the other nucleus. As the atoms approach each other the interchange of electrons becomes more rapid, and at the equilibrium distance between the two hydrogen nuclei in the H_2 molecule (0·74 Å) the frequency of the interchange is of the order of 10^{18} per sec. This is very much more rapid than the frequencies of the vibrations of the relatively heavy atoms, which are of the order of 10^{12}

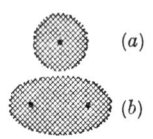

(a)

(b)
[*Courtesy Prentice-Hall, Inc.*]

FIG. III.31. Diagrammatic representations of H_2 molecule.

The electron cloud is axially symmetrical about the line joining the two nuclei, and Fig. III.31(a) represents a cross-section perpendicular to this axis, whilst (b) represents a cross-section containing the line which joins the nuclei. The details of the electron-cloud density can be obtained from Fig. III.32.

per sec. The interchange between the two configurations is thus so rapid that for most purposes we are no longer justified in thinking of the electrons as changing from one atom to another, but have rather to think of the average electron distribution of the system as a whole. Interchange or resonance of this kind has the effect of lowering the energy and reducing the interatomic distances, and these effects are greater as the energies of the two configurations are more nearly equal. In the hydrogen molecule the interchange of electrons between the two atoms produces systems which are indistinguishable, and hence the conditions for strong resonance bonding are present. The electrons in the H_2 molecule must therefore no longer be regarded as associated with individual atoms, but rather as forming a new configuration in which they are associated with the system as a whole, and this new configuration may have properties which are different from those of the configurations between which resonance occurs.

The electron density in the H_2 molecule can be represented by diagrams analogous to those used for free atoms, and Fig. III.31 shows the general form of the electron-cloud pattern of the (σ_g 1s) state in sections perpendicular to and passing through the line joining the centres of the two atoms. Higher energy states of the H_2 molecule can also be

Assemblies of Atoms 145

formed, and these possess nodal surfaces or planes analogous to those in higher states of the free atom.

Fig. III.32 shows the electron-cloud density in a section through the two nuclei. The contour lines are lines of constant electron density, and they tend to crowd together in the space between the nuclei in agreement with the general characteristic of symmetric functions.

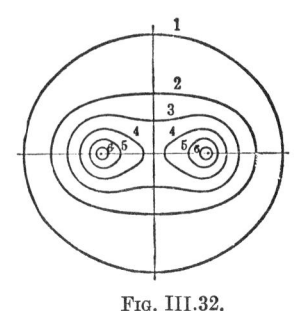

FIG. III.32.

[*Courtesy Prentice-Hall, Inc.*

In this diagram the electron-cloud density in the space surrounding a hydrogen H_2 molecule is shown for a section passing through the two nuclei; the distance between the nuclei is 0·74 A. The curves are curves of equal probability density, and the numbers give the relative values of the density. It will be seen that the high electron density (contours 3, 4, 5, 6) tends to be concentrated in the region between the nuclei. The electron cloud is axially symmetrical about the line joining the nuclei, so that the energy contours in sections perpendicular to this line are circles whose spacing can be determined by drawing vertical lines in Fig. III.32, and seeing where these intersect the different contours. Fig. III.32 thus shows the detail inside Fig. III.31(*b*), whilst Fig. III.31(*a*) could be elaborated into a series of diagrams with circular contours depending on the position of the section across the internuclear axis. The contours of low electron density (contour 1) are more nearly spherical, and when dealing with problems which involve mainly the outermost part of the electron cloud, the hydrogen molecule can sometimes be regarded as approximately spherical.

The Heitler–London method in its simple form does not give exactly the correct value for the energy of the H_2 molecule. The reason for this is that the form of the wave function does not take into account the possibility of both electrons being near the same nucleus at the same time. When this occurs, the molecule has a positive charge at one end, and a negative charge at the other, and is said to be in a *polar state*. This effect can be allowed for by a perturbation calculation in which a contribution is regarded as being made by the state H^+H^-, and when this is done a good agreement with the observed energy of the molecule is obtained.

It will be seen that, at large interatomic distances, the simple Heitler–London method is a better approximation than the simple method of molecular orbitals, whereas at small internuclear distances the reverse is true. It can be shown that when the two simple methods are corrected in full detail they lead to identical results. In a sense, in each approach the H_2 molecule is regarded from an extreme viewpoint, and the pictures presented become increasingly inaccurate as one departs from the point of view assumed. Both methods can be corrected so as to give the same result, but as pointed out by Slater * care must be

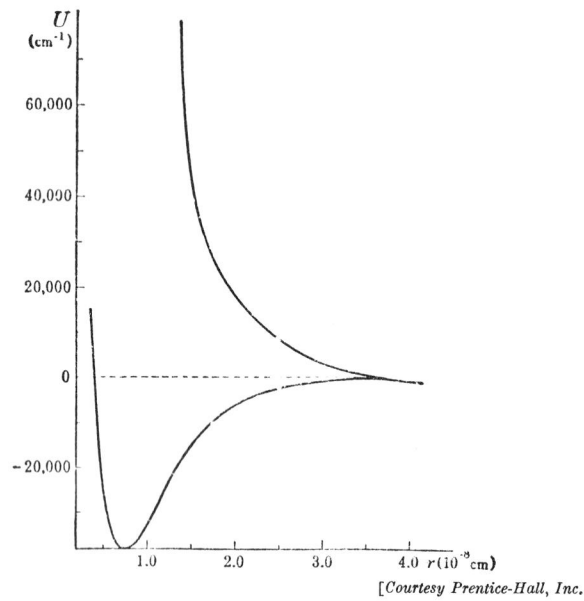

[Courtesy Prentice-Hall, Inc.

FIG. III.33. Potential curves of the two lowest states of the hydrogen molecule. The curves show how the internuclear distance affects the total energy of the molecule when the electrons are in the repulsive orbital (upper curve), and the attractive orbital (lower curve) respectively.

taken in speaking of the polar properties of a co-valent bond as is done in the Heitler–London method. For a particular co-valent bond it may be necessary to introduce a contribution of, say, $x\%$ of an ionic form into the calculation. The value of $x\%$ is obtained by correcting the approximate and over-simplified wave functions of the simple Heitler–London theory, and it may be incorrect to suggest that the actual bond contains $x\%$ of an ionized form.

Fig. III.33 shows the relation between the energy and the internuclear distance for the H_2 molecule when the electrons are in the repulsive

* J. C. Slater, "Quantum Theory of Matter." 1951: New York and London (McGraw-Hill).

Assemblies of Atoms

orbital (upper curve) and the attractive orbital (lower curve). These curves are partly theoretical, and have then been adjusted to agree with experimental data.

The Formation of Molecules.

In the description of the theory of the H_2^+ and H_2 molecules, we have seen that if we regard the two nuclei as approaching so closely that they coalesce, the wave functions pass into those of He^+ and He atoms. This is an example of a general method of considering the formation of diatomic molecules, by imagining the two atoms to approach each other so closely that they coalesce and form what is called a "united atom." The united atom may then be regarded as gradually separating into two atoms, and it can be shown that the first stages of this separation affect the energy levels of the united atom in the same way as the application of a strong electric field. The interest of this method is that when combined with Pauli's Principle, it gives an indication of the way in which the energy levels of the diatomic molecule are related to those of the free atoms. We may suppose, for example, that two lithium atoms, each containing 3 electrons, unite to form a carbon atom containing 6 electrons. In a free lithium atom, the electronic configuration is $(1s)^2(2s)^1$, but since the $(2s)$ level can contain not more than 2 electrons, the united atom will have some of its electrons in $(2p)$ states. In this way it is possible to draw a diagram such as that of Fig. III.34, in which the right-hand side shows the electron levels of the free atoms and the left-hand side the levels of the united atom, and the connecting lines show how the states of the one pass into those of the other as the interatomic distance is varied. If, on passing from right to left, a line slopes upwards, it implies that the energy increases as the distance between the two atoms decreases, and this corresponds to a repulsive or anti-bonding orbital; whilst if the line slopes downwards it implies that the orbital is attractive or bonding. Pauli's Principle indicates that each σ sub-group of the molecule can contain not more than 2 electrons, and each π or δ sub-group can contain not more than 4 electrons.

If Fig. III.34 is applied to the case of two hydrogen atoms, it will be seen that the $(1s)$ level is shown as splitting into the $(\sigma_u\ 1s)$ and $(\sigma_g\ 1s)$ levels, and the second of these slopes downwards, and hence indicates the bonding orbital, which can contain the two electrons. The line for the $(\sigma_u\ 1s)$ level slopes upwards, and thus indicates an anti-bonding orbital, and hence a combination of two helium atoms has two electrons in bonding, and two in anti-bonding orbitals, and there is no stable molecule.

148   Atomic Theory

In the case of lithium there are six electrons to be accommodated in a diatomic molecule. Of these the two K electrons of each atom enter the (σ_u 1s) and (σ_g 1s) orbitals of the pair of atoms, and these may be regarded as taking no part in the bonding process, since the K electrons

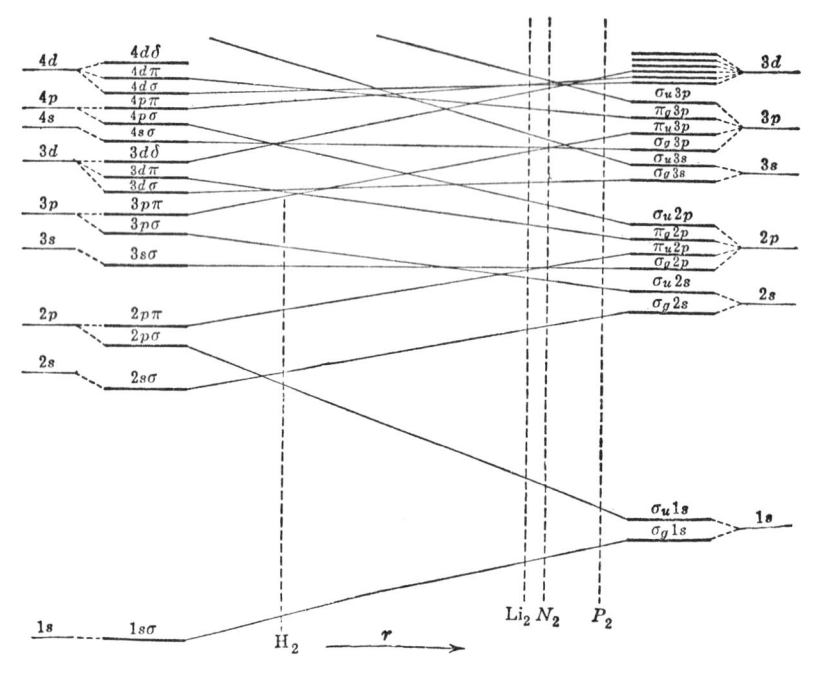

[Courtesy Prentice-Hall, Inc.

FIG. III.34. Correlation of electronic orbitals in a two-centre system for equal nuclear charges.

To the extreme left and the extreme right are given the orbitals in the united and separated atoms, respectively, and, beside them, those in the molecule for very small and very large nuclear separations, respectively. The region between corresponds to intermediate nuclear separations. The vertical broken lines give the approximate positions in the diagram that correspond to the molecules indicated. It should be noticed that the scale of r is by no means linear, but becomes rapidly smaller on the right-hand side. Owing to this change of scale, the straight connecting lines between right and left do not correspond to a linear energy change with r, but to one that is closer to reality, even though it is an admittedly poor approximation.

occupy the inner part of the atom and so are affected appreciably only when the atoms are very close together, and in any case they only give rise to equal numbers of bonding and anti-bonding orbitals. The (2s) valency electrons of the free atoms can enter the (σ_g 2s) orbital

Assemblies of Atoms 149

of the pair of atoms, and since the line for this slopes down from right to left in Fig. III.34, a stable Li_2 molecule is formed.

Fig. III.34 is purely diagrammatic: the horizontal scale of distances is not uniform, and if it were uniform, the lines joining the various levels would no longer be straight. Further, the relative position of the energy levels of the individual atoms and the united atom will vary according to the element concerned. But when used in conjunction with Pauli's Principle, diagrams such as that of Fig. III.34 are extremely useful in showing how, when two atoms approach one another, bonding and/or anti-bonding orbitals may be formed. It will be noted further that for relatively large distances of separation, the splitting of the groups shown in Fig. III.34 results in the groups of sub-levels derived from the $(1s)$, $(2s)$, $(2p)$, &c., levels of the free atoms remaining separated from one another, although at closer distances of approach they overlap or intersect. This is entirely analogous to the overlapping of the energy bands in metals referred to on p. 90, and Fig. III.34 is really only showing for a diatomic molecule what Fig. III.3 shows for the immense molecule formed by the crystal of a metal. Fig. III.34 refers to the interaction between two like atoms, but analogous diagrams can also be drawn for the processes which occur when two unlike atoms approach one another, and in principle similar diagrams could be drawn for more complicated systems. Although a proper discussion of these effects lies outside the scope of the present book, it should be mentioned that, when a molecule has been formed, there are low-energy excitations that correspond to setting up rotations and vibrations of the molecule, as well as the rather high-energy excitations of electrons between the orbitals of Fig. III.34.

Exchange Repulsion.

In the above description (p. 143) we have seen how the use of particular combinations of wave functions may be interpreted as giving rise to interchange of the electrons between the two atoms of a diatomic molecule, and that these exchange effects affect the energy of the molecule. Exchange effects are also important in connection with the repulsion produced when the electron clouds of two ions overlap. In the above description we have frequently referred to atoms or ions being drawn together until the attractive forces are balanced by the repulsion caused by the overlapping of the electron clouds of the ions. If the electron clouds were spherical, static, and undeformable, repulsion would occur only when the atoms were so close together that the nucleus of one was within the electron cloud of the other. If, however, we are dealing with completed groups of 2 (the helium configuration), 8, or 18 electrons, the ion contains only completely filled shells, and the exchange effects imply that comparatively slight overlapping of the electron clouds must result in an enormously increased energy. This is because when two ions approach so closely that their electron clouds overlap and they begin to form a joint system, the energy levels of the original ions split up to form groups of levels as in Fig. III.34. If the ions contain completed groups of 8 or 18 electrons, the Pauli Principle prevents the electrons from redistributing themselves so as to obtain a lower energy as the atoms approach, and the energies of the states whose electrons give rise to the outer part of the ions increase rapidly as the atoms approach, *i.e.*, they correspond to lines sloping upwards from right to left in Fig. III.34.

In some cases this kind of repulsion may be superimposed on an attraction of the van der Waals type. In a crystal of copper, for example, the valency electrons are to be regarded as drawing the atoms together, and the electrons in the outermost parts of the ions then move in harmony with one another, and so give rise to additional attractive forces of a van der Waals type. But at the same time as the electron clouds of the ions begin to overlap, the repulsion resulting from the effects of the Pauli Principle comes into play. In a sense, therefore, we may say that the electrons in the outermost parts of the ions of copper, silver, and gold are responsible both for increased attraction, and hence higher melting points, and increased repulsion, and hence lower compressibilities.

SUGGESTIONS FOR FURTHER READING.

C. N. Hinshelwood, "The Structure of Physical Chemistry," Part IV. **1951**: Oxford (Clarendon Press).
C. A. Coulson, "Valence." **1952**: London (Oxford University Press).
W. Heitler, "Elementary Wave Mechanics," 2nd edn. **1956**: Oxford (Clarendon Press).

7. DIRECTED VALENCIES.

In the preceding section we have described some of the electronic processes which result in the formation of the H_2 hydrogen molecule. These processes involve the $(1s)$ electrons of the free hydrogen atoms, and as the electron-cloud patterns of these are spherically symmetrical, there is no reason to expect any directional effects. We have seen (p. 51) that on passing along the series of elements preceding an inert gas, the electronic structure builds up in such a way that first of all the $(ns)^2$ sub-group is built up with two electrons of opposite spins, and then as the electrons enter the $(np)^6$ sub-group, they take up parallel spins as

Assemblies of Atoms 151

long as this is possible. Since the $(np)^6$ sub-group contains three orbitals, each of which can contain two electrons (of opposite spin), this implies that a normal atom of Group VB has one electron in each p orbital. In the theory of simple superposition of functions, this would result in an electron cloud of spherical symmetry, but as explained on p. 75 the effects of electron correlation are not completely known.

In Group VIB, each atom in its normal state has one p orbital completely occupied by two electrons of opposite spins, whilst the other two p orbitals contain one electron each. Similarly, in Group VIIB

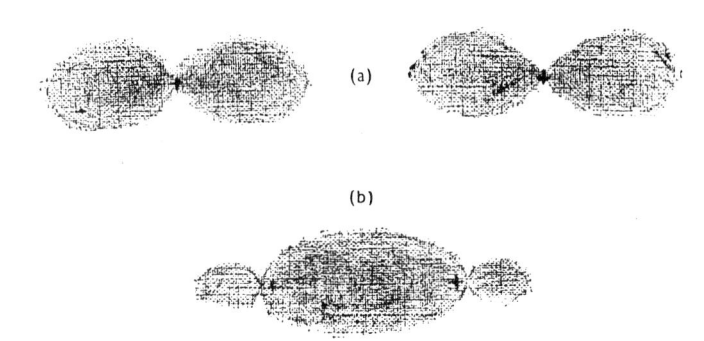

FIG. III.35.

(a) Half-filled p orbitals of the free atoms approaching one another. Each p orbital contains one electron, and these must have opposite spins if bonding is to occur.

(b) The same two atoms after bonding has occurred, and the electron cloud is drawn between two atoms. The bond is formed by two electrons of opposite spin.

the atom in its normal state has two of its p orbitals completely filled, whilst the third contains one electron only.

We have already seen that a co-valent bond is formed by two electrons of opposite spin, and we can understand, therefore, why a fully occupied orbital does not take part in normal bond formation, since its two electrons have already paired their spins. The co-valent bonds of these elements are therefore formed by electrons in the orbitals which contain only one electron, or in other words by the unpaired electrons of the free atom.

When a bond is formed, the electron cloud is drawn into the space between the two atoms, so that if the electron-cloud patterns of p states in two atoms are as shown in Fig. III.35(a), the cloud pattern after the bond is formed is more like Fig. III.35(b).

In Part II we have seen that the electron clouds of the three p orbitals are mutually perpendicular, and we might therefore represent a

L

152   *Atomic Theory*

normal atom of selenium by a diagram such as that of Fig. III.36(a). Here the electron clouds of the inner electrons and of the two (4s) electrons are shown as a shaded region, and the electron cloud of the fully occupied (4p) orbital is shown with cross hatching, and since this contains

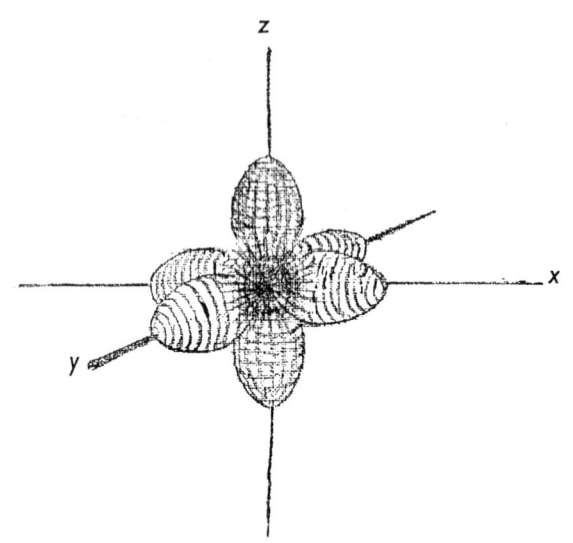

Fig. III.36(a). Diagrammatic representation of an atom of selenium.

The inner electrons are shown as a dark area. The $(4s)^2$ sub-group is not shown; its electron cloud is spherically symmetrical with 3 spherical nodes. The 4p orbital extending along the z axis is shown with double markings and contains two electrons of opposite spins. The other two p orbitals, which contain one electron each, extend in the direction of the x and y axes respectively, and are shown with single markings. The figure is highly diagrammatic and is shown to emphasize the directions of the orbitals. In the actual atom the total electron-cloud density is the sum of the densities of the three orbitals. In the free atoms of N, P, As, Sb, and Bi each p orbital contains one electron, and in the simple theory the total electron-cloud density is spherically symmetrical as explained on p. 79. In selenium, where one p orbital contains two electrons, the electron density will be greater in the direction of this orbital, i.e., greater in the direction of the z axis in Fig. III.36(a).

two electrons of opposite spin, it does not give rise to bonding. The two remaining (4p) electron clouds contain one electron each, and as the electron clouds are perpendicular, it is easy to see how a series of these atoms unite, as shown in Fig. III.36(b), to form the spiral chain of the selenium structure shown in Fig. III.9 (p. 101). The bonds in the crystal are more properly described as sp hybrids, in which the p function is predominant. The actual angles in the structures of selenium and tellurium are 105° and 102°, so that the exact right angles have been slightly

Assemblies of Atoms 153

distorted. Similarly, in the Group V elements, where the free atoms have three unpaired p electrons, we can see how these will tend to bind together to form the antimony structure of Fig.III.10. In this case the angles between the bonds are much more nearly right angles, being 97°, 96°, and 94° for arsenic, antimony, and bismuth respectively. We can thus see that the wave-mechanical picture of mutually perpendicular p orbitals gives a very satisfactory account of the crystal struc-

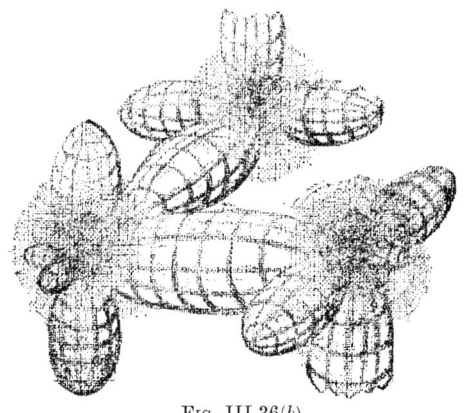

Fig. III.36(b)

This figure shows diagrammatically how the selenium atoms of Fig. III.36 unite to form the spiral chains in the crystal of selenium shown in Fig. III.9. The inner electrons of each atom are shown by a shaded region. The two unpaired p orbitals of the valency electron of each selenium atom coalesce with those of adjacent atoms to form completely filled orbitals, each containing two electrons of opposite spin.

The electron clouds of the shared electrons in each co-valent bond are, of course, not really bounded by sharp surfaces but are diffuse. The surfaces are drawn as shown to illustrate the axial symmetry of the electron cloud about the direction of the bond, and about four-fifths of the charge would be contained within the volumes shown.

tures of arsenic, antimony, bismuth, selenium, and tellurium. It is also to be noted that the co-valent bonds formed by p orbitals are stronger than those formed by s orbitals.

The co-valent bond in the H_2 molecule (p. 144) and the p bonds referred to above have their maximum electron densities along the lines joining the two nuclei; such bonds are called σ bonds. If, however, two atoms approach with p orbitals orientated as in Fig. III.37(a), bonding may occur with the formation of what is called a π orbital, whose general shape is illustrated in Fig. III.37(b). Here the electron cloud has a zero density in the plane shown, which passes through the two nuclei, and although two electrons are being shared, their electron cloud has a minimum density along the line joining the two nuclei. In an analogous way the d states of atoms may produce bonding in which the shared electrons give rise to molecular orbitals of the σ type, or of other types. The details of these lie outside the present book,* but it is important to note that co-valent bonding does not necessarily result in an electron cloud whose density is greatest along the line joining the nuclei.

The electron configuration of the normal state of the carbon atom is $(1s)^2(2s)^2(2p)^2$, in which the two $(2s)$ electrons form a pair with opposite spins, whilst the two $(2p)$ electrons are unpaired. In this state the

Fig. III.37(a).

Fig. III.37(b)

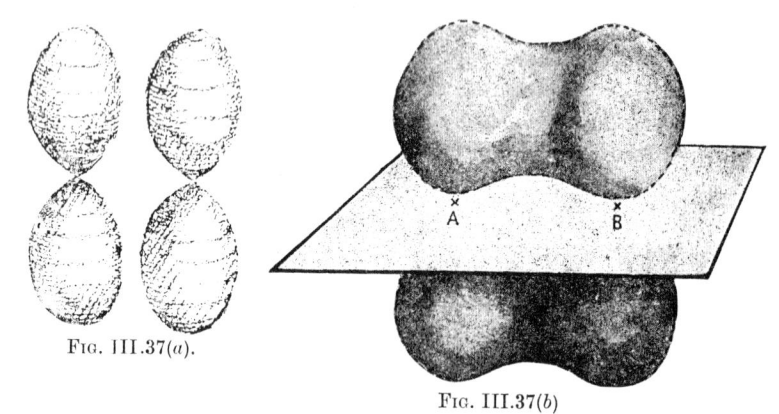

Figs. III.37(a)–(b). Bonding with formation of a π orbital.

carbon atom would have a co-valency of two, since there are only two unpaired electrons. The normal co-valency of four which is shown by the carbon atom is the result of the fact that the electron cloud of the $(2s)^2$ group in the free atom is still near the surface of the atom. In the free atom the configuration $(1s)^2(2s)^2(2p)^2$ gives the lowest possible energy, but in an assembly of atoms, more stable bonds can be formed by an electronic rearrangement which may be regarded as giving rise to bonds derived from the $(1s)^2(2s)^1(2p)^3$ state of the free atom. In this state, the free atom has four unpaired electrons, and can consequently give rise to four co-valent bonds. It might at first be thought that these would consist of three mutually perpendicular bonds derived from the three $2p$ states, together with a fourth bond at some other angle, resulting from the $2s$ state. Actually this is not so, and the process of hybridization (p. 90) occurs, as a result of which the $2s$ and $2p$ states combine to form a set of hybrid orbitals, and it can be shown that these

* See C. A. Coulson, "Valence." **1952**: Oxford (Clarendon Press).

Assemblies of Atoms 155

are directed tetrahedrally; * they are very strong bonds, in the sense that there is a high concentration of electrons in the region between two atoms which are close neighbours. This, of course, is possible only if the two electrons in a bond have opposite spins, and all co-valent bonds of this type involve two electrons of opposite spin. The formation of co-valent bonds of this type is not confined to cases where the atoms are similar, and in this way the wave-mechanical theory is able to explain the tetrahedral structure of crystals such as those of silicon, germanium, and grey tin, and also the wurtzite and zinc-blende structures, where the tetrahedral arrangement is again found provided that the two kinds of atom contain on the average 4 valency electrons per atom.

Hybridization of the (sp^3) type becomes increasingly difficult as the energy difference between the ns and np levels increases. In the B sub-groups, it is a general principle that this difference increases with increasing atomic number, and hence it increases on passing from C → Si → Ge → Sn → Pb in Group IV. According to Cohen and Heine,† this may be the reason why the diamond structure formed by carbon, silicon, and germanium is stable only at low temperatures in the case of tin, and does not appear at all for lead. In the extreme case where the energy of the s level is depressed sufficiently, this may become an atomic level as is assumed in the earlier concept of incomplete ionization for lead (p. 112).

The co-valent bonds in structures such as those of the diamond involve exchange forces similar to those described above (p. 144) for the H_2 molecule, and we have again to imagine the electrons concerned as undergoing a rapid interchange between two atoms. Since each atom has four neighbours, the valency electrons undergo a continual interchange between the different atoms in the crystal, and we must not regard a given pair of electrons as permanently associated with a given bond. The whole swarm of electrons is continually circulating through the crystal, although the electron-cloud density round an atom is greatest in the direction of the four co-valent bonds joining the atom concerned to its four neighbours. Claims have been made to detect this concentration of electrons in the directions of the co-valent bonds, by means of refined X-ray crystal analysis. At first sight it might be imagined that this continual circulation of the swarm of electrons through the crystal would make the latter a conductor, but as we shall see later (Part V) this difficulty has been resolved

---

* When hybridization takes place in carbon, the tetrahedral angles are not necessarily exactly equal, and may in some cases (*e.g.*, CH_3Cl) be a few degrees different from the angles of a regular tetrahedral arrangement.

† M. H. Cohen and V. Heine, *Advances in Physics*, 1958, 7, 395.

satisfactorily by the theory of Brillouin zones. Here we may content ourselves with saying that electrical conductivity implies the possibility of a resultant flow of electrons in one direction, and this necessitates the existence of unoccupied electron states into which the electrons may be transferred by the action of an external field. If all the electronic states of a system are occupied, the application of an external field * can produce no alteration in the velocity distribution of the electrons as a whole, and it can be shown that in the case of the diamond the valency electrons completely fill one band of energy states, and that a very high external e.m.f. is necessary to excite electrons into the electron states of the next band. In this way the concept of a swarm of electrons in continual motion through the crystal can be reconciled with the absence of a resultant flow in any direction, and hence with the properties of an insulator.

Hybridization between s and p orbitals cannot give rise to more than four bonding orbitals to each atom, and consequently the highest covalency of the elements in the first two Short Periods is usually four, since more than this number of bonds would necessitate the use of the d orbitals and these are of considerably higher energy. We have, however, seen that in the first Long Period, the $3d$ and $4s$ states of the free atoms have nearly the same energies, and in compound molecules of these elements hybridization can occur between the $3d$, $4s$, and $4p$ orbitals with the production of very stable bonding orbitals, which may assume different configurations. Thus, in stable complex ions such as $Co(NH_3)_6^{+++}$, the six NH_3 groups are situated at the corners of a regular octahedron, and are bound to the central cobalt atom by hybrid s, p, d orbitals. In the $(PdCl_4)^{--}$ ion the four chlorine atoms are arranged at the corners of a square. Until recently, directed bonds of this kind have not been of much interest in connection with the structure of metals and alloys, but according to Pauling the fact that hybridization may occur between the ns, np and $(n-1)d$ orbitals of transition elements † is of great interest in connection with the cohesive forces in the crystals of these metals. The s, p, and d groups contain 1, 3, and 5 orbitals respectively, so that as many as nine orbitals per atom are available for bond formation in the transition elements, and although the evidence suggests that in metals these are not all used, the high melting points and strong cohesion of the crystals of transition metals are connected with the larger number of orbitals from which bonds may be formed.

* We are here ignoring the effects of very strong fields which break down insulators.
† Here n is the quantum number of the valency electrons.

Assemblies of Atoms

In recent years it has been suggested by several authors that directed bonds of the hybrid type may play an important part in determining the crystal structures of the transition metals, and some of these ideas are described later.

8. Resonance Bonding and the Metallic Linkage.

In the description of the hydrogen molecule we have seen that it is necessary to take into account the fact that two configurations of identical energy are produced if an arrangement with electrons No. 1 and 2 near nuclei A and B respectively changes to one with electron No. 1 near to nucleus B, and electron No. 2 near to nucleus A. We have seen further that this produces a rapid interchange or resonance between the two electronic configurations, with a resulting lowering of energy and decrease in the interatomic distance. This is an illustration of a general principle which may be summarized by saying that if a molecule can exist in more than one structure, the most stable state does not correspond to the wave function of any one structure, but to the superposition of wave functions of the different structures. If, for example, $\Psi_1, \Psi_2, \Psi_3 \ldots$ are the wave functions for a given molecule in a number of structures 1, 2, 3 . . ., then a general function of the form

$$\Psi = a\Psi_1 + b\Psi_2 + c\Psi_3 \ldots$$

where, $a, b, c \ldots$ are arbitrary numerical coefficients, is a possible wave function of the system, and the lowest possible energy of the system will correspond to a definite ratio $a:b:c$. . . . If one structure, say Ψ_2, is relatively much more stable than the others, then the ratios $a:b, c:b, \ldots$ will be small, and the most stable state will resemble that of Ψ_2 more closely than any other state, although there will be a slight mixture or superposition of the other states. If, however, the stabilities of the different structures are more nearly equal, the lowering of the energy produced by resonance is relatively greater. It must be emphasized that in this phenomenon it is not a question of molecules of two or more structures existing in equilibrium with one another. The resonance vibration is so rapid that for most purposes we have to consider the electron as in a new state characteristic of the resonance process, and the resonance structure may have properties which are different from those of the states between which resonance occurs.

Resonance of this kind can occur only if the alternative structures contain the same number of unpaired electrons. Now in our description of the hydrogen molecule we have seen that, in the Heitler-London theory the simple wave functions have to be corrected to take into account the possibility of both electrons being near to the same atom. At any instant when this occurs the molecule has a dipole moment, and it is customary to say that the co-valent H — H structure is to be regarded as in resonance with the ionized structures $H_A{}^+H_B{}^-$ and $H_A{}^-H_B{}^+$, and that about 5% of the bond energy results from resonance with the ionic forms. As explained above (p. 146), this method of expression is not really satisfactory, since the figure of 5% is obtained by reference to the admittedly incorrect oversimplified wave function of the simple Heitler–London theory. It is, however, a general principle that if the approximate wave functions are corrected in this way, the correction becomes larger as the stability of the hypothetical ion increases.

In the H_2 molecule, the co-valent form is very much the more stable of the two, and so the actual molecule has the properties of the co-valent rather than the ionic form, and since the two forms have very different stabilities, the resonance has comparatively little effect on either the bond energy or the interatomic distance. On the other hand, in the case of the molecule of hydrogen fluoride, HF, the ionized molecule

$$H^+F^-$$

and the co-valent molecule

$$H - F$$

have energies which are much more nearly equal, and hence the conditions are favourable for resonance between the ionized and co-valent forms, and the lowering of energy and decrease in interatomic distance are correspondingly great.

Resonance of this kind can occur not only in simple diatomic molecules, but between different possible forms of complicated molecules. In the case of benzene, C_6H_6, for example, it has long been known that the six carbon atoms form a closed ring, but the distribution of the four valencies of each carbon atom in this structure was for a long time a matter of dispute. Fig. III.38 shows some of the structures that were proposed, and each of these seemed to explain some of the properties of the substance, but none would account for all the facts. From the point of view of the wave mechanics, the actual structure of benzene is the result of resonance between these different structures. It can be shown that the pairing of electrons represented by the Armstrong–Baeyer structure (Fig. III.38(c)) is equivalent to a combination of the pairings of the five structures represented by Figs. III.38(a) and (b). The actual structure of benzene is thus the result of resonance between the three Dewar and the two Kekulé structures, and of these the Kekulé

Assemblies of Atoms

structures play the most important part. The resonance effect greatly stabilizes the molecule, so that the actual substance benzene has properties which are more than the sum of those to be expected from the mere addition of the properties of the different structures. It must again be emphasized that it is not a question of molecules of the different structures existing in equilibrium with one another. Some writers would say that the structures $a, b, c \ldots$ represent the end points of the resonance vibrations, so that the molecule spends a short fraction

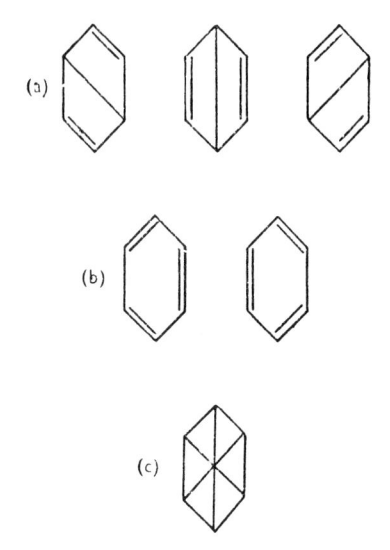

Fig. III.38. Older structural models for benzene (C_6H_6).
(a) Dewar, (b) Kekulé, and (c) Armstrong–Baeyer.

of its time in these structures. This, however, may be misleading, and it is better to regard the molecule as existing in a form which is neither one structure nor the other, but is characteristic of the resonance. The frequency of the resonance process is so high (of the order of 10^{18} per sec) that the resonance vibrations affect the bonding electrons only, and the relatively heavy atoms are not to be regarded as oscillating with this high frequency. The atoms in molecules, and in crystals, do undergo vibration for other reasons, but the frequencies concerned are very much lower (ca. 10^{12} per sec).

In the molecule of benzene, there are 30 valency electrons, 24 of which come from the carbon atoms and six from the hydrogen atoms. The resulting structure may be described by saying that 12 electrons are concerned in binding the six hydrogen atoms by co-valent bonds to the carbon atoms of the ring, whilst of the remaining 18 electrons 12 may be regarded as giving rise to single bonds within the ring, and 6 are associated with the benzene ring as a whole, and move freely in this ring in a configuration resulting from the resonance between the different structures of Fig. III.38.

In the above description the term " resonance " has been used both in connection with the exchange effect in the co-valent bond (*e.g.*, the H_2 molecule, p. 144), and for the resonance that occurs between different structures such as those of the benzene molecule (p. 158). This is the accepted practice, and is the result of the two processes being mathematically similar, but it would undoubtedly be better if the term resonance were used for the second type of process alone. In the formation of the co-valent bond, the indistinguishability of electrons means that we cannot even in principle distinguish between the configuration with electrons Nos. 1 and 2 near to atoms A and B respectively, and that with electron No. 1 near to atom B and No. 2 near to atom A. Although the process is one of resonance, it would be clearer if it were called the exchange effect. The term resonance could then be used for the second type of phenomenon, in which resonance occurs between different structures such as those of the benzene molecule. In this case the different structures contain co-valent bonds in which the exchange effect has already taken place. Further, although the actual substance benzene always possesses the resonance structure, the different modifications (Dewar, Kekulé) could in principle be distinguished if one imagined them isolated.

The molecules whose structures have so far been described have been mainly those in which the atoms contain one, two, three, or four electrons fewer than the configuration of the next inert gas. In such cases, from the viewpoint of the simple Lewis theory, there are sufficient electrons present for it to be possible to build up stable octets (or a stable pair in the case of hydrogen) by means of normal electron sharing in which single, double, and triple bonds mean that two, four, or six electrons respectively are held in common between two atoms. In terms of the wave-mechanical theories, these elements contain sufficient electrons to fill completely all the bonding orbitals derived from the states of the valency electrons in the normal atoms. In the crystals of the elements which obey the $(8-N)$ rule, the normal valencies are satisfied by each atom sharing one electron with each of its $(8-N)$ neighbours. It is clear that this process can proceed as far back as Group IV, but no further. In the diamond structure, each atom contains four valency electrons, and uses all of these to exert its maximum co-valency of four,

Assemblies of Atoms 161

and so acquires an octet. In the elements of Group III there are only 3 valency electrons per atom, and consequently only 3 co-valent bonds can be formed, whereas the atom must acquire 5 additional electrons if it is to build up an octet. It is thus impossible to build up an octet of electrons by means of co-valent bonds, and the same difficulty applies to an even more marked extent in the case of the elements of Groups I and II, where there are only one and two valency electrons per atom respectively. It can readily be shown that unstable structures would result if these elements formed ionic structures. For example, a structure in which $Cu^{-------}$ ions were surrounded by seven Cu^+ ions which had given up their valency electrons to the central ion, would be most unstable, because the positive charge on the copper nucleus is not great enough to enable it to hold an octet of 4-quantum electrons firmly outside a $(1s)^2(2s)^2(2p)^6(3s)^2(3p)^6(3d)^{10}$ core.

We can thus readily see that on passing backwards from Group IVB, some new kind of atomic binding process will have come into being, because there are not sufficient electrons present to enable the atoms to build up octets by means of co-valent bonds. Reference to the Periodic Table will show that it is at this stage that the metallic elements * are found, and we can thus understand these as being the result of the constituent atoms containing too few electrons to build up stable groups by means of simple co-valent bonds. From the diagram of Fig. III.34 it might at first be thought that the main characteristic of the elements of Group I would be the formation of diatomic molecules, since the lines for the (σ_g ns) state slope downward from right to left and thus indicate bonding orbitals. Actually, molecules of formulæ Li_2, Na_2, and K_2 have been observed, but these are not very stable, and all trace of them is lost in the metallic crystal, where each atom is surrounded perfectly symmetrically by the eight neighbours characteristic of the body-centred cubic structure.

In general, the metallic elements are characterized by the formation of monatomic vapours and of crystal structures of high co-ordination number, *i.e.*, structures in which each atom is surrounded by a large number of equidistant or almost equidistant neighbours. Under these conditions there are insufficient electrons to hold the atoms together by means of electron-pair or single-electron bonds, and it has long been recognized that in metallic cohesion, the valency electrons † are shared between the atoms, so that a metallic crystal may be regarded as

---

* Boron in the First Period is a semi-conductor.
† In the transition elements the cohesion involves not only the ns valency electrons, but also the outermost $(n-1)d$ electrons of the ion.

Atomic Theory

consisting of an array of positive ions held together by attractions to the common system of negatively charged electrons. It has been suggested by Pauling that this metallic bonding may be regarded as an extension of the resonance process referred to above. In the case of the alkali metals, for example, the unit cell of the body-centred cubic structure has been shown in Fig. III.12. This unit contains two atoms,* and consequently two valency electrons are available to bind the central atom to its eight neighbours. Pauling's suggestion is that the resulting condition of affairs is analogous to a resonance between all the different structures which might be obtained by allotting the electrons to definite one-electron or two-electron bonds between the various atoms. From some points of view this is merely another way of expressing the old idea that the valency electrons are shared between the atoms, but Pauling's method of approach to the problem serves to emphasize two important features. In the first place the interatomic distances in metallic crystals are usually of the same order as those to be expected for resonating, co-valent, or one-electron bonds. In the development of the theory of co-valency a large amount of data has been accumulated regarding the interatomic distances in crystals involving co-valent and resonance bonds, and in this way it has been found possible to deduce a set of atomic radii which account for the interatomic distances in this class of compound. These radii are approximately the same as those required to account for the interatomic distance in metallic crystals,† and this clearly suggests that the bonding process is essentially of the same nature.

In the second place, Pauling's views have served to explain why the metallic state of the elements extends to higher Groups of the Periodic Table in the elements of the A Sub-Groups than in those of the B Sub-Groups; for this purpose we consider the first two Short Periods as analogous to the B Sub-Groups. In the elements of the B Sub-Groups the orbitals available for bond formation are limited to four per atom, namely the one s orbital and the three p orbitals of the valency electrons. We can therefore understand why metallic structures cease at carbon, silicon, and germanium in the first three Periods, and at antimony and bismuth in the fourth and fifth Periods.‡ For when we have four

---

* The atoms at the eight corners of the unit cell are shared between eight units, and so may be regarded as each contributing $\frac{1}{8}$th of an atom to the cell under consideration. The atom at the centre of the cell belongs to this cell alone, and so counts as one atom. The total number of atoms associated with the cell is therefore two.

† There are slight corrections for variation in the co-ordination numbers in different structures, but these do not affect the general conclusion.

‡ We have already explained (p. 108) why lead does not crystallize in the diamond type of structure.

Assemblies of Atoms 163

electrons or more we have the possibility of filling the orbitals and building up stable groups of electrons. In the transition elements of the A Sub-Groups, we have already seen (p. 60) that the tendency is for the $(n-1)d$ electrons of the ions to have lower energies than the np states of the valency electrons. For these elements, therefore, there are five d orbitals of the ion available for bond formation as well as the s and p orbitals of the valency electrons. With four electrons per atom there is no possibility of filling all these orbitals, and consequently the metallic state continues past Group IV in the transition elements. This aspect of Pauling's theory is dealt with in detail in Part VI (p. 360).

In many of the earlier treatments of the subject, the reader will find it stated that the electrical conductivity of metals indicates that the electrons are free to move in metallic crystals, but not in crystals of other substances. This is, however, only partly true, and is sometimes a very misleading statement. In crystals of salts (including normal oxides),* the units of structure are the positive and negative ions, and there is little or no interchange of electrons from one ion to another. On the other hand, in co-valent crystals such as the diamond, the exchange effects (p. 155) mean that the electrons are continually moving from one atom to another, and the whole swarm of electrons is to be regarded as in continual circulation through the crystal. The difference between an insulator, such as the diamond, and a conductor is not that the electrons are bound in the one case and free to move in the other, but rather that in the conductor an external field can produce a resultant flow in a given direction. This, as we have already explained, means that the conductor must have unoccupied electron states into which the electrons may pass when the external field is applied. We can therefore begin to understand why the elements with comparatively few valency electrons per atom are those which show metallic conductivity. For if there are only a few electrons per atom, the available orbitals of the assembly (which are derived from those of the free atoms) will not be completely filled, and there will be unoccupied states in those orbitals into which the electrons can pass under the influence of an electric field.

After what has been said of the difficulties of solving a simple problem such as that of the H_2 molecule, it will readily be appreciated that the problem of the motion of electrons in a metallic crystal is far too complicated to be solved in any detail. In the following chapters we shall describe some of the attempts which have been made to deal with the problem, and in these certain clear stages can be recognized.

* There are some oxides which show what is called defect conductivity. In this process if an electron is missing from one O^{--} ion, an electron may jump from the next O^{--} ion. Here exchange forces may be present.

The first and most simplified treatment is to consider the behaviour of an assembly of electrons confined in an enclosed space, which is assumed to contain an atmosphere of uniformly distributed positive electricity so that the whole is electrically neutral. This free-electron theory or "electron box" model is described in Sections IV.1–3 and is, of course, far too simplified to bear much resemblance to the state of affairs in most actual metals, but the theory is of great interest because it shows very clearly how an assembly of electrons at the concentrations concerned in solid metals has properties which are very different from those of an assembly obeying the classical gas laws. For some purposes the theory is a satisfactory approximation to the state of affairs in the alkali metals, but for other metals the main value of the free-electron theory is to provide a simplified background against which the characteristics of the more complicated theories can be seen most clearly.

The next stage in the theory consists in examining the behaviour of an assembly of electrons moving in the periodic fields of the various crystal structures. This theory is due to Bloch, and is described in Section V.1. In this treatment the electrons are regarded as moving in a simple periodic field which has the periodicity of the lattice, and the quantitative details of the field are not considered. This theory of electrons in a periodic field has been brilliantly successful, and has shown that when the periodicity of the field is of the order of 10^{-7} to 10^{-8} cm (*i.e.*, the size of the unit cell of the crystal) the behaviour of electrons according to wave mechanics may be entirely different from that predicted by classical mechanics. In particular, it is possible to show that in a periodic field there are both permitted and forbidden energy states for the electrons, and in this way an explanation is obtained of the existence of the three main classes of substance—metals, semi-conductors, and insulators.

In the final stage of the theory, attempts are made to work out the theory for particular metals in much greater quantitative detail. This leads to the introduction of electrons in s states, p states, &c., in the crystal, and is the mathematical development of the ideas shown in simple form in Figs. III.3 and III.34. It must be emphasized that this part of the theory, which is described in Section V.3, requires rather involved mathematical computations. It has shown that some of the conclusions of the theory of the simple periodic field require modification when the problem is attacked in greater detail, and in recent years some new ideas have yielded important results, but in many cases the problems have become involved in an immense mass of mathematics from which a solution can be obtained only by the introduction of so many approxi-

Assemblies of Atoms 165

mations that the final result is really of little quantitative value. The non-mathematical reader must be warned that if he attempts to draw conclusions from papers of this class, it is nearly always necessary to seek advice from a professional mathematical physicist in order to appreciate the real significance of the calculated results. It is unfortunate that in many cases the results of numerical calculations are given to 3 or 4 figures even where the second figure is uncertain, and there is a singular reluctance on the part of some mathematical physicists to publish estimates of the uncertainty in the results which they claim to calculate.

Suggestions for Further Reading.

C. S. Barrett, " The Structure of Metals. Crystallographic Principles, Techniques, and Data," 3rd edn. New York and London: **1956** (McGraw-Hill).

M. Born, " Atomic Physics." Translated by J. Dougall. 6th edn. London and Glasgow: **1957** (Blackie and Son).

W. Hume-Rothery, R. E. Smallman, and C. W. Haworth, "The Structure of Metals and Alloys." London: **1969** (Institute of Metals).

F. Seitz, " The Physics of Metals." New York: **1943** (McGraw-Hill Book Co., Inc.).

C. Kittel, " Introduction to Solid-State Physics," 2nd edn. New York: **1956** (John Wiley and Sons, Inc.).

A. J. Dekker, " Solid-State Physics." London: **1958** (Macmillan).

PART IV. THE FREE-ELECTRON THEORY OF METALS.

1. THE FERMI–DIRAC STATISTICS AND THE ELECTRON GAS.

IN Part III we have seen that when atoms are brought into close contact, the general effect is for each sharp energy level of the free atom to be broadened into a band which becomes wider as the atoms approach each other more closely. The exact details of the process in a crystalline solid constitute a problem of great complexity, and only approximate solutions are now available. Of these theories, the most simple is the so-called free-electron theory, in which the structure of the solid is ignored, and the metal is regarded as containing a number of electrons which move about freely as though they constituted particles of an electron gas. The electrostatic repulsion between the electrons is also ignored, and the model really involves the assumption that the

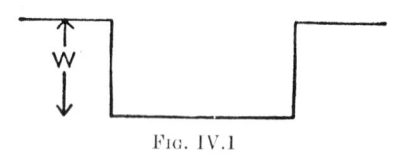

FIG. IV.1

metal contains a uniformly distributed positive charge which neutralizes the charge on the electrons and enables the latter to be treated as ordinary gas particles of very small mass. It need scarcely be said that such an assumption is a great over-simplification of the problem, and, as we shall see later, the free-electron theory is a fair approximation to the state of affairs existing in the alkali metals, but is a very poor approximation for nearly all other metals. The theory is retained largely because it is a convenient approximation which lends itself to comparatively easy mathematical treatment, and also because it provides a simplified background against which the complexities of the more complete theories can be seen most clearly.

The free-electron theory assumes that the potential within the metal is constant, and that there is a sharp rise in the potential at the surface of the metal, which prevents the electrons from escaping. In problems of electronic emission, the exact form of the potential curve at the surface is of great importance, but for many other problems these details are unimportant, and it is customary to draw a potential diagram of the type shown in Fig. IV.1, where W represents the potential step at the boundary. Comparison with experimental data suggests that W is of the order 10 eV for most metals. For many purposes it is customary to assume the zero of potential energy to be that of an electron at rest outside the metal, and in this case the energy of an electron within the metal is:

$$- W + \tfrac{1}{2}mu^2 \quad \ldots \ldots \quad \text{IV (1)}$$

since the kinetic energy of a free electron is $\tfrac{1}{2}mu^2$, where u is the velocity. In comparing the conclusions of the theory with the experimental results for actual metals, the question of the variation of W has always to be considered. In more elementary developments, it is customary to assume that for a given metal W is independent of both temperature and volume. As regards the dependence of W on volume, this is quite unjustified, and some of the discrepancies between the conclusions of the theory and the experimental data for electronic emission are to be ascribed to the fact that the simple theory ignores the change in W resulting from the thermal expansion. It is less certain whether W is affected by temperature alone, apart from the effect due to thermal expansion.

For many purposes it is unnecessary to consider W, and it is often customary to take the zero of potential energy to be that of an electron at rest within the metal, in which case the energy of an electron is purely kinetic. This is the procedure adopted in the well-known text-book of Mott and Jones, but if this convention is used it must always be remembered that there is really a potential energy term which may be variable under some conditions. In their discussion of the subject, Fowler and Guggenheim * conclude that a calculation of the variation of W with temperature and volume is prohibitively difficult, and this prevents the free-electron theory from being used to explain phenomena connected with changes of pressure or volume.

One of the most useful features of the free-electron theory is that it shows in a simplified way some of the main differences between a " gas " of electrons at the densities at which electrons exist in solid metals, and an ordinary gas obeying the laws of the Classical Theory. We have first to consider how the state of a gas is to be described, and here if we adopt the older viewpoint, we may imagine the state of a gas molecule to be defined by its position (x, y, z) relative to three mutually perpendicular axes, and by the three components of momenta (p_x, p_y, p_z) parallel to the same three axes. If then we have a gas containing N molecules, the state of the whole gas at any instant can be represented

---
* R. H. Fowler and E. A. Guggenheim, " Statistical Thermodynamics ", Sections 1107 and 1118, London: **1939** (Macmillan & Co., Ltd.).

by a single point in a so-called *Phase Space* of $6N$ dimensions, these dimensions being (x, y, z, p_x, p_y, p_z) for each molecule. We could also describe the condition of a gas by an assembly of points in a *six-dimensional Phase Space*, where the dimensions are the three co-ordinates of position x, y, z and the three co-ordinates of momenta (p_x, p_y, p_z), and each point refers to one molecule of the gas. This six-dimensional phase space is called a *Molecular Phase Space*. We can also describe the momentum characteristics of a gas by an assembly of points in a so-called *Momentum Diagram* in which the three co-ordinates are the components of momenta (p_x, p_y, p_z) parallel to the three mutually perpendicular directions (x, y, z) in real space. In this case in the older statistics each particle is represented by a point P, such that the length OP (Fig. IV.2(a)) is equal to the momentum, and the direction OP gives the direction of motion of the particle concerned. The momenta of all

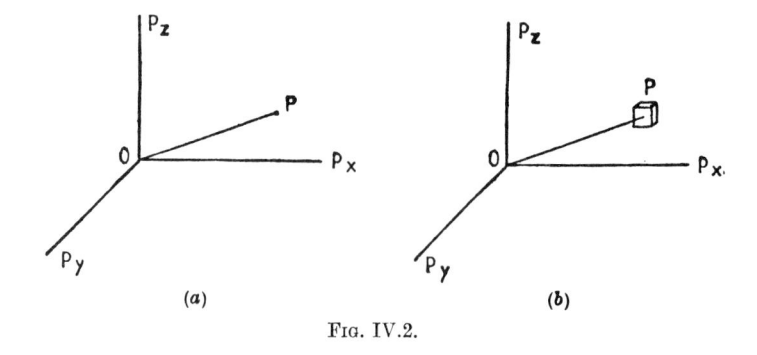

FIG. IV.2.

the particles of a gas will then be represented by a dust or cloud of points in the momentum diagram, and the density distribution of this dust can indicate the energy of the assembly, a point near to the origin corresponding to a particle of low momentum, and hence of low velocity and small kinetic energy.

It will be appreciated that the momentum diagram by its very nature tells us nothing about the position of a molecule, but refers solely to the momentum. In the older statistics, where the mean energy of a gas particle was proportional to the absolute temperature, diminishing temperature resulted in the dust of points becoming more closely concentrated about the origin (*i.e.*, the velocities became smaller), and at the absolute zero all the dust points were concentrated with infinite density at the origin of Fig. IV.2(a).

In the new quantum statistics a momentum diagram analogous to that of Fig. IV.2(a) is still used, but the restrictions imposed by the

The Free-Electron Theory of Metals 169

Principles of Heisenberg and Pauli have to be considered, and for this purpose we may imagine the gas to be contained in a cube of side L, and volume L^3. Since each particle is now restricted to a volume V, its momentum can no longer be specified exactly. The effect of this is to divide up the momentum diagram into little cells of volume h^3/V, since, if the particle is confined to a cube of side L, the uncertainty in its momentum must be at least h/L in each dimension, and so its momentum cannot be defined more precisely than as being associated with a volume h^3/V in the momentum diagram; it can be shown that this volume is independent of the shape of the metal. These cells become smaller as the volume of the metal increases, and each cell represents a quantum state. For an electron gas, in accordance with Pauli's Principle, each state can be occupied by not more than two electrons, and the statistical theory based on this assumption is known as the Fermi–Dirac statistics.

The Heisenberg Principle refers to the *minimum* uncertainties, and consequently when we come to consider an individual electron in a volume V, we can no longer regard it as definitely associated with one cell in the momentum diagram. We have instead to say that its momentum is most probably that of a particular cell, but that there are certain probabilities of its momentum being that of an adjacent cell, these probabilities diminishing as the values of the momentum differ from the most probable value. In the free-electron theory, the electrons are regarded as undergoing continual collisions with the atoms (or more properly the thermal vibrations of the lattice), as a result of which an electron is in a given energy state for only a very short time. According to Heisenberg's Principle, we can know the energy exactly only if the time is completely indeterminate, and consequently for electrons in a metal each energy state must have a corresponding width or uncertainty, and for any solid appreciably larger than atomic dimensions the width of the energy states is greater than the distance between them. This fact, together with the uncertainty in the momentum and the smallness of the constant h, means that it is justifiable to treat the energy states as forming a continuum, although Pauli's Principle still applies and there cannot be more than two electrons in each volume h^3/V of the momentum diagram.

The above remarks are based on the fact that we know the electron is within the volume V, but if we attempt to localize its position inside the volume, then the uncertainty in its momentum becomes correspondingly greater. If, for example, we consider the electron as travelling about inside the solid, we have to consider it as a probability wave-packet (see p. 43), and the more we localize its position the wider is

170   Atomic Theory

the range of cells in the momentum diagram in which the momentum of the electron may lie. We have already seen (p. 42) that the position of the electron in real space can only be represented by a probability wave-packet, which extends over a larger range, the more precisely we define the momentum of the electron. Similarly, the position of the point representing the momentum of the electron in the momentum diagram can be defined exactly only if the position of the electron is completely indeterminate. As soon as we place restrictions on the position of the electron, e.g., by saying that it is in the volume or part of the volume V, there is an uncertainty in its momentum, and the representative point in the momentum diagram has to be replaced by a probability wave-packet, the amplitude of which at any place is a measure of the probability of the electron having the momentum concerned. In all ordinary problems the electron cells of volume h^3/V in the momentum diagram which represent the possible electron states are so small that the region in which the probability momentum wave-packet has an appreciable amplitude extends over many cells, and the more we localize the position of the electron, the larger is the volume of momentum space in which the representative point may lie. In the free-electron theory this point is not often of importance, but in the more complex theories where the atomic structure is considered the situation may be very different. If, for example, we consider an electron as moving between a given pair of atoms, we are localizing its position to within a distance of the order of 10^{-8} cm, and under these conditions its momentum will be correspondingly uncertain, and its point in the momentum diagram may lie in any one of a large number of cells.

As explained above, Pauli's Principle states that each cell in the momentum diagram can contain not more than two points, corresponding with electrons of opposite spins. At high temperatures and low densities this restriction on the number of points in each cell of the momentum diagram has little effect, since the dust of points is widely scattered (i.e., the momenta and energy are large). But at low temperatures the restriction is of great importance, because it implies that at the absolute zero the dust of points can no longer contract to the origin. Since each cell of volume h^3/V can contain only two dust points, representing two electrons in the same state but with opposite spins, an assembly of N electrons must occupy at least $N/2$ cells, and hence a volume of momentum space equal to at least $Nh^3/2V$. At the absolute zero of temperature, the cells closest to the origin are each occupied by two dust points, and the lowest energy will clearly be when the distribution is symmetrical, so that the occupied states constitute a sphere in the

The Free-Electron Theory of Metals

momentum diagram, the volume of this sphere being given by the equation:

$$\frac{4}{3}\pi p^3_{\max.} = \frac{Nh^3}{2V}. \qquad \text{IV (2)}$$

where $p_{\max.}$ represents the momentum of the highest occupied state. If we adopt the convention that the potential energy of an electron at rest inside the metal is zero, the total energy of each electron will be kinetic, and will be equal to $\frac{1}{2}mu^2$, which equals $\frac{p^2}{2m}$, where p is the total momentum. It follows, therefore, that at the absolute zero, where the occupied states form a sphere round the origin of the momentum diagram, the energies of the electrons extend over a range from the almost zero value* of the lowest state, to a maximum value, $E_{\max.}$, given by:

$$E_{\max.} = \frac{1}{2m}p^2_{\max.} = \frac{h^2}{2m}\left(\frac{3N}{8\pi V}\right)^{2/3} = \left(\frac{3}{\pi}\right)^{2/3}\left(\frac{N}{V}\right)^{2/3}\frac{\pi^2\hbar^2}{2m}. \qquad \text{IV (3)}$$

where $\hbar$ is an abbreviation for $\frac{h}{2\pi}$.

The free-electron theory does, therefore, satisfy the first requirement of a satisfactory theory in that, even at the absolute zero, the electronic energies are spread over a range of values, which increases as the volume becomes smaller. It will be appreciated that as the volume of the metal becomes larger, the volume of the cells in momentum space (h^3/V) becomes continually smaller, so that the difference in the energies of successive states becomes smaller, and the magnitude of h is such that for any piece of metal of reasonable size, the difference between successive states or energy levels is so small that they may be regarded as forming a continuum. $E_{\max.}$, however, depends not on V, but on N/V, and consequently for any given metal, $E_{\max.}$ is independent of the size of the specimen, since if we double the volume we also double the number of electrons present. If the numbers of electrons present are of the order one per atom, the calculated values of $E_{\max.}$ for most metals are of the order of 1 to 10 eV. Thus Mott and Jones† give the following values of $E_{\max.}$ for the univalent metals on the assumption of one electron per atom, using the accepted value of Avogadro's number, and the experimentally determined densities and atomic weights.

Element	Li.	Na.	K.	Rb.	Cs.	Cu.	Ag.	Au.
$E_{\max.}$, eV	4·74	3·16	2·06	1·79	1·53	7·10	5·52	5·56

For the alkalis, the method is a rough approximation, and the steady

---

* As explained in Section I.3 (p. 21), the Heisenberg Uncertainty Principle prevents us from thinking of an electron in a finite volume as being at rest.
† "The Theory of the Properties of Metals and Alloys," p. 54. Oxford: **1936**.

decrease in $E_{max.}$ with increasing atomic number is a result of the steady increase in atomic volume.

It will be seen, therefore, that the picture of an electron gas at the absolute zero of temperature, given by the Fermi–Dirac statistics, is very different from that of the older statistics, since even at the absolute zero there are electrons of high energy, and it can be shown that the total energy of an assembly of N electrons at the absolute zero is $3/5 N E_{max.}$, so that the mean energy of one electron, which is called the *mean Fermi Energy*, is $3/5 E_{max.}$. For convenience we may replace $E_{max.}$ by the symbol ε^*, and this quantity is of great importance, because many of the equations for the properties of the electron gas can be expressed comparatively simply in terms of ε^*. The fact that many of the equations involve the constant π can be readily understood if the concept of a sphere of occupied states in the momentum diagram is borne in mind. The high energy of the electron gas at the absolute zero is the direct result of the Pauli Exclusion Principle, which prevents more than two electrons from occupying the same state. Other ways of expressing this principle are that the density of states in momentum space cannot exceed $2V/h^3$, and that the density of states in 6-dimensional phase space cannot be greater than $2/h^3$.

If, at the absolute zero of temperature, we introduce electrons into a box of definite volume V, an argument similar to the above shows that the number of electronic states with energy less than a given value E, is:

$$\frac{V}{6\pi^2}\left[\frac{2m}{h^2} \cdot E\right]^{3/2}.$$

Consequently, as the electrons are introduced into the box, their number varies as $E^{3/2}$, where E is the energy of the highest occupied state.

The free-electron model thus requires the energies of the electrons to be spread over a range, and the properties of the assembly will clearly depend on how the electrons are divided among the possible energy states. We have explained in Section III.1 how the $N(E)$ curves indicate the distribution of the energy states over the different ranges of energy. In the free-electron model it is clear that $N(E)$ will increase continuously with E, since the greater the energy the larger is the sphere in momentum space which corresponds to the value of E concerned, and hence the greater the number of states in the range E to $E + dE$. The exact expression is given by differentiating the above expression with respect to E, so that:

$$N(E)dE = \frac{V}{4\pi^2}\left(\frac{2m}{h^2}\right)^{3/2}\sqrt{E} \cdot dE. \quad . \quad . \quad \text{IV (4)}$$

The relation between $N(E)$ and E is thus parabolic, as shown in Fig. IV.3, and at the absolute zero of temperature the assembly of N

The Free-Electron Theory of Metals

electrons occupies the $N/2$ lowest electron states extending up to $E = \varepsilon^*$ and represented by the shaded area in Fig. IV.4. From the definition of

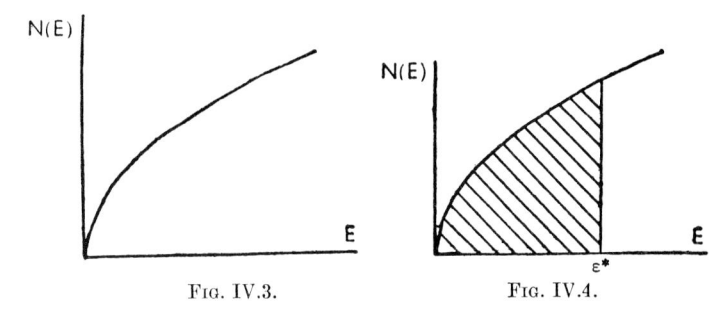

Fig. IV.3.   Fig. IV.4.

$N(E)$ it follows that at the absolute zero the number of occupied electronic states in unit volume of the metal is equal to the shaded area, and the number of electrons is twice this value, since each state contains two electrons.

The preceding remarks refer to the absolute zero of temperature, where the electrons occupy the $N/2$ lowest energy states, which form a sphere round the origin of the momentum diagram in Fig. IV.2(b). The surface of this sphere may be called the *Fermi Surface*, and the electrons in states on the Fermi Surface have energy ε^* at the absolute zero. We have next to consider the effect of rising temperature, and how this affects the distribution of points among the cells of the momentum diagram. Since increasing temperature produces increasing energy, we shall clearly expect a tendency for the occupied states to move outward from the origin of the momentum diagram as the temperature is raised, and it is here that the Pauli Exclusion Principle produces one of its most striking effects. In the older statistics, a rise in temperature could produce an increase in the energy of any of the electrons, without any restrictions, but in the Fermi–Dirac statistics this is no longer so. Each cell in the momentum diagram of Fig. IV.2(b) can contain only two points, and consequently when the temperature is first raised above the absolute zero, an electron in a low energy state (*i.e.*, one near to the origin of Fig. IV.2(b)) cannot be raised to a state of slightly higher energy, because all these states already contain two electrons, which is the maximum allowed by Pauli's Principle. Electrons of low energy could therefore only undergo a large excitation, carrying them to the Fermi surface where unoccupied states exist, but the probability of such a large excitation is very small. Hence the first rise in temperature affects only those electrons in states near to the Fermi surface, since these can undergo excitation into adjacent states which are unoccupied.

When, therefore, we first raise the temperature from the absolute zero, the electrons of low energy are left unaffected, and only electrons with energies near ε^* are excited. As the temperature becomes higher and higher, continually greater numbers of electrons are excited, and at very high temperatures the excitation would affect the whole of the electrons, but at the electron densities concerned in metals these temperatures are far above the melting points. These general considerations apply also to the more complete electron theories, in which the first rise in temperature above the absolute zero again affects only the electrons near the Fermi surface. This statement does not, of course, imply that an individual electron in a low energy state always remains in that state. The statistical representation always allows interchange between electrons, so that electron No. 1 in state A and electron No. 2 in state B may interchange to give electron No. 2 in state A and No. 1 in state B. Since electrons are indistinguishable, it is doubtful whether this has any real meaning, and in any case the final and initial distributions are the same. The thermal excitation which we have been discussing involves the transition of an electron from one state to another with the production of a new distribution.

It remains to express these general considerations in a more precise form, and for the purpose we require a function showing the probability of finding an electron in a given state, or the probability that a given cell in the momentum diagram or the 6-dimensional phase space is occupied. This may be expressed by saying that the number of electrons in an element of 6-dimensional phase space $d\Phi$ is $2fd\Phi$, where f is the so-called distribution function, and the factor 2 arises from the fact that the electrons may have opposite spins. Since each volume h^3 of the 6-dimensional phase space corresponds with one state, this is equivalent to saying that the number dn of electrons per unit volume with velocities between the limits ξ, η, ζ, and $\xi + d\xi$, $\eta + d\eta$, and $\zeta + d\zeta$ is:

$$dn = 2\left(\frac{m}{h}\right)^3 f \cdot d\xi \cdot d\eta \cdot d\zeta \quad \ldots \quad \text{IV (5)}$$

and in the Fermi–Dirac statistics the function f is given by:

$$f = \frac{1}{\exp[(E-z)/k\theta] + 1} \cdot \quad \ldots \quad \text{IV (6)}$$

where $E = \dfrac{mu^2}{2} = \dfrac{m}{2}(\xi^2 + \eta^2 + \zeta^2)$ and z is the thermodynamic potential of one molecule of gas; z is thus given by:†

$$z = U - \theta S + pv \quad \ldots \quad \text{IV (7)}$$

† In this equation U is the total internal energy, θ the absolute temperature, S the entropy, p the pressure, and v the volume.

The Free-Electron Theory of Metals

and is a function of temperature. At the absolute zero of temperature $z = \varepsilon^*$, so that at the absolute zero the thermodynamic potential is equal to the energy of the electrons on the Fermi surface. It can be shown that the variation of z with temperature is given by:

$$z = \varepsilon^* - \frac{\pi^2 \, k^2\theta^2}{12 \, \varepsilon^*} \qquad \ldots \ldots \quad \text{IV (8)}$$

for the range in which the behaviour of the gas approximates to that at the absolute zero. When in this condition the gas is said to be degenerate, and the conditions for degeneracy are described on p. 176. We have already seen that ε^* is of the order of 1 to 10 eV, and since at $1000°$ C, $k\theta$ is only about $\frac{1}{10}$ eV, it can be seen that z is approximately equal to ε^* at ordinary temperatures. At the absolute zero of temperature, since $E = \varepsilon^*$, it can readily be seen from equation (6) that if $E < \varepsilon^*$, $f = 1$, and that if $E > \varepsilon^*$, $f = 0$. If, therefore, we plot f against E, we obtain a curve of the type shown in Fig. IV.5, No. 1, with a sharp fall at $E = \varepsilon^*$. The meaning of this is that up to $E = \varepsilon^*$ all the states are occupied by two electrons, since $f = 1$, and $2f = 2$ in Fig. IV.5, whilst

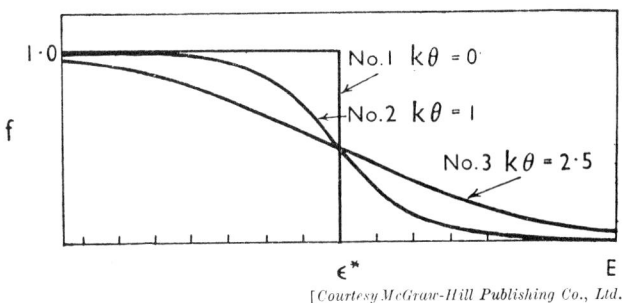

[Courtesy McGraw-Hill Publishing Co., Ltd.]
Fig. IV.5.

for energies greater than ε^* no state is occupied, the Fermi surface being perfectly sharp. At higher temperatures the corresponding curves take the form shown in Fig. IV.5, Nos. 2 and 3. These curves have the characteristic that they pass through the point $f = \frac{1}{2}$, $E = \varepsilon^*$, and are symmetrical with change of sign about this point. From the relative magnitudes of ε^* and $k\theta$, it will be appreciated that it is only at very high temperatures that there is any considerable difference from the curve at the absolute zero. The meaning of these curves is simply that as the temperature is first increased above the absolute zero, the electrons of low energy are left unchanged (i.e., $f = 1$) and the states of low energy continue to be fully occupied, whilst some of those of higher energy are excited. This excitation means that there is a probability of finding electrons in states with $E > \varepsilon^*$, so that the value of f for these states is now greater than zero. This implies that the states for which E is slightly less than ε^* are no longer completely occupied, and so the value of f for these states is less than unity. If we consider the electrons of highest energy for which $E - \varepsilon^* \gg k0$, the number of electrons per energy range dE is, from the above equations:

$$2N(E) \exp\left[-(E-z)/k0\right] \quad \ldots \quad \text{IV (9)}$$

This is the equation for a distribution of a Maxwellian type, so that if $E = z$ is taken as the zero of energy, the electrons of highest energy have a Maxwellian distribution.

The curves shown in Fig. IV.5 indicate the probability of states of given energy being fully occupied at different temperatures.† We have next to consider the conditions which determine whether the state of a gas approximates to that at the absolute zero (curve No. 1 in Fig. IV.5). Gases in this state are said to be degraded or degenerate, and it can be shown that the condition for almost complete degradation‡ is that:

$$\frac{Nh^3}{2V}(2\pi mk0)^{-3/2} \gg 1 \quad \ldots \quad \text{IV (10)}$$

This condition is satisfied by all metals at ordinary temperatures, and even at 2000° K the deviation is very slight. It will be noted that degradation is favoured by the small mass of the electron, and by large values of N/V, i.e., by a high concentration of electrons in a given volume. Roughly speaking, a gas of electrons at ordinary temperatures at the densities concerned in metals is degraded to an extent which would correspond with that of a gas such as helium at a few degrees above the absolute zero, if it obeyed the Fermi–Dirac statistics. Actually helium obeys the Einstein–Bose statistics, and so the comparison cannot be made.§ For many purposes, therefore, electrons in metals at ordinary temperatures can be considered as completely degraded, and the equations for the behaviour at the absolute zero can be used, with a great simplification of the mathematical treatment. The first application of the Fermi–Dirac statistics to the problem of electrons in metals was made by Pauli∥ in connection with the paramagnetic properties of the alkali metals (see p. 191). The whole subject

---

† Since there are two directions of spin, a state for which $f = 1$ contains $2f = 2$ electrons.
‡ The German word for the phenomenon is *Entartung*.
§ For information on the difference between the Fermi–Dirac and Einstein–Bose statistics, the reader may consult J. C. Slater's "Introduction to Chemical Physics." New York: **1939** (McGraw-Hill Book Co., Inc.).
∥ *Z. Physik*, 1926, **41**, 81.

was then developed by Sommerfeld† in a remarkable series of papers which stimulated work in many countries. Although the simple theory of Sommerfeld is open to several objections, the application of the Fermi–Dirac statistics in this way represents one of the great advances in the theory of metals. The earlier work was usually subject to the simplifying assumption that the electron gas was completely degraded. At higher temperatures, where the electron gas is partly degraded, the treatment becomes more complicated, and the equations have been discussed by Stoner.‡ At extremely high temperatures or very low densities where:

$$\frac{Nh^3}{2V}(2\pi mk\theta)^{-3.2} \ll 1 \quad \ldots \quad \text{IV (11)}$$

the Fermi–Dirac statistics reduces to the ordinary classical Maxwell–Boltzmann statistics,§ and the function f takes the form:

$$f = \exp[(E-z)/k\theta] \quad \ldots \quad \text{IV (12)}$$

The mathematical treatment then again becomes simple, but this condition is never satisfied by electrons within a metal, although it is sometimes satisfied by an electron atmosphere outside a metal, and in some problems of electron emission we are really concerned with an equilibrium between a degenerate Fermi–Dirac gas of electrons inside a metal and a gas of electrons obeying the Classical Laws outside the metal.

The different curves of Fig. IV.5 are, of course, closely related to the energy of the electron gas, since they indicate the distribution of points in the cells of the momentum diagram. For the degraded gas of electrons, the energy may be expressed in the form:

$$E = N\left[\frac{3}{5}\varepsilon^* + \frac{\pi^2}{4}\frac{(k\theta)^2}{\varepsilon^*}\right] \quad \ldots \quad \text{IV (13)}$$

The relation between energy and temperature is of the form shown in Fig. IV.6, and a metal at ordinary temperatures corresponds with the

† *Z. Physik.*, 1928, **47**, 1.
‡ *Phil. Mag.*, 1938, [vii], **25**, 899; 1939, [vii], **28**, 257.
§ It may be noted that this condition is readily related to the Uncertainty Relation. Omitting the small numerical factors, equation IV (11) is equivalent to $\left(\frac{V}{N}\right)^{1/3}(mk\theta)^{1/2} \gg h$. Now $\left(\frac{V}{N}\right)^{1/3}$ is of the order of the average distance between two particles, and in classical mechanics $(mk\theta)^{1/2}$ is the average momentum. The product $\left(\frac{V}{N}\right)^{1/3}(mk\theta)^{1/2}$ is therefore of the same order as the product of the uncertainties in the momentum and position, and equation IV (11) states that this must be very much greater than h if classical mechanics is to apply.

178   *Atomic Theory*

almost horizontal part of the curve, so that the energy of the degraded electron gas is almost independent of the temperature. This is in marked contrast to the behaviour of an electron gas obeying the older statistics which required the energy to be directly proportional to the absolute temperature. Since the energy of a degraded electron gas is almost independent of the temperature, its specific heat is almost zero,

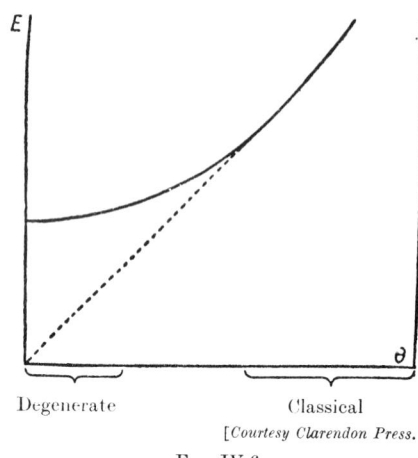

Degenerate   Classical
[*Courtesy Clarendon Press.*]
Fig. IV.6.

and in this way we can understand why the specific heat of a metal can be explained by considering only the atomic vibrations, even though free electrons are present whose number is of the order one per atom. This is one of the great triumphs of the free-electron theory, and the general conclusion applies also to the more complete theories.

The specific heat of a degraded electron gas is obtained by differentiating equation IV (13) with respect to the temperature, and, neglecting the small variation of ϵ^* with temperature, this is given by:

$$c_v = \frac{\pi^2 k^2}{2\epsilon^*} N\theta \quad . \quad . \quad . \quad . \quad \text{IV (14)}$$

so that the electronic specific heat is proportional to the absolute temperature. At ordinary temperatures the specific heat of the electrons is negligible compared with that of the atoms, but at very low temperatures where the atomic heat has become small, the electronic specific heat is relatively much greater, and for some metals it has been possible to express the results of specific-heat measurements in the form of the sum of two terms, one representing the normal atomic heat, and the other being proportional to θ, and identified as being the electronic

The Free-Electron Theory of Metals

specific heat. These remarks refer to normal metals, but the transition elements often show abnormal specific heats for reasons which are described later (p. 354).

Fig. IV.5 shows the effect of temperature on the relations between f and E, and the corresponding effect on the $n(E)$ curve is shown in Fig.

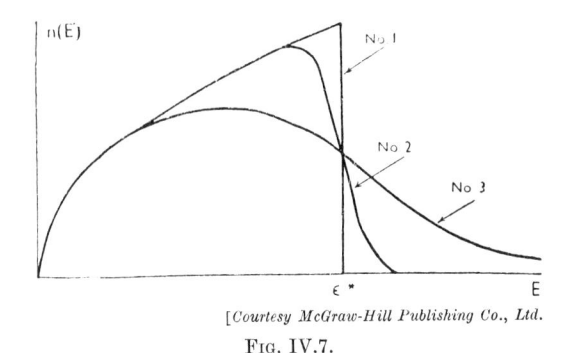

[Courtesy McGraw-Hill Publishing Co., Ltd.
Fig. IV.7.

IV.7, in which the vertical line showing the limit of the electronic energies at the absolute zero is replaced by curves of the form shown.

SUGGESTIONS FOR FURTHER READING.

Much of interest in connection with this chapter will be found in " Introduction to Chemical Physics," by J. C. Slater. 1939: New York and London (McGraw-Hill), and "The Theory of the Properties of Metals and Alloys," by N. F. Mott and H. Jones. 1936: Oxford (Clarendon Press); Dover Reprints, New York. Though often out of date as regards experimental work this book remains one of the best introductions to the subject.

2. THE MODELS OF WAVE MECHANICS.

In the above description we have referred to the Fermi–Dirac statistics, with the Pauli restriction of not more than two electrons in any one state, and we have accepted this as an arbitrary constraint which the electron gas obeys. It remains now to consider how this concept is related to the wave mechanics, and how the wave-like characteristics of an electron affect the electron gas. When we deal with the observed properties of metals such as electronic specific heat, electrical conductivity, thermal conductivity, electronic emission, &c., we are always concerned with the behaviour of a large number of electrons, and we are never able to follow the individual electron along its path. The requirements of a satisfactory theory are therefore that it accounts for the behaviour of large numbers or assemblies of electrons. The complete problem of the motion of electrons in an actual crystal where there is a periodic field disturbed by the thermal vibrations of the lattice is at present too difficult for solution, and even the problem of a gas of electrons which interact with one another (Coulomb repulsion, exchange forces, &c.) is very complicated; it is therefore necessary to begin by considering first the behaviour of one electron enclosed in a volume V, and to derive the wave-functions and the equations for the energy, &c. From this the theory goes on to consider the behaviour of an assembly of N electrons in a volume V, and it is assumed that the possible electron states are the same as those in the one-electron problem, subject only to the Pauli restriction that not more than two electrons are in any one state. So long as the assumption of non-interacting particles is accepted, this part of the theory is satisfactory, but it need scarcely be said that as regards the actual state of affairs inside a metal the treatment is a very great over-simplification of the problem, and to regard complete ψ patterns inside an actual metal as identical with those for one electron in an empty box of the same dimensions, raises as many difficulties as some of the inconsistencies of the older theories. Actually, as we shall see later, the free-electron model is usually a fair approximation to the condition of affairs in the alkali metals, but is a very bad model in nearly all other cases, and the theory is retained largely as a matter of mathematical convenience, and because it forms a simplified background against which the complexities of the more detailed theories can be viewed.

For convenience it is usual to assume that the metal exists in the form of a cube of side L, and it is necessary to make some assumption as to the conditions which exist at the surface. Since the electrons exist within the volume $V = L^3$, the most simple boundary condition is clearly that ψ vanishes at the boundary. The wave-function which satisfies this condition is of the form:

$$\psi = \sin \frac{\pi l_1 x}{L} \cdot \sin \frac{\pi l_2 y}{L} \cdot \sin \frac{\pi l_3 z}{L} \qquad \text{IV (15)}$$

where l_1, l_2, l_3 are positive integers. If the zero of potential energy is that of an electron at rest outside the metal, and $-W$ is the potential inside the metal, the energies of the possible electron states are given by the relation:

$$E = -W + \frac{h^2}{8mL^2}(l_1{}^2 + l_2{}^2 + l_3{}^2) \qquad \text{IV (16a)}$$

It is a general assumption of the theory that W is independent of the temperature, and the limitations of this have been discussed (p. 167).

The Free-Electron Theory of Metals

If we regard the potential inside the metal as zero, the expression for the energy, which is then purely kinetic, becomes:

$$E = \frac{h^2}{8mL^2}(l_1{}^2 + l_2{}^2 + l_3{}^2) \qquad . \quad . \quad \text{IV (16}b\text{)}$$

which is the form used by Mott and Jones. The energies of the possible electronic states are thus characterized by the squares of three positive integers. Since $E = \tfrac{1}{2}mu^2$, and $\lambda = h/mu$, the wave-length of the lowest state is $2L/\sqrt{3}$ and corresponds with the combination $l_1{}^2 = l_2{}^2 = l_3{}^2 = 1$. The interval between successive energy levels is $h^2/8mL^2$, and since h is of the order of 10^{-27} erg-sec and m is of the order of 10^{-28} g, the interval between successive levels becomes appreciable only when L is of atomic dimensions. As we have already explained (p. 168), when L is large compared with atomic dimensions, the energy levels are so close together that the Uncertainty Principle prevents us from saying that an electron is definitely in any one state, and the states may be regarded as forming a continuum, although the Pauli restriction still applies, and there cannot be more than two electrons to each state. An assembly of N electrons in a volume V, at the absolute zero of temperature, will thus occupy the $N/2$ lowest energy states given by equation IV (15), and this number $(N/2)$ will be equal to the number of sets of positive integers such that:

$$\frac{h^2}{8mL^2}(l_1{}^2 + l_2{}^2 + l_3{}^2) < E_{\max}. \qquad . \quad . \quad \text{IV (17)}$$

where $E_{\max.}$ is the energy of the highest occupied state. It can be shown that this is given by:

$$\frac{N}{2} = \frac{\pi}{6}\left[\frac{8mL^2}{h^2} \cdot E_{\max.}\right]^{3/2} = \frac{V}{6\pi^2}\left[\frac{2m}{h^2} \cdot E_{\max.}\right]^{3/2} \quad . \quad \text{IV (18)}$$

This expression is the same as that on p. 172, so that the box model gives the Fermi–Dirac relation between E, V, and N. The ψ patterns given by the above equations may be regarded as those which fit conveniently into the cube of side L. They represent each electron state as a standing wave reflected from the sides of the box, but these standing waves are not to be interpreted as stationary electrons, but rather as indicating that there is an equal probability of finding the electron moving in one of two opposite directions. If the motion is referred to rectangular axes parallel to the three sides of the cube, then l_1, l_2, and l_3 are the direction cosines of the motion of the electron in the state defined by the integers l_1, l_2, and l_3.

The simple condition that ψ vanishes at the boundary of the box is not really sufficient, because it does not prevent an electron from moving into the region outside the box. To prevent this, as explained on p. 46, we use a function in which ψ dies away exponentially* outside the box, and Fig. I.5 (p. 46) shows the general form of the wave-function for the stationary state whose wave-length is equal to $2L/5$, where L is the length of a one-dimensional box. It can be shown that the exact form of the boundary condition assumed does not affect the behaviour of the swarm of electrons as a whole, for properties which do not involve the surface. Thus the energy content and specific heat of the assembly of electrons are not affected by the exact boundary conditions, although the latter may determine the exact form of the equation for, say, thermionic emission.

For problems such as conductivity it is convenient to use a model in which the electron states are represented by travelling waves, and for this purpose what is called a " cyclical metal " is assumed. This may be regarded as an infinite metal divided up into small cubes of side L, and the boundary condition assumed is that ψ is triply periodic with period L, i.e., that ψ repeats itself in each small cube so that the waves travel on continuously through the infinite crystal. The wave-functions then assume the form:

$$\psi = \exp[2\pi i(\chi_1 x + \chi_2 y + \chi_3 z)/L) \quad . \quad . \quad \text{IV (19)}$$

where χ_1, χ_2, χ_3 are whole numbers which may be either positive or negative, including zero. This model refers to an infinite metal with the internal periodicity given by the division into the small cubes of side L. To apply the argument to a metal of finite size, it is assumed that this consists of a cube of side L which can be treated with the same boundary condition, so that ψ is triply periodic with the period of the whole piece of metal of side L, but no condition is imposed to restrict the electron to the volume considered. Fortunately it can be shown that this does not affect many of the conclusions for the behaviour of the assembly of electrons as a whole, and the model is satisfactory as long as we are not concerned with properties near a surface.

The energies corresponding with equation IV (19) are:

$$E = -W + \frac{h^2}{2mL^2}(\chi_1^2 + \chi_2^2 + \chi_3^2) \quad . \quad . \quad \text{IV (20)}$$

where the χ integers are either positive or negative, including zero. If W is assumed to be zero, the energy and wave-length of the lowest state are zero and infinite, respectively, as compared with the values $h^2/8mL^2$ and $2L/\sqrt{3}$ given by the box model. The fact that the integers in equation IV (17) are positive only, whilst those in equation IV (20) are positive and negative, results in the two models giving the same values for the

---

* Examples of the fitting of exponential tails in this way are given in Chapter III of " Elementary Quantum Mechanics," by R. W. Gurney. **1934**: Cambridge (University Press).

number of energy states per unit energy range, and for all values of the properties of the swarm of electrons as a whole. The properties of the whole assembly of electrons are thus independent of the boundary conditions assumed, although the latter affect profoundly the energies, wavelengths, &c., of the individual states. This point sometimes causes great confusion, since there is a natural tendency to ask whether the electron states in a metal are " really those " of equation IV (17), where the states are at intervals of $h^2/8mL^2$, or those of equation IV (20), where the intervals are four times as great. To answer this question we have to ask what we mean by reality, and if we adopt the attitude that this implies something measurable, then any method of measuring the energy will involve a certain error ΔE, and if the metal is large enough, *i.e.*, if L is large compared with atomic dimensions, the intervals between the successive energy states of the above equations will be smaller than ΔE. In this sense it may be said that the individual states of these equations have no real existence, although the energy of the assembly of electrons has a real existence and can be measured. The difference between successive energy states in equation IV (20) is four times as great as those in equation IV (17), but the existence of both positive and negative integers in equation IV (20) means that the density of states is four times as great as in equation IV (17), where the integers are positive only, and consequently both equations give the same results for the assembly as a whole, and both give the same result for the " reality " which we can measure; so that for a solid of ordinary size it is merely a matter of mathematical convenience which model is used. If, on the other hand, we were dealing with a very small solid, of dimensions of a few Ångstrom units, then the two equations would not be interchangeable, and on the whole we might expect the equations for the box model to be the more satisfactory, since this model contains the all-important condition that the electron is confined to the volume under consideration.

In the cyclic model, where the electron state is represented as a travelling wave, the individual electron must again be thought of as a wave-packet, and the mean velocity v associated with a state is given by the equation for the group velocity (p. 40):

$$v = \frac{d\nu}{d\left(\frac{1}{\lambda}\right)} \quad \ldots \ldots \quad \text{IV (21)}$$

Since the energy E is equal to $h\nu - mc^2$, this is equivalent to:

$$v = \frac{1}{h} \cdot \frac{dE}{d\left(\frac{1}{\lambda}\right)} \quad \ldots \ldots \quad \text{IV (22)}$$

in problems where the velocity is small compared with c. For free electrons, the energy $E = h^2/2m\lambda^2$ and the mean velocity v associated with a state is given by $v = h/m\lambda$ and is equal to the root mean square velocity u. In the more complicated theories of Brillouin zones, equations IV (21) and (22) still apply, but the two velocities are in general not the same (p. 45).

3. Applications of the Free-Electron Theory.

In Section IV.1 we have seen how the specific heat of the degenerate electron gas is quite different from that of an ordinary gas obeying the classical statistics, and we have explained that these general ideas are carried over into the more complete electron theories. We may now consider briefly some other physical properties in which the general ideas of the free-electron theory apply to the more complete theories, even though the details require considerable modification.

(a) *Electrical Conductivity.*

For simplicity in drawing we shall consider a hypothetical two-dimensional metal, and in Fig. IV.8 we show a momentum diagram in

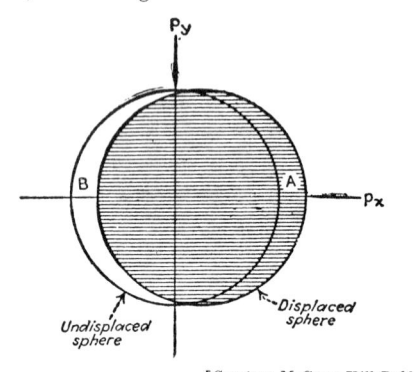

[Courtesy McGraw-Hill Publishing Co., Ltd.
Fig. IV.8.

which the two axes show the co-ordinates of momenta p_x and p_y. The full circle in Fig. IV.8 represents the Fermi surface; at the absolute zero of temperature, in the absence of an external field, this will be a sharp surface and perfectly symmetrical about the origin, and at a higher temperature the surface will be symmetrical but diffuse. If an electric field is now applied in the direction of the x-axis, the whole assembly of electrons will be accelerated in this direction, and the effect

The Free-Electron Theory of Metals

of this is to shift the Fermi surface bodily as shown by the shaded circle. This process would proceed indefinitely were it not for the " collisions with the atoms." The actual physical process is more properly described as an interaction between the electron and the thermal vibrations of the lattice. This is often called a " collision with an atom," because in the earlier electron theories of metals the process was regarded as a collision between an electron, considered as a particle, and an atom; in some treatments it was even assumed that both the electrons and the atoms were spherical. It is now realized that any such precise picture is quite misleading, and that the real process involves the interaction between the thermal vibrations of the lattice and the wave system of the electron. It is, however, a convenient abbreviation* to speak of a " collision," and this is justifiable provided that the real nature of the process is borne in mind. The free-electron theory then makes the following assumptions regarding the collisions:

(1) That the collisions are nearly elastic, and that after each collision the electron sets out with a velocity whose direction is completely random. If the average time between two collisions is 2τ, it is customary to call τ the *time of relaxation*, and during the time 2τ, an electron has an acceleration in the direction of the field given by:

$$\frac{d^2x}{dt^2} = \frac{eF}{m} \quad \ldots \quad \text{IV (23)}$$

The average component of velocity in the direction of the field is thus zero immediately after a collision, and $2eF\tau/m$ after a time 2τ, when a further collision occurs, and the direction of motion is again quite random.

(2) That the changes in the velocities of the electrons produced by the field are small compared with their average velocities as particles of an electron gas.

The question then arises as to what restrictions are imposed on the collisions of the electrons on account of the degraded nature of the electron gas. Since the collisions are assumed to be nearly elastic, an electron can be scattered only into a state of nearly the same energy, *i.e.*, into a state at approximately the same distance from the origin of the momentum diagram of Fig. IV.2(*b*). In a degraded electron gas, all the states of low energy are fully occupied, and the conclusion is, therefore, that electrons of low energy cannot undergo collisions to cause a resistance. The electrons of low energy occupy the close-packed states round the origin of the momentum diagram, and this

---

* This avoids the continual use of expressions such as " the interaction between the electron and the thermal vibrations of the lattice."

group of electrons produces no resultant flow in any direction, since for each two electrons with velocity $+v$ there are two with velocity $-v$, and since the states are fully occupied no change in the distribution can be produced by collisions. Two electrons in low energy states may, of course, interact with the lattice vibrations in such a way that if electron No. 1 is in state A before the collision, and electron No. 2 in state B, then after the collision electron No. 2 is in state A, and electron No. 1 is in state B. Since electrons are indistinguishable, it is a matter of definition whether this has any physical meaning, and in any case it does not cause a resistance, because the final state is the same as the first state. On the other hand, electrons at the edge of the Fermi surface, *i.e.*, those in the two crescents of Fig. IV.8, can undergo collisions, the general effect of which is to carry the point in the momentum diagram from the right-hand to the left-hand crescent, and so to create a resistance, and to set up a steady state in which the movement of the crescent to the right-hand of Fig. IV.8 is balanced by the throwing back of the representative points to the left. The condition of affairs is, therefore, one in which all the electrons are accelerated in the direction of the electric field, and all take part in the conductivity process, but only those at the edge of the Fermi surface can produce a resultant flow in any direction, or undergo collisions which produce a resistance.

The expression deduced by Sommerfeld for the electrical conductivity of a degraded electron gas is of the form:

$$\sigma = \frac{8\pi e^2 l}{3h} \left(\frac{3N}{8\pi V}\right)^{2/3} \quad \ldots \quad \text{IV (24)}$$

Hence l is the mean free path of the electrons *at the edge of the Fermi zone*,* so that if $\bar{u}$ is the velocity of these electrons, l is defined by the relation:

$$l = 2\tau\bar{u} \quad \ldots \quad \ldots \quad \text{IV (25)}$$

The variation of conductivity with temperature has then to be ascribed to the variation of l, which cannot be calculated. The theory is unsatisfactory, because when numerical values are inserted the mean free path is of the order of several hundred times the interatomic distances, and a free path of this magnitude is incompatible with the details of the calculation which involve the assumption of free gas particles colliding with spherical atoms. The long free paths can be understood, since wave mechanics permits an electron to move unimpeded through a

* It should be noted that the mean free path l of the Sommerfeld theory has not the same meaning as that of the older theories. In the Sommerfeld theory $l = 2\tau\bar{u}$, where $\bar{u}$ refers to the electrons at the edge of the Fermi surface; whereas in the older theories $l = 2\tau u$, where u is the root mean square velocity of all the electrons.

The Free-Electron Theory of Metals

perfectly regular lattice, but this involves the more detailed theories described later.

Just as a difference of electric potential causes a drift of electrons, so when a difference of temperature is set up between two parts of a metal, the electron gas carries energy from the hot to the cold parts, and in this way the thermal conductivity is accounted for. The expression obtained by Sommerfeld is:

$$K = \frac{8\pi^3}{9} \cdot \frac{lk^2\theta}{h} \left(\frac{3N}{8\pi V}\right)^{2/3} \qquad \ldots \quad \text{IV (26)}$$

where θ is the temperature on the absolute scale. The fact that the thermal conductivity is almost independent of temperature at normal temperatures has again to be ascribed to the appropriate variation of l with θ.

The *Wiedemann–Franz Ratio*, i.e., the ratio of thermal to electrical conductivity, is thus, according to Sommerfeld:

$$\frac{K}{\sigma} = \frac{\pi^2}{3}\left(\frac{k}{e}\right)^2 \theta \qquad \ldots \quad \text{IV (27)}$$

This expression is independent of l, N, and V, and the theory thus predicts that K/σ has a constant value for all metals at a given temperature (the Wiedemann–Franz Law), and that the ratio varies in direct proportion to the absolute temperature (Wiedemann–Franz–Lorentz Law), and not merely are these predictions in general agreement with the facts, but the numerical value is very nearly correct. Actually the theory proves a little too much, since there are real differences between the values of K/σ for the different metals, even when allowance is made for the part of the thermal conductivity due to the atoms, as distinct from the electronic thermal conductivity.

The free-electron theory is by its very nature unable to deal with any of the problems involving the dependence of electrical conductivity on direction relative to the crystal axes. As a quantitative theory its value is not very great, because although the correct value of the Wiedemann–Franz ratio is obtained, this is only by the cancelling of the l terms, which are recognized as having magnitudes incompatible with the mathematical treatment. The general conclusion that the electrons at the edge of the Fermi surface are those effective in determining the resistance is, however, quite correct, and is developed in greater detail in the more complete theories.

(b) *Electronic Emission Phenomena.*

In dealing with emission properties it is usual to denote the potential energy of an electron at rest outside the metal as zero, and in the free-electron model, the potential inside the metal has a constant value — W, as shown in Fig. IV.9. At the absolute zero of temperature, the electrons occupy the $N/2$ lowest energy states extending up to the value ε^*. The minimum amount of energy required to expel an electron is therefore $(W - \varepsilon^*)$, in contrast to the picture presented by the classical theory in which, at the absolute zero, the amount of work required to expel an electron was W, since the electrons were at rest. The experimental results then indicate that W is of the order of 10 eV, and $(W - \varepsilon^*)$ is usually of the order 2-5 eV.

We may consider first the process of photoelectric emission in which the electron absorbs a quantum of energy $h\nu$, where ν is the frequency of

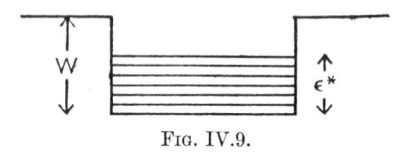

Fig. IV.9.

the radiation. It is clear from Fig. IV.9 that, at the absolute zero, emission will only take place if $h\nu$ is greater than $(W - \varepsilon^*)$, so that there will be a sharp threshold frequency given by:

$$h\nu = (W - \varepsilon^*) \quad \ldots \quad \text{IV (28)}$$

Further, since ε^* is approximately constant at low temperatures, the threshold frequency should be almost independent of temperature, and this is in agreement with the facts. To a higher degree of accuracy, as the temperature is raised, the Fermi surface is no longer completely sharp, and the threshold frequency should no longer be completely sharp; this is in agreement with the experimental data, although up to room temperatures the effect is very small. At higher temperatures the Fermi surface becomes more and more diffuse, and the process of photoelectric emission becomes increasingly complicated and becomes mixed with that of pure thermionic emission.

It is, of course, well known that electrons are emitted from metals at high temperatures, quite apart from any photoelectric effects. This process of thermionic emission has been studied from the viewpoint of the free-electron theory by two different methods. On the one hand, equilibrium between the electrons inside a metal and those in the surrounding space is treated from the viewpoint of the new statistics. On the assumption that the number of electrons emitted per second is equal to the number striking the surface per second, the theory leads to an equation of the form:

$$I = A\theta^2 e^{-B/\theta} \quad \ldots \quad \text{IV (29)}$$

where I is the emission current, and A and B are constants, B being known as the thermionic work-function. This differs from the corresponding equation of the classical theory, which was of the form:

$$I = C\theta^{\frac{1}{2}}e^{-D/\theta} \quad \ldots \quad \text{IV (30)}$$

and this difference is the result of the fact that the new statistics requires the specific heat of the electrons within the metal to be zero. The two equations are in almost equally good agreement with the experimental data, although the equation involving θ^2 is the more correct.

In the second line of approach, the problem of penetration of the surface by electrons is considered from the standpoint of wave mechanics. It can be shown that only the energy associated with the component of velocity perpendicular to the surface need be considered, and the problem becomes one of the passage of wave trains (representing the electrons) through changes of potential occurring at the surface. When an electron moves away from the surface, it induces a charge on the metal and this gives rise to an image potential which attracts the electron back towards the metal. This image potential is equal to $-e^2/4x$, where x is the distance from the surface, and e is the electronic charge. The result of this is that the simple potential step of Fig. IV.10(a) becomes modified to that of Fig. IV.10(b). It can be shown

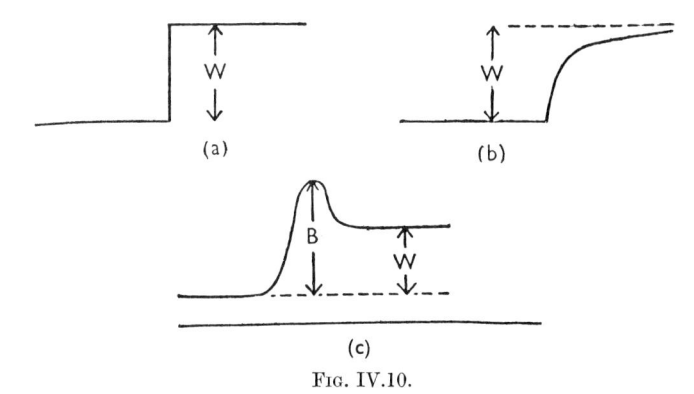

Fig. IV.10.

that only the component of the velocity normal to the emitting surface need be considered, and if E_n is the kinetic energy due to the velocity component normal to the surface, the emission current may be expressed in the form:

$$I = e\int_0^\infty N(E_n)D(E_n)dE_n \quad \ldots \quad \text{IV (31)}$$

Here the potential inside the metal is taken to be zero, and $N(E_n)$ is the number of electrons with energy E_n which strike unit area of the surface in unit time, and $D(E_n)$ is the fraction transmitted. The Sommerfeld theory leads to the relation:

$$N(E_n)dE_n = \frac{4\pi m}{h^3} k\theta \log (1 + e^{-(E_n - Z)/k\theta})dE_n . \qquad \text{IV (32)}$$

where Z is the thermodynamic potential of an electron. In the older theories, where the electron has only the properties of a charged particle, we shall expect the electron to remain within the metal if E_n is less than W, and to escape if E_n is greater than W. The wave-like properties of the electron produce a more complicated state of affairs in which the electron is still unable to escape if E_n is less than W, but in which the value of D for the electrons which do escape depends on the exact form of the potential boundary. For the potential step of Fig. IV.10(a), the mean value of D for $E_n > W$ is about $\frac{1}{2}$, whilst for the more correct picture of Fig. IV.10(b) the mean value of D for $E_n > W$ is nearly unity.

On substituting equation IV (32) into IV (31) and integrating, we obtain for the model of Fig. IV.10(b), where $D = 1$:

$$I = \frac{4\pi mek^2}{h^3} \theta^2 e^{-(W-Z)/k\theta} . \qquad \ldots \qquad \text{IV (33)}$$

$$120 \; \theta^2 e^{-(W-Z)/k\theta} \; \text{amp/cm}^2$$

This is of the correct form if $(W - Z)$ is taken to be constant, except that the factor 120 is about double the correct value. This discrepancy may be ascribed to the fact that the treatment ignores the volume expansion of the metal which affects the potential energy. It is interesting to note that the over-simplified model of Fig. IV.10(a) gives almost the correct value for the thermionic constant, but this is fortuitous.

The above remarks refer to clean metals, but if a surface film is present, the wave-like properties of an electron produce very interesting results. If, for example, the potential step at the surface is of the form shown in Fig. IV.10(c) the older theories would allow an electron to escape only if its kinetic energy were greater than B, where B is the height of the potential hill due to the surface film. But according to wave mechanics, if the film is not too thick, the electron has a chance of escaping if its kinetic energy normal to the surface is greater than W, so that electrons can escape which could not do so according to the older theories. In this case the emission is still given by an equation of the type IV (29), but with reduced values of A and B.

(c) *Magnetic Properties.*

The simple free-electron theory does not deal with ferromagnetism or with the abnormal paramagnetism of transition elements such as

The Free-Electron Theory of Metals

platinum. For simple metals such as the alkalis, the free-electron theory indicates two effects, a paramagnetic effect due to the electron spin, and a diamagnetic effect due to the translational motion. In comparing the theoretical conclusions with experimental results, allowance has to be made for the fact that the theory deals only with the effects of the free electrons, whereas the observed results include also the effect of the ions. All ions with complete outer electron groups of 8 or 18 electrons are diamagnetic, whilst some of the ions of transition elements are paramagnetic. In comparing theory and experiment great care is necessary to ensure that the observed magnetic properties are not affected by the presence of ferromagnetic impurities.

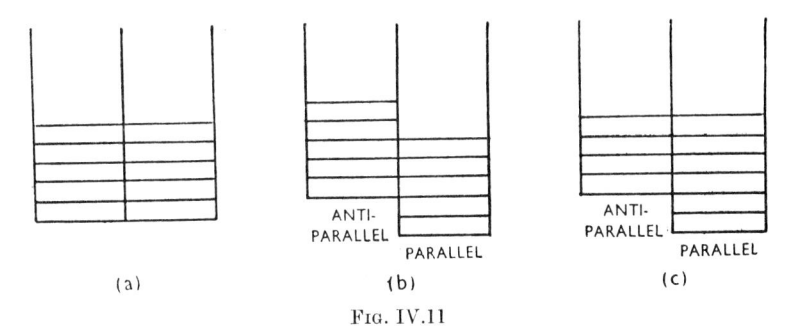

FIG. IV.11

We may consider first the paramagnetism and imagine that we have an electron gas containing N electrons which at the absolute zero in the absence of a magnetic field occupy the $N/2$ lowest energy levels as shown in Fig. IV.11(a). We may now imagine that a magnetic field of strength H is applied, and we may consider the electrons with spins parallel and anti-parallel to the field, respectively. The energy of each electron with parallel spin will then be diminished by an amount $H\mu$, where μ is the Bohr Magneton (p. 62) given by:

$$\mu = \frac{eh}{4\pi mc} \qquad \ldots \qquad \text{IV (34)}$$

Similarly, the energy of the electrons with anti-parallel spins will be increased by $H\mu$, and the condition of affairs will be as illustrated in Fig. IV.11(b). It is clear that the energy of the system can now be diminished if some of the electrons with anti-parallel spins change to those of the parallel type. This change produces a decrease in magnetic energy of $2H\mu$ for each electron concerned, but since a given state can contain only one electron of a given spin, the electrons undergoing this reversal must enter states lying above those already occupied by the electrons

192 Atomic Theory

with parallel spins, and there is thus an increase in kinetic energy. This process will continue until an equilibrium state is reached in which the increase in kinetic energy during a transfer is equal to the decrease in magnetic energy so that the total energy (magnetic + kinetic) is unaffected during a transfer (Fig. IV.11(c)). There will thus be an excess of electrons with parallel spins, and hence a paramagnetic effect. The fact that the observed paramagnetism is almost independent of the temperature is merely the result of the electron gas being almost completely degraded at ordinary temperatures. To a first approximation the expression for the paramagnetic volume susceptibility is:

$$\varkappa = \frac{3}{2} \cdot \frac{N\mu^2}{\varepsilon^*} \left[1 - \frac{\pi^2}{12} \left(\frac{k\theta}{\varepsilon^*} \right)^2 \right] \quad \ldots \quad \text{IV (35)}$$

where the second term in the square bracket is extremely small. The numerical constants are such as to make $\varkappa$ of the order of 10^{-6}, which is in agreement with the experimental data. Equation IV (35) is in marked contrast to the conclusions of the classical theory, which required the susceptibility to vary inversely as the absolute temperature.

Equation IV (35) requires the susceptibility to become very large when ε^* is very small. Equation IV (3) (p. 171) shows that ε^* becomes smaller and smaller as the electron gas becomes more dilute, and it is one of the defects of the free-electron theory that it would require a very dilute electron gas to be ferromagnetic. In some applications of the theory it has, for example, led to the conclusion that the valency electrons in caesium, which has a very large atomic volume, should be ferromagnetic, in contradiction to the facts. These difficulties have been removed by the considerations of electron correlation discussed on p. 298.

According to Classical Mechanics, the diamagnetic susceptibility resulting from the translatory motion of the particles of an electron gas is zero. In quantum mechanics this is no longer the case, because the electron moves in circular orbits round the direction of the field, and these orbits have to satisfy the quantum conditions. The theory works out in such a way that the diamagnetic susceptibility of the electron gas is numerically one-third of that given by the equation IV (35) for the paramagnetic susceptibility due to the spin. Consequently, the total paramagnetic susceptibility of an electron gas is two-thirds of that indicated by equation IV (35).

(d) *Free-Electron Theory of Liquid Metals.*

In the previous sections, some of the consequences of a free-electron model for electrons in a metal have been reviewed, and in Part VI it will

The Free-Electron Theory of Metals

be shown that this model is not too unsatisfactory a basis for discussion of the alkali metals. In most other solid pure metals the role of the periodic crystalline potential cannot be ignored and the considerations of Part V must be taken into account. The possibility has, however, often been entertained that, with the destruction of long-range periodicity on melting, a free-electron model may be appropriate for a wide range of liquid metals, and experimental evidence is now available in a number of cases.

A detailed theory of a liquid metal is inevitably complicated* and the significance of measured quantities should not be over-emphasized. For the best-known property, the electrical conductivity, the theory must attempt to describe the scattering of the electrons by the disordered atomic arrangement, and it is not easy to judge whether the assumption of a free-electron description of the *electronic* structure is justified. In theoretical calculations, such as that of Ziman,† the conduction electrons are assumed to form a free-electron gas which is scattered by the atoms of the liquid metal. As a first approximation (the gas-like limit), the scattering from different atoms is treated independently, and the predictions of the theory depend mainly on the potential at each atom site that scatters the electrons. (For alkali metals, estimates of these potentials are available from band-structure calculations for the solid state.) The corrections that result from the interference of scattering from different atoms can be made, using experimental evidence on the structural character of the liquid as expressed in a radial probability distribution, and these lead to a reduction in the calculated resistivity. Unfortunately, in sodium, the first (gas-like) estimate is already too small, and an extra source of resistance has to be sought, and is assigned to density fluctuations.

In polyvalent metals, which are considered in a later paper,‡ this last complication can probably be ignored, but the atomic scattering potential is much less certain. If the complications in the electronic structure can also be ignored, the electrical resistivity of the liquid can be treated as a source of information about this potential. The resultant values for polyvalent metals seem rather small, but it is by no means clear that the use of a non-free-electron description of the electrons would increase them.

It should be emphasized that, in this kind of theory, the assumption

---
* A survey of the theoretical and experimental situation is given in the Proceedings of the 1966 Brookhaven Conference on Liquid Metals, published in *Advances in Physics*, 1967, **16**, Parts 62–64.

† J. M. Ziman, *Phil. Mag.*, 1961, **6**, 1013.

‡ C. C. Bradley, T. E. Faber, E. G. Wilson, and J. M. Ziman, *Phil. Mag.*, 1962, **7**, 865.

of free-electron character made for the electrons is in fact only the assumption that the maximum momentum of the current-carrying electrons is that which a free-electron gas involving all the outer electrons of the metal in question would have. There is no need to assume that the relationship between energy and momentum over the whole energy range of these electrons is the same and has a free-electron character. The success or otherwise of these theories for any transport properties cannot be used as an argument for a particular " valency " in the metallic state.

It is remarkable that the Hall coefficients (see p. 260) of a large number of liquid metals, including Ga, Ge, and Bi, which are violently non-free-electron-like in their solid-state properties, agree closely with the values calculated on a free-electron model, assuming that all electrons outside closed atomic electron shells are free.

(e) *The Free-Electron Theory and Electron Correlation.*

The great defect of the free-electron theory is that the wave-functions ascribed to the different electrons in an assembly are the same as those which are possible for a single electron in an empty box. No allowance is made for the effect of the different electrons on each other's motion, in spite of the high densities at which they are present in metals. These effects, which are known as correlation effects, are of different kinds. Coulomb repulsion between electrons means that two of them can never be very close together, and this of course is not allowed for, if simple one-electron wave-functions are used. Apart from this, the Pauli Exclusion Principle and the Exchange Forces affect the motions of the electrons in an analogous manner to the effects in the H_2 molecule described on p. 141. These complications have now been examined mathematically and this has led to some quite new concepts as regards the behaviour of an assembly of electrons. This new work has been tested by calculating the physical constants of the alkali metals, and is described on p. 295.

SUGGESTIONS FOR FURTHER READING.

N. F. Mott and H. Jones, " The Theory of the Properties of Metals and Alloys." **1936:** Oxford (Clarendon Press); Dover Reprints, New York. Though often out of date as regards experimental work this book remains one of the best introductions to the subject.
A. Sommerfeld and H. Bethe, " Elektronentheorie der Metalle." ("Handbuch der Physik, Zweite Auflage," Band XXIV, Zweiter Teil.) **1933:** Berlin.

PART V. THE BRILLOUIN-ZONE THEORY OF METALS.

1. The Simple Theory of Brillouin Zones.

The simple free-electron theory imagines the electrons to move in a field of constant potential, and the next development of the theory consists in applying wave mechanics to the motion of an electron in a periodic field. If we were to take a line passing through the centres of the atoms in a metal, the potential energy would vary with the distance as shown in Fig. V.1(a), where there is a singularity at each atomic nucleus. The variation of potential along a line parallel to the first, but not passing through the centres of the atoms, would be as shown in Fig. V.1(b), whilst Fig. V.1(c) shows the variations of potential

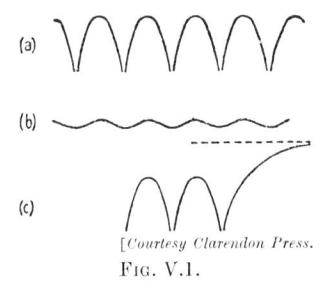

[Courtesy Clarendon Press.
Fig. V.1.

at the surface. The valleys in the potential curves are at distances of the order of 10^{-8} cm, and consequently, as explained before (p. 46), electrons of low energy have a probability of passing from one valley to the next even though their energies, according to classical mechanics, would not be sufficiently great to enable them to surmount the peaks. Electrons with energies greater than that of the peaks in Fig. V.1(a) can, of course, move freely through the crystal in both classical and wave mechanics, but it is only in wave mechanics that the electrons of low energy are free to move.

The theory of electrons in a potential field was first developed in its most general form by Bloch, and different simplified models were then examined by Brillouin, Peierls, Mott, Jones, and other investigators. We shall describe first the case where the variation in potential is small compared with the total kinetic energy of the electrons, and shall indicate which conclusions apply to the more general case. For simplicity, we shall adopt an approach in which the conclusions are directly related to the diffraction of electrons by the crystal lattice, which we shall assume to be cubic. As explained later (p. 247) this is not completely satisfactory, but it is by far the easiest way to understand the underlying ideas. For convenience, we shall deal first with normal crystal structures in which all the lattice sites are occupied by atoms, and we shall describe later (p. 204) some of the complications met with when defect structures are considered.

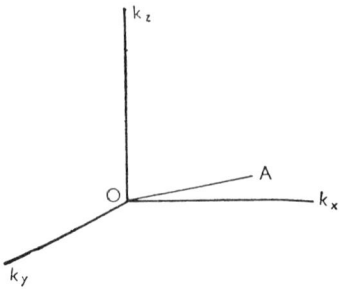

FIG. V.2.

In Section IV.1 we have described the use of a momentum diagram. For a free electron, the wave-length, λ, is given by:

$$\lambda = \frac{h}{mu} \qquad \ldots \ldots \quad \text{V (1)}$$

where u is the velocity, so that the wave number, or the number of waves per unit length, is given by:

$$\frac{1}{\lambda} = \frac{mu}{h} \qquad \ldots \ldots \quad \text{V (2)}$$

Instead of a momentum diagram, we could therefore have used a wave-number diagram, since the wave number $1/\lambda$ is directly proportional to the momentum. For the purpose of the present theory, it is often convenient to multiply the wave number $1/\lambda$ by 2π, and the resulting quantity is termed the wave number k, defined by:

$$k = \frac{2\pi}{\lambda} = \text{(for free electrons)} \ \frac{2\pi mu}{h} \quad \ldots \quad \text{V (3)}$$

This convention is, however, not universally accepted, and the symbol k (or $\boldsymbol{k}$, when we are concerned with its vector character) is often used to denote the simple wave number $1/\lambda$.

We may thus construct a diagram in k-space exactly in the same way that we used a diagram in momentum space. If Fig. V.2 represents three-dimensional k-space, the three axes k_x, k_y, k_z correspond to wave motion parallel to the x, y, and z axes of real space. In general, x, y,

The Brillouin-Zone Theory of Metals

and z will be chosen so that they are related to the axes of the crystal, and in the cubic system, for example, k_x, k_y, and k_z will correspond with waves moving in the directions of the sides of the unit cell. The distance from the origin O of k-space to a point such as A represents the value of k for the electron state concerned, and the direction OA is the direction of the waves associated with the state $\boldsymbol{k}$. In the free-electron theory, OA is also the direction of motion of the electron itself, considered as a wave packet or wave group built up from a number of states in the region of A. In the periodic field this is not necessarily so, and in general the group velocity of the waves of an electron in states near to A is not in the direction OA. That is to say, individual states in the region of the point A in k-space correspond with wave motion in the direction OA, but these waves combine together to form a wave group or wave packet which, in general, does not move in the direction OA. Much of the difficulty in understanding the theory lies in a failure to realize this distinction between the direction of motion of the waves and of the wave group, and the reader must appreciate that in a periodic field it is quite possible for waves moving in one direction to build up a wave packet which moves in another direction.

In many descriptions the reader will find k or $\frac{h}{2\pi}k$ referred to as the momentum. The reason for this is that when a field of force is applied, the rate of change of $\frac{h}{2\pi}k$ is equal to the force, so that k satisfies Newton's definition of momentum. In the free-electron theory k is proportional to mass × velocity, but in the periodic field, as we shall see later, this is no longer so, and we may have a high value of k with a zero velocity. It therefore seems better not to call k or $\frac{h}{2\pi}k$, the momentum, because there is a natural tendency to visualize momentum as mass × velocity.

In our description of the free-electron theory, we employed the concept of a Fermi surface of occupied states. In the free-electron theory the Fermi surface is spherical, and we can imagine a number of concentric spheres in momentum space, or in k-space, the surfaces of which represent states of equal energy, and may therefore be called energy contours. The direction of motion of an electron in the free-electron theory is then perpendicular to the energy contour at the position of the point in k-space concerned.* In the same way, in the theory

---

* The motion of the electron is, of course, in real space, but since the axes in k-space are chosen to correspond with directions of motion of waves in real space, it is customary to speak of motion parallel to a direction in k-space. As explained on p. 247, this is justifiable for cubic crystals only.

of the periodic field, we may again construct energy contours, or surfaces of constant energy in k-space. In general, these surfaces are not spherical (see p. 202), but the direction of motion of the electron (*i.e.*, the group velocity of the waves) is again perpendicular to the energy contour at the point concerned in k-space. Just as in a momentum diagram a point near to the origin represents a state of low momentum, so in k-space a point near the origin represents an electron in a state of low wave number. In the free-electron theory, the kinetic energy of an electron E and its wave number k are related by the simple equation:

$$E = \frac{h^2 k^2}{8\pi^2 m} \qquad \cdots \cdots \quad \text{V (4)}$$

which is independent of the direction of k; but the effect of a periodic field is to make the relation between E and k more complicated, and dependent upon the direction of k.

The theory is developed on the assumption of the cyclic model already described in connection with the free-electron theory, and is therefore able to deal with problems involving the flow of electrons in a definite direction. The theory assumes further that in considering the motion of any one electron the remaining electrons may be regarded as smoothed out throughout the crystal, so that the one electron under consideration moves in a strictly periodic field. Subject to these assumptions, Bloch showed that the wave function for the motion of an electron in a periodic field is of the form:

$$\psi = e^{i(k \cdot r)} \phi_k(x, y, z) \quad \cdots \cdots \quad \text{V (5)}$$

where ϕ is a function which depends in general on k, and has the threefold periodicity of the lattice. This equation is general, and applies to all regular periodic variations in potential, and it may be compared with equation IV (19) of the free-electron theory. Equation V (5) represents a plane wave of wave-length $2\pi/k$ travelling in the direction of the vector k, but the wave is modulated in the rhythm of the lattice. Examination of this equation by the methods of wave mechanics then shows that there are only certain values of the energy for which k is real, and that for other values of the energy k would be imaginary. From this it is concluded that if we consider electron states corresponding with any one direction in k-space, the energies will not be a continuous function of the wave numbers, but will show gaps of forbidden energies.

The general conclusions of the theory are best understood by considering first a hypothetical one-dimensional metal, in which the potential undergoes a regular periodic variation as shown in Fig. V.1(*b*). We may consider first the relation between the energy E and the wave number k. If the potential were uniform, these would be related by the

The Brillouin-Zone Theory of Metals 199

simple equation V (4), so that the curve connecting E and k would be of the form shown in Fig. V.3(a). In the presence of a periodic potential, the corresponding curve is of the type shown in Fig. V.3(b), where the vertical portions a_1b_1, a_2b_2, ... represent the forbidden ranges of energy. If electrons are introduced with continually increasing values of k, then when k reaches the critical value corresponding to a_1, there is an abrupt increase in the energy on passing to the next highest wave number k. The lowest portion of the curve in Fig. V.3(b) closely resembles that of the free-electron curve of Fig. V.3(a), so that on introducing electrons into the one-dimensional model with periodic potential, the first electrons behave very like those of the free-electron model; but at the higher energies there is a series of breaks in the curve, and near

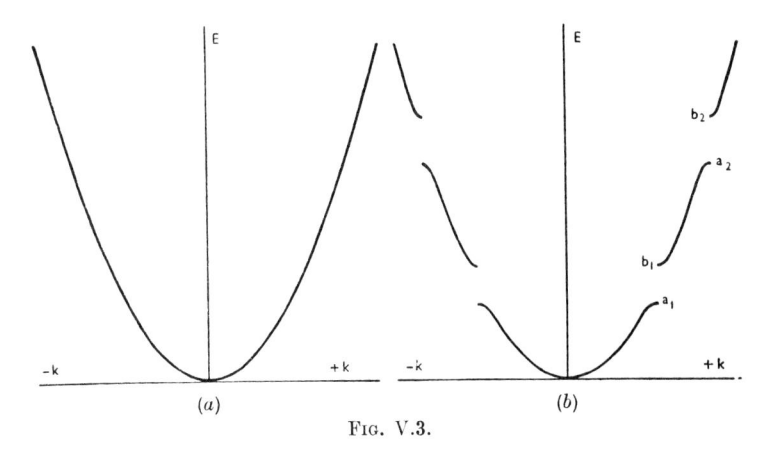

FIG. V.3.

to these the behaviour is very different. Each portion of the curve in Fig. V.3(b) may be called a band, and the curves are horizontal at the top and bottom of each band. Near to the bottom of each band the curves are parabolic if the energy is measured from the bottom of the band, and the curves are also parabolic at the top of a band, with the curvature in the opposite direction. The curves are of such a nature that d^2E/dk^2 is positive in the lower portion of a band and negative at the upper portion. These general characteristics apply to all regular periodic variations of potential, but the exact assumptions made determine the magnitudes of the energy gaps a_1b_1, a_2b_2, &c., and affect also the finer details of Fig. V.3(b), such as the exact positions of the points a_1b_1 ... relative to the simple curve of Fig. V.3(a).

We may now turn to a two-dimensional crystal and suppose that Fig. V.4(a) represents a simple square lattice of atoms, and that Fig. V.4(b)

shows the corresponding two-dimensional wave-number diagram in which the axes k_x and k_y are parallel to the sides of the square. If we consider any line such as OA, then the passage from O to A represents a series of electron states with increasing values of k, all referring to the same direction of the wave motion.* If we confine our attention to any one direction such as OA, the theory indicates that the relation between the energy E and the wave number k is given by a curve of the same general form as that in Fig. V.3(b), so that as k increases there are breaks in the curve corresponding to forbidden ranges of energy. The physical significance of this is readily understood in view of the wave-like characteristics of an electron. Suppose that in Fig. V.4(c) the lines p_1, p_2, p_3 represent a series of rows of atoms in the two-dimensional crystal, and that OA is a line whose direction is the same as that of OA in Fig. V.4(b). Then, in general, an electron whose state is described by a wave number on OA in the wave-number diagram of Fig. V.4(b) will pass through these lines of atoms with only slight scattering. But if the electronic wavelength satisfies the Bragg equation:

$$n\lambda = 2d \sin \theta \quad . \quad . \quad . \quad . \quad . \quad \text{V (6)}$$

where d is the atomic spacing of the lines, and n is a whole number, then strong reflection takes place, and the electron cannot travel through the lattice. It can be shown that at this critical value of λ, there is an abrupt increase in energy on passing to the next highest wave number. If therefore in the wave-number diagram of Fig. V.4(b) we consider wave numbers lying on the line OA, there will be a series of points a_1, a_2, a_3 . . . for which the Bragg equation is satisfied, and the energy will increase continuously on passing from O to a_1, from a_1 to a_2, a_2 to a_3 . . . &c., but there will be sudden and discontinuous increases in energy at a_1, a_2, a_3 . . .

The Bragg equation involves the angle, θ, between the direction of the waves and the scattering lines p_1, p_2, p_3 . . . If therefore, as in Fig. V.4(d), we consider a whole series of lines $OA_1, OA_2, \ldots$ in the wave-number diagram, the points corresponding to a_1 in Fig. V.4(b) will lie at distances from the origin which depend on the direction of OA. These points are represented as a_{11}, a_{12}, a_{13} . . . in Fig. V.4(d), and it is a consequence of the theory that such points lie on straight lines which form the boundaries of a polygon known as the *First Brillouin Zone* of the two-dimensional structure. Similarly, the point a_2 in Fig. V.4(b) is one of a series of points lying on the sides of a second polygon, the *Second Brillouin Zone*, these points corresponding to wave-lengths which

---

* As explained above, although the individual states at A correspond with motion in the direction OA, they combine together to form a wave group which in general does not move in the direction OA.

The Brillouin-Zone Theory of Metals 201

satisfy the Bragg equation for reflection from another set of atomic rows in the two-dimensional structure. Each crystal structure gives rise to its own characteristic Brillouin zones, and the first two zones for a simple square lattice are shown in Fig. V.4(e). The meaning of this diagram is that inside the square $ABCD$, the energy varies continuously with the wave number, but that an abrupt increase in energy occurs on passing

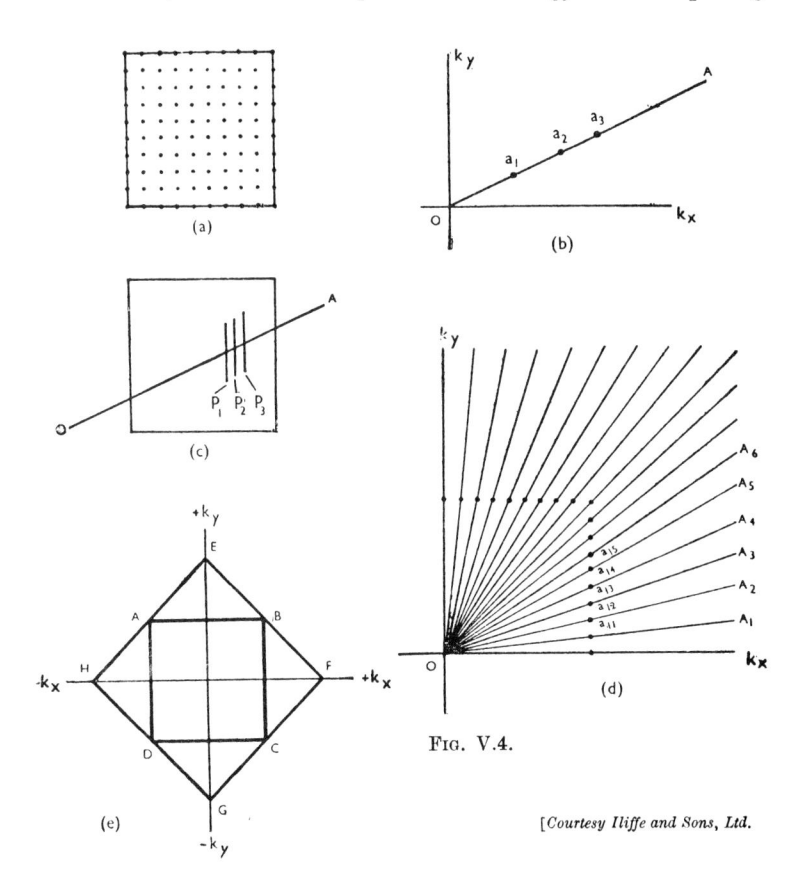

Fig. V.4.

[*Courtesy Iliffe and Sons, Ltd.*

through the sides of the square. Similarly, within the regions AEB, BFC . . . the energy varies continuously with the wave number, but an abrupt increase occurs on passing through the lines AE, EB, BF . . .

For the reasons described on p. 253, some writers now use the expression *Jones Zones* for the Brillouin zones referred to above, and use the latter term in a slightly different sense. At the moment of writing

202 Atomic Theory

(1969), this is not universal practice, and we have therefore retained the name Brillouin zone in most of the present book.

For a three-dimensional crystal structure the same principles apply. Instead of the two-dimensional wave-number diagram of Fig. V.4(b) we construct a three-dimensional wave-number diagram, and if we consider energy states lying along any one direction in this diagram, there will be critical values of the wave number for which the Bragg equation is

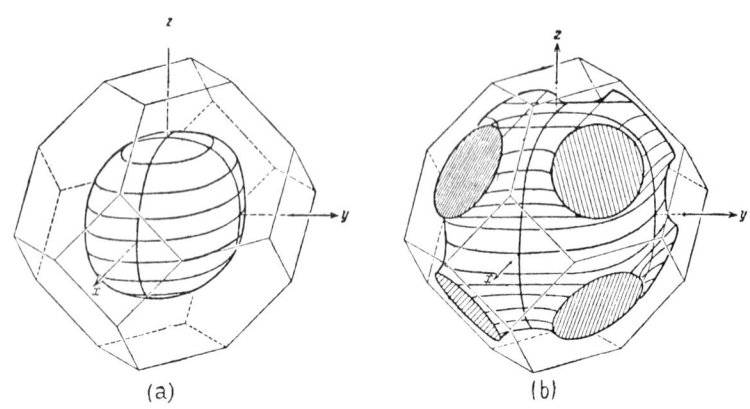

(a)   (b)

[*Courtesy Clarendon Press.*]

FIG. V.5. The First Brillouin Zone of the face-centred cubic structure. The zone is bounded by planes parallel to the cube and octahedral faces. Within the zone the energy varies continuously with the wave number, but on reaching the zone face there is an abrupt increase in energy on passing into the next zone. In Fig. V.5(a) the number of electrons per atom is small, and the ruled surface shows the limit of the occupied states; this corresponds with the condition of affairs in some univalent metals, and it will be seen that the ruled surface is nearly spherical. The ruled surface is a surface of constant energy (see p. 198), and, as explained on pp. 212 and 213, the direction of motion of an electron in a group of states on the ruled surface will be perpendicular to the ruled surface at the place considered. In Fig. V.5(b) the number of electrons is greater, and the Fermi surface of occupied states has reached the zone boundary. It will be seen that near the zone boundaries the ruled surface of constant energy is very different from a sphere, and bends round so as to be perpendicular to the zone faces at the points where they meet the ruled surface. This is an expression of the fact described on p. 211 that for electron states on the zone surface the component of velocity of the electron perpendicular to the zone face is zero.

satisfied for strong reflections from planes of atoms in the three-dimensional crystal. The positions at which these critical points occur will depend on the direction considered, and the critical points lie on planes which build up the three-dimensional *Brillouin Zones*. Thus, Fig. V.5 shows the first Brillouin zone of the face-centred cubic structure. In this figure k_x, k_y, and k_z are drawn parallel to the sides of the cube, and the zone is bounded by two sets of faces, corresponding to the satisfying

The Brillouin-Zone Theory of Metals 203

of the Bragg equation for reflection from the (111) octahedral and (200) cubic planes of the crystal. The meaning of this diagram is that if we start at the origin of k-space and consider a series of states with continually increasing k, then E varies continuously with k as long as k lies within the zone, but when k reaches the surface of the zone there is an abrupt increase in energy. For any one direction in k-space the relation between E and k is of the same general form as that shown in Fig. V.3(b), but the positions and magnitudes of the gaps $a_1 b_1$, $a_2 b_2$ . . . depend on the direction considered. If, therefore, we consider the position of the gap $a_1 b_1$ on the surface of the first zone—this is also called the top of the first band—this will depend on the direction, and there will be a series of curves such as those of Fig. V.6(a), where each broken curve such

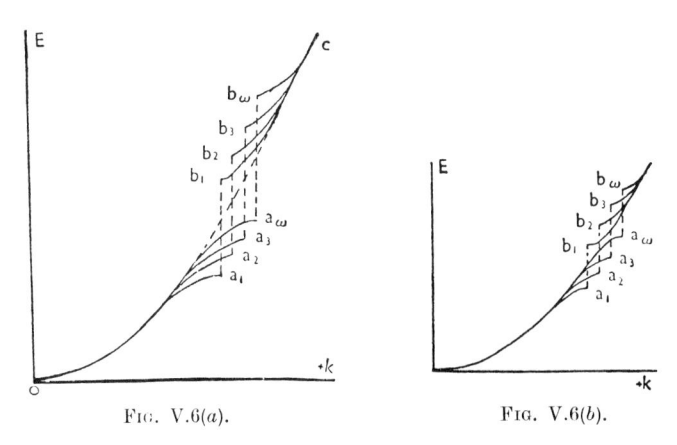

Fig. V.6(a).   Fig. V.6(b).

as $Oa_1 b_1 c$ refers to one direction in k-space. In general, these curves are not horizontal at points such as a_1 or b_1, but for some directions they are horizontal (see p. 211). Exactly similar considerations apply to the positions of the energy gaps at the top of the later bands, and these again vary with the direction considered. Diagrams such as Fig V.5 do not give the magnitude of the energy gaps at different points on the surface of the zone, but they show clearly the dependence on direction of the points at which the breaks occur in the curves of Fig. V.6(a).

In view of frequent misunderstandings, it must be emphasized that diagrams such as those of Fig. V.5 are not pictures of the crystal in real space. They are diagrams in wave-number or k-space, and show the critical wave numbers at which abrupt increases in energy occur on moving outwards from the origin of the k-space diagram.

If the value of k at which the energy gaps occurred were independent of direction, the Brillouin zone would be a sphere, but as we shall find later the shapes of some Brillouin zones are very different from the spherical. It will be seen, therefore, that there is an essential difference between the E/k relations in the free-electron theory and in the theory of the periodic field. In the free-electron theory, for all directions of the wave number, E increases continuously with k, and as electrons are introduced into the box model, the energy increases smoothly and continuously. In the theory of the periodic field, for each direction of the wave number there is a series of breaks in the E/k curve, and the values of k at which these breaks occur lie on the surfaces of the Brillouin zones.

The Bragg reflections are determined by the arrangement of the atoms in the unit cell of the crystal lattice, and in Brillouin-zone theory the fundamental quantity is, therefore, the number of electrons per unit cell of the crystal structure. In normal structures where every lattice point is occupied, the theory may be described in terms of the number of electrons per atom, since this is a simple fraction of the number per unit cell. In certain alloys, such as those described on p. 325 there are defect structures in which some lattice points are unoccupied and for these the description must be in terms of electrons per unit cell. It is, however, common practice to describe the theory in terms of the number of electrons per atom, which is also called the electron concentration, and we shall adopt this policy for the present, but the reader must realize that the number of electrons per unit cell is the really fundamental quantity.

In the free-electron theory we have seen how the occupied states form a sphere in the momentum diagram, the states on the surface of the sphere having the energy $E_{max.}$. Electron states having energies proportional to 1, 2, 3, 4 . . . (in arbitrary units) lie on a series of spherical surfaces of radii proportional to $\sqrt{1}$, $\sqrt{2}$, $\sqrt{3}$ . . ., because $E \propto k^2$ in the free-electron theory. Such surfaces may be called *energy contours*, and in the free-electron theory the contours are always spherical. In the theory of the periodic field energy contours may again be drawn as surfaces in k-space, and for electrons of low energy for which the E/k relations of Fig. V.6(a) resemble those of free electrons, the contours are approximately spherical. Thus, in Fig. V.5(a) the ruled surface shows the limit of the occupied states when the number of electrons per atom is relatively small, and the occupied states lie well within the zone boundaries. This corresponds to the condition of affairs in some univalent metals, and the ruled surface is an energy contour or surface of constant energy. In Fig. V.5(b) the number of electrons per atom is sufficiently great for the occupied states to have reached the surface of the zone, where the abrupt increases in the energy at the zone surfaces prevent electrons from entering the next states in the directions con-

The Brillouin-Zone Theory of Metals 205

cerned. The ruled surface in Fig. V.5(b) is again an energy contour, and near the zone boundaries the shape is no longer spherical.

In Fig. V.5(a) the Fermi surface of occupied states is not very greatly distorted from the spherical surface of the free-electron theory. This, however, is not necessarily the case, and in some metals the Fermi surface is greatly distorted. In the case of copper, for example (p. 304), the distortion is so great that, even though there is only one valency electron per atom, the Fermi surface of occupied states has almost certainly touched the face of the zone.

The relation between the energy contours and the E/k curve is readily seen from Fig. V.7, which is taken from Raynor's " Introduction to the Electron Theory of Metals ".* In this, the central square represents the first Brillouin zone of the two-dimensional square lattice, and is analogous to the zone $ABCD$ in Fig. V.4(e). For the free-electron theory the energy contours would be circles, analogous to the spheres described above, and the contours of low energy in Fig. V.7 are drawn as circles which become closer and closer together as k increases. Near the boundaries of the zone the circles become deformed, and the diagram shows how the spacing of the contours is related to the bends in the E/k curve below the places where the sudden increases in energy occur at the zone boundary.

In considering a Brillouin zone such as that of Fig. V.5 two possibilities arise. In the first case, which is that illustrated in Fig. V.6(a), the energies of states at the top of the first band may *for all directions* be less than the energy of any state for any direction at the bottom of the second band. In this case, if we introduce increasing numbers of electrons into the crystal,† the first electrons will fill up the states in k-space almost as in the free-electron model, the Fermi surface (see p. 173) will be almost spherical, and the $N(E)$ curve (see p. 207) will be parabolic as in the free-electron model. With increasing numbers of electrons the values of k for the highest occupied states will begin to approach the surface of the zone in certain directions, and when the surface of the zone is reached, it will no longer be possible to add electrons whose states are in the directions concerned, since this would involve an abrupt increase in energy. When this stage is reached, the $N(E)$ curve will show a sharp fall, because there will no longer be

---

* G. V. Raynor, *Inst. Metals Monograph and Rep. Series* No. **4**, 1949.

† When we speak of introducing increasing numbers of electrons into the crystal it is, of course, always assumed that a corresponding positive charge is present, so that the whole is electrically neutral. In actual alloys, the most usual way of changing the electron concentration is to alloy a given metal with another which forms a simple primary solid solution, but is of different valency. In this way the number of valency electrons per atom is changed, but the structure is unaltered.

206   *Atomic Theory*

possible energy states for some directions of the wave number. The resulting $N(E)$ curve is of the form shown in Fig. V.8(b), the point a_1

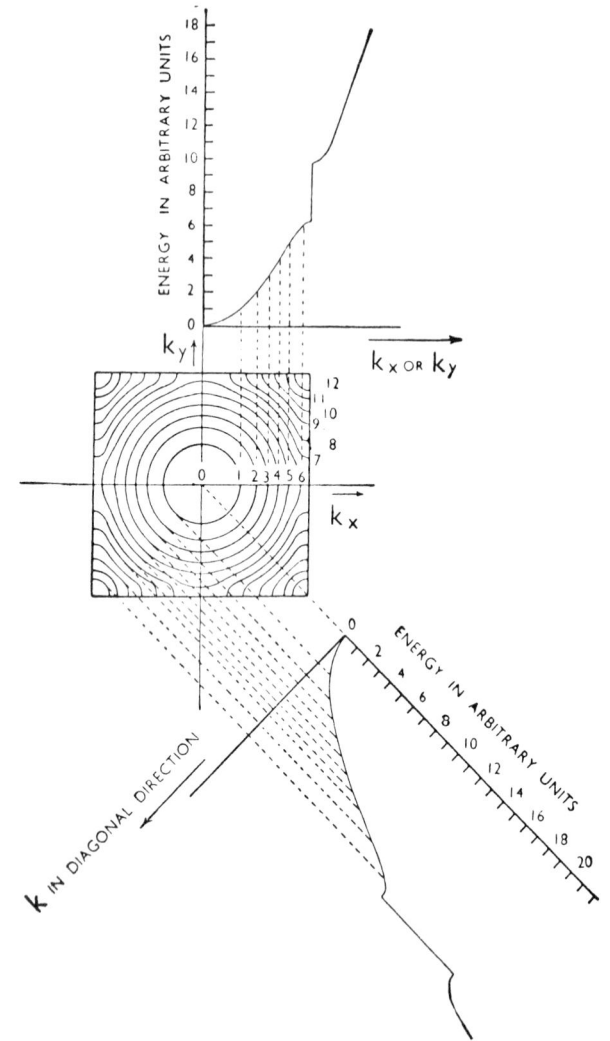

Fig. V.7. Derivation of energy contours inside a zone.

representing the stage at which the surface of the zone is first touched, corresponding with the point a_1 on the lowest curve of Fig. V.6(a). It will be noted that just before a_1 the $N(E)$ curve rises above that for the

The Brillouin-Zone Theory of Metals

free-electron model; this rise is a natural consequence of the decrease in the slope of the E/k curve as a zone boundary is approached. In this region the Fermi surface is no longer spherical since, as one can see in Fig. V.7, the numerical magnitude of k for a given energy in this region will be larger for those directions in k-space for which this decrease in dE/dk has begun. On adding still more electrons, the values of k increase, and more and more directions are affected, until eventually all the states in the first Brillouin zone are occupied; at the same time the $N(E)$ curve sinks to zero, as shown in Fig. V.8(b), where the point a_ω represents the top of the last curve for the first band in Fig. V.6(a). When the first Brillouin zone is filled, the next electron state will involve a sudden increase in the energy, this increase being equal to the difference

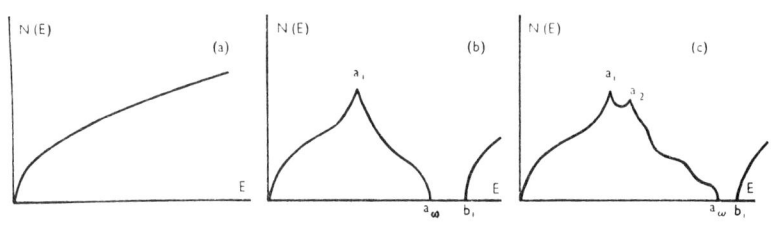

Fig. V.8(a) shows the form of the $N(E)$ curve in the free-electron theory; here $N(E) \propto E^{\frac{1}{2}}$.

Fig. V.8(b) shows the corresponding curve in the Brillouin-zone theories; the example is one in which the first and second zones do not overlap, and the zone has only one type of face, and one kind of corner.

Fig. V.8(c) shows the $N(E)$ curve for a zone with two types of face, and three different kinds of corner.

between the energies of the point b_1, representing the lowest energy of the group of curves for the second band in Fig. V.6(a), and the point a_ω at the top of the highest curve of the first band. The second and third Brillouin zones may be separated in the same way, so that on introducing increasing numbers of electrons the second zone is completely filled before any electrons enter states in the third zone.

In the description of Fig. V.8(b) we have assumed that the first Brillouin zone is bounded by faces of one type only, the point a_1 representing the stage at which the Fermi surface of occupied states first touches the faces of the zone, all of which are at the same distance from the origin. In many cases the first zone is bounded by faces of more than one type. Thus, as shown in Fig. V.5, the first zone for the face-centred cubic structure is bounded by a combination of octahedral and cube faces, and these lie at different distances from the origin. In this case the $N(E)$ curve will rise to a first peak at the stage where the Fermi surface first touches the octahedral faces of the zone, and will then fall and rise to a second peak when the Fermi surface touches the cube faces of the zone. It can readily be seen that there is a rapid fall in the $N(E)$ curve as the Fermi surface of occupied states fills the corner of a zone, because here the zone restrictions reduce the number of states available. In the general case, a zone will have more than one kind of corner, and there will be rapid falls in the $N(E)$ curve at the stages where each set of corners is filled. Fig. V.8(c) shows a $N(E)$ curve for a hypothetical zone with two types of face, and three kinds of corner. It will be noted that the valley between the two peaks is rounded and smooth.

The second general kind of energy relationship is that in which the curves of Fig. V.6(a) are so related that, although for any one direction

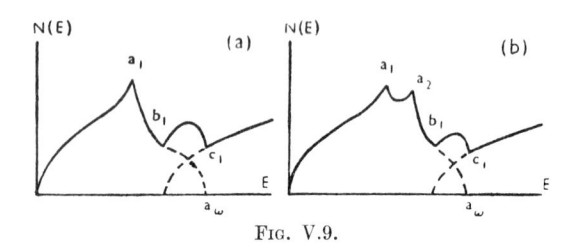

FIG. V.9.

there is an energy gap at the top of the first band, some of the states at the bottom of the second band have lower energies than the highest states at the top of the first band. This condition of affairs is shown in Fig. V.6(b), and the resulting $N(E)$ curve in Fig. V.9(a). In this case, on adding increasing numbers of electrons to the crystal, the first electrons enter the first Brillouin zone, as in the preceding case, and the $N(E)$ curve is of the form shown in Fig. V.9(a), the sharp fall at a_1 corresponding with the stage at which the Fermi surface of occupied states first touches the surface of the first Brillouin zone, i.e., the point a_1 on the lowest curve in Fig. V.6(b). The point b_1 in Fig. V.9(a) represents the stage at which the electrons begin to enter the second Brillouin zone, and corresponds with point b_1 on the lowest curve of the second zone in Fig. V.6(b). At this stage the states in the first zone are not completely filled, because the energy of the point a_ω on the highest of the curves for the first zone in Fig. V.6(b) is greater than the energy for the point b_1. As more electrons are added, they enter the remaining states of the first zone and the lowest states of the second zone, and the resulting $N(E)$ curve in Fig. V.9(a) is given by the sum of the ordinates for the two zones where these overlap. The sudden rise in the $N(E)$ curve after b_1 is the result of the opening up of a series of new states when the second zone is entered. Finally, the point a_ω represents the stage at which the first zone is completely filled, and further electrons continue to fill up the

The Brillouin-Zone Theory of Metals

second zone. With increasing numbers of electrons, a second peak occurs when the Fermi surface reaches the surface of the second Brillouin zone, and a second kink in the curve occurs if the second zone overlaps the third. Later zones produce similar effects, and, if the zones overlap, the typical $N(E)$ curve shows a whole series of peaks and valleys, the positions of which depend upon the crystal structure, and on the magnitude of the energy gaps at the surface of a zone. Some of the more interesting of these curves are discussed in Part VI.

The description of Fig. V.9(a) assumes that the zone has only one type of face, the point a_1 representing the stage at which the Fermi surface of occupied states first touches the faces of the zone, all of which are at the same distance from the origin. If the zone has two types of face at different distances, the $N(E)$ curve (Fig. V.9(b)) will have a double peak for the reasons described in connection with Fig. V.8(c), and it is necessary to distinguish between the rounded valleys due to two types of face on a zone, and the sharp valleys at b_1 and c_1 (Fig. V.9(b)) due to the zone overlap. If the zone has more than one kind of corner, the $N(E)$ curve of Fig. V.9(b) will be further complicated by inflections analogous to those in Fig. V.8(c).

The Velocities of Electrons.

The cyclic model is able to deal with problems involving a flow of electrons in a definite direction, and, as in the free-electron theory, the moving electron must be thought of as a wave packet, which travels with the group velocity v. We may deal first with the one-dimensional model referred to on p. 198. The group velocity v is given by:

$$v = \frac{d\nu}{d\left(\frac{1}{\lambda}\right)} \quad \ldots \ldots \quad \text{V (7)}$$

which for electron waves is equivalent to:

$$v = \frac{2\pi}{h} \frac{dE}{dk} \quad \ldots \ldots \quad \text{V (8)}$$

From the form of the curves in Fig. V.3(b), it follows that the group velocity is zero for electrons at the top and bottom of each band. In the free-electron theory, increasing wave number or increasing energy means a continually increasing velocity, but in the periodic field this is no longer so, and, on starting from the bottom of the first band, increasing energy first results in an increase in v, but as the top of the band is approached dE/dk diminishes, and eventually becomes zero where the curve is horizontal at the point a_1 in Fig. V.3(b). The energies in the range a_1b_1 are forbidden energies, and on reaching b_1 in the second band the velocity is zero; on passing up the second band the velocity, v, at first increases and then decreases. The relation between velocity and wave number is thus one in which the velocity in each band is zero at the bottom of the band, rises to a maximum, and then sinks to zero at the top of the band.

The fact that the velocity, v, is zero at points such as a_1 and b_1 in Fig. V.3(b) is not to be taken as implying that the electron is at rest. The wave functions at points such as a_1 and b_1 represent standing waves, but since the group velocity of the wave packet is zero, these standing waves cannot be interpreted as representing equal probabilities of the electrons moving in opposite directions up and down the length L of the one-dimensional model, as was the case for the standing waves in the box model of the free-electron theory. In the periodic field, the standing waves at points such as a_1 and b_1 in Fig. V.3(b) may still be regarded as indicating an equal probability of finding the electron moving in opposite directions, but since the velocity of the wave packet as a whole is zero, the process concerned is within the length of the wave packet, and cannot therefore be visualized in the form of a definite trajectory. Strictly speaking, we cannot do more than say that the velocity of the wave packets is zero for electrons in states such as a_1 and b_1 of Fig. V.3(b). If we try to look into the process further, however, we may perhaps say that the velocity $v = 0$ is to be interpreted as meaning that the electron is travelling first in one direction and then in the opposite direction; but these reversals of motion occur within the length of the wave packet, so that we cannot visualize the process or draw a definite trajectory. The energy of electrons in these states is thus neither entirely potential energy, nor to be visualized as the kinetic energy of long-range translatory motion, but is in a form which is characteristic of the wave-mechanical behaviour of an electron in a periodic field.

We may now turn to the three-dimensional model * for which the relation between energy and wave number is of the form shown in Fig. V.6(a), where each curve refers to one direction of wave number, and the breaks in the different curves occur at the values of k which lie on the surface of a Brillouin zone, such as that of Fig. V.5. The velocity of an electron can be resolved into components parallel and perpendicular to the surface of a zone, and the picture presented is that an electron in a state on the surface of a zone has a zero component of velocity in a direction perpendicular to the surface concerned. The

---

* The formal expression for the velocity in the three-dimensional case (replacing V(8)) is $v = \dfrac{2\pi}{h} \operatorname{grad}_k E$.

The Brillouin-Zone Theory of Metals 211

physical process associated with this is that the electron has a wave-length such that it undergoes a Bragg reflection within the crystal. Let us suppose, for example, that Fig. V.10 shows parallel planes of atoms within the crystal, and that we allow a monochromatic beam of electrons travelling in the direction AB to impinge on the crystal from outside. Then, in general, an electron with wave number k, corresponding to

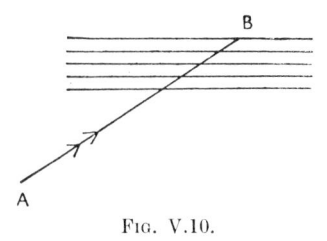

Fig. V.10.

motion in this direction, will be transmitted with comparatively little scattering. If, however, we consider a series of electrons with increasing k, a stage will be reached at which the wave-length satisfies the Bragg equation V(6) and then strong reflection takes place so that the electrons do not penetrate far into the lattice. During this reflection, the component of velocity parallel to the reflecting plane is unaltered, whilst the component perpendicular to the reflecting plane is reversed in direction. As explained above, it is the electron states on the surface of a zone whose wave-lengths satisfy the conditions for a Bragg reflection inside the crystal. Electrons in these states have a zero component of velocity perpendicular to the face of the zone concerned, and if we are asked what picture we can give of the motion inside a crystal of an electron whose wave number lies on the surface of a Brillouin zone, we may perhaps answer that, since in a Bragg reflection the angle of incidence is equal to the angle of reflection, the electron may be regarded as undergoing repeated Bragg reflections inside the crystal, so that the mean component of velocity perpendicular to the zone face is zero, but successive reflections occur within the length of the wave packet, and consequently we cannot draw a definite trajectory or form any picture of the process in terms of the classical mechanics of " ordinary particles."

From the above description it will be appreciated that if the corner of a zone involves the intersection of three mutually perpendicular planes, the mean velocity associated with an electron state whose wave number lies at this corner is zero, although as in the one-dimensional case discussed above, this is to be interpreted not as a stationary electron, but as equal probabilities of motion in opposite directions, the reversals of motion occurring within the length of a wave packet, so that no definite picture can be formed.

The above considerations indicate clearly that there is an intimate relation between the shapes of the Brillouin zones and the X-ray or electron-diffraction patterns characteristic of the structure concerned. Thus, the first Brillouin zone shown in Fig. V.5 for the face-centred cubic structure is bounded by octahedral planes, cut off by planes parallel to the faces of the cube, and this corresponds with the fact that the first two strong reflections for the face-centred cubic structure are those from the (111) octahedral and (200) cubic sets of planes. This correspondence is the foundation of many of the methods for obtaining the shapes of the Brillouin zones of the different structures, but these lie outside the scope of the present work (see p. 253).

For an electron state at the surface of a Brillouin zone, the mean

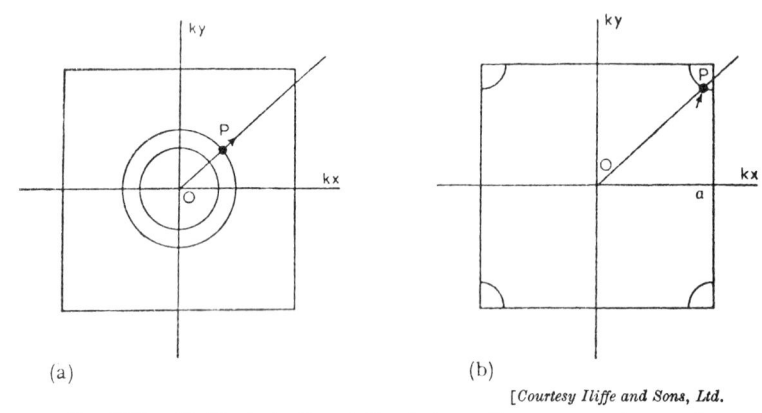

(a) (b)

[*Courtesy Iliffe and Sons, Ltd.*]

FIG. V.11. To illustrate the relation between the energy contours and directions of motion of the electron in a cubic crystal.

velocity v associated with the state has a component perpendicular to the zone face equal to zero. From this it will be appreciated that if we consider the relation between E and k for states along one direction of k-space, such as the line OA in Fig. V.2, then the curves in Figs. V.6(a) and (b) will be horizontal at the points a and b if OA in Fig. V.2 is perpendicular to a zone face, or if OA passes through a corner formed by mutually perpendicular planes. In the general case, however, the curves will not be horizontal, because although the component of v perpendicular to the zone face vanishes, the component parallel to the zone face is unaffected.

Each component of the group velocity v of the wave packet is proportional to dE/dk, and consequently the direction of the velocity of the electron is perpendicular to the energy contours in the region of the group of states from which the wave packet is built up. If in a two-dimensional crystal the energy contours near the centre of the zone are circles, as in Fig. V.11(a), and we consider a small group of states in the region of P, the group velocity, as in the free-electron theory, will be in the direction OP. If the contours are distorted, as in Fig. V.11(b), then OP gives the direction of the waves from which the wave packet is built up, but the group velocity of the wave packet, and hence the velocity of the electron, is in the direction of the arrow which is perpendicular to the energy contour. The same principle applies in three dimensions, where the group velocity is perpendicular to the energy-contour surface.

Brillouin Zones and Superlattices.

The above description has been concerned mainly with the Brillouin zones of pure metals. Here each crystal structure gives rise to its characteristic Brillouin zone, although the energy gaps at the surface of a zone may vary greatly from one metal to another. An entirely random solid solution will give rise to a Brillouin zone of the same general type as that formed by the same structure with all the atoms identical. If, however, a random structure undergoes a transformation to an ordered structure or superlattice, then just as new lines occur on X-ray-diffraction films, so new planes of energy discontinuity are formed in wave-number space. The result of this is that, for a superlattice, the zones of the main structure are retained but are crossed by new planes of energy discontinuity at wave-numbers corresponding to the superlattice reflections of the electrons.

Thus Fig. V.12, due to Nicholas,* shows the first Brillouin zone of the ordered Ag_3Mg superlattice lying inside the first zone of the random face-centred cubic structure (Fig. V.5). In such a case the superlattice zone produces a new peak on the $N(E)$ curve inside the first peak of the main structure (see Fig. V.13), and this may stabilize the superlattice for reasons similar to those discussed on p. 309. In most superlattices, the stabilization results from the reduction in strain energy accompanying the ordered arrangement, but in some cases (*e.g.*, CuPt and Ag_3Mg) the lowering of Fermi energy is probably more important. In the general case both factors must be considered. Certain types of *Long-Period* superlattice, the best known of which is the CuAu II structure, have been studied intensively by Sato and Toth† and they have given

* J. F. Nicholas, *Proc. Phys. Soc.*, 1953, [A], **66**, 201.
† H. Sato and R. S. Toth, " Alloying Behaviour and Effects in Concentrated Solid Solutions " (edited by T. B. Massalski), p. 295. **1965**: New York (Gordon and Breach).

convincing arguments in favour of the suggestion that the superlattice structure is governed by the stabilizing influence of a Fermi surface/Brillouin zone interaction.

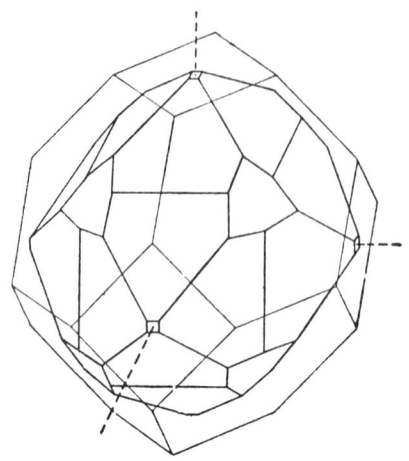

[Courtesy Physical Society.]
FIG. V.12. First Brillouin Zone of ordered Ag_3Mg. Thin lines show the first zone of the disordered structure. (After J. F. Nicholas.)

Direct evidence of an interaction between the Fermi surface and *superlattice*-zone boundaries is provided by the transport properties of Cu_3Au. The Hall effect (p. 260) changes, on atomic ordering, from

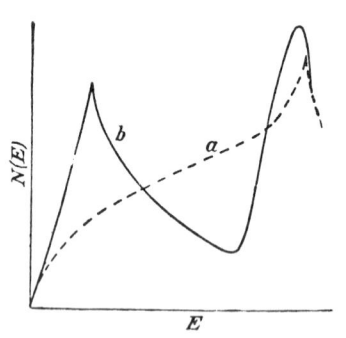

[Courtesy Physical Society.]
FIG. V.13. Schematic diagram of the $N(E)$ curves for: *a* the disordered structure; *b* the ordered structure. (After J. F. Nicholas.)

a negative value similar to that of pure copper to a positive one, as found in divalent metals; in the latter, Fermi-surface interactions with

The Brillouin-Zone Theory of Metals 215

ordinary zone boundaries are known to be responsible, and are usually discussed in terms of the effective mass of the conduction electrons (p. 225).

SUGGESTIONS FOR FURTHER READING.

N. F. Mott and H. Jones, " The Theory of the Properties of Metals and Alloys." **1936:** Oxford (Clarendon Press); Dover Reprints (New York).
F. Seitz, " The Modern Theory of Solids." **1940:** New York (McGraw-Hill).
"Handbuch der Physik," Vol. XIX. **1956:** Berlin (Springer-Verlag). This volume contains a series of articles on the electron theory of solids, including one by J. C. Slater on the electronic structure of solids.
J. M. Ziman, " Electrons in Metals." **1963:** London (Taylor and Francis).
S. Raimes " The Wave Mechanics of Electrons in Metals." **1961:** Amsterdam (North-Holland Publishing Co.).

2. INSULATORS, SEMI-CONDUCTORS, AND METALS.

Before dealing with the more complete theories of Brillouin zones, we shall describe the general bearing of this kind of theory upon the problem of electrical conductivity, since this can be understood in terms of the simple zone theories. From the point of view of electrical conduction, substances may be divided into the following main classes : insulators, semi-conductors, electronic conductors, and electrolytic conductors. Most pure metals are normal electronic conductors and their " ideal " electrical resistances (the values corrected for impurity and defect scattering) tend towards zero as the temperature approaches absolute zero. Their resistances increase with increasing temperature and increasing impurity content, while the different classes of semi-conductors (p. 228) in general have resistances that decrease with an increase in either temperature or impurity content. We shall not consider electrolytic conductors, since they are seldom met with in the study of solid metals and alloys, although it should be noted that if the constituent elements differ greatly in their electrochemical nature, a certain amount of electrolytic conduction may take place when an electric current is passed through a liquid alloy, and electrolytic migration in solid alloys has also been observed. It is one of the great triumphs of the Bloch theory that it is able to give a satisfactory explanation of the remaining three classes of substance, and in particular to explain the enormous difference between the electrical resistance of a substance such as a diamond and that of a metal. Differences in the strengths of the binding forces between electrons and atoms might affect the resistance by a factor of 10 or 100, but the actual resistance of the diamond is so great that some entirely new factor must be at work, and the Brillouin-zone theory provides the required explanation.

P

216 *Atomic Theory*

In the description of the simple free-electron theory, we saw how, in the absence of an electric field, the velocity distribution of the electrons was perfectly symmetrical, so that no resultant flow occurred in any one direction, whilst in the presence of a field, a slight displacement of the velocity distribution took place, until a steady state was set up in which the acceleration of the electrons in the direction of the field was counterbalanced by the collisions with the atoms, or more properly the thermal vibrations of the lattice. In the zone theories, the same line of approach may be used, but the conclusions are very different owing to the ranges of forbidden energy which correspond with wave numbers k on the surfaces of the Brillouin zones. We may note first that wave mechanics indicates that an electron can move freely and without resistance in a perfectly periodic field, and in this way the very long free paths indicated by experiment can be understood. If we look at the matter from the point of view of the free-electron theory, an electron may be scattered by interaction with (*a*) lattice vibrations (phonons), (*b*) crystal imperfections or impurity atoms, and (*c*) collisions with other conduction electrons. It is easy to show that (*c*) is relatively unimportant at all ordinary temperatures. In an electron/electron collision, momentum and energy are conserved, and an electron in the Fermi surface cannot collide with all the $(N-1)$ other electrons, but only with NkT/ε^* of them. The energy exchange does not exceed a few kT, so that the mean free path is not a few Å but varies as $(\varepsilon^*/kT)^2$, being of the order of 10^{-4} cm at room temperature, and as much as 10 cm at 1° K. We may, thus, regard transport phenomena as involving particles that undergo collisions with the lattice or with imperfections.

We may begin first by considering the behaviour of one single electron in a periodic field, and for simplicity in drawing we may consider a hypothetical two-dimensional solid with a periodic field for which the first Brillouin zone is a square, as shown in Fig. V.14(*a*). Let us suppose that the electron is in the state represented by the point P. If the energy contours are circles, the velocity of the electron will then be in the direction OP, whilst if the energy contours are of a different form the velocity of the electron will be perpendicular to the contour at the value of k concerned. We may now suppose that an electric field is applied in the direction shown. In the presence of the field the wave number of the electron tends to increase in the direction of the field, and the effect of the field is to make the electron pass through a series of states p_1, p_2, p_3, ... until the zone boundary is reached, and a break occurs in the E/k curve. Calculation then shows that even if the gap in the energy curve is quite small, say 1 eV, the probability of the electron passing to a state in the next zone is very small. On reaching the state a in Fig.

The Brillouin-Zone Theory of Metals   217

V.14(a) the electron, therefore, does not pass into a state in the next zone, but undergoes a Bragg reflection in the lattice, as a result of which the component of velocity perpendicular to the zone face is reversed, and the electron state jumps from the point a to the point b in k-space. Under the influence of the electric field, the electron then passes through a succession of states b_1, b_2, b_3, . . . until it reaches the point c on the

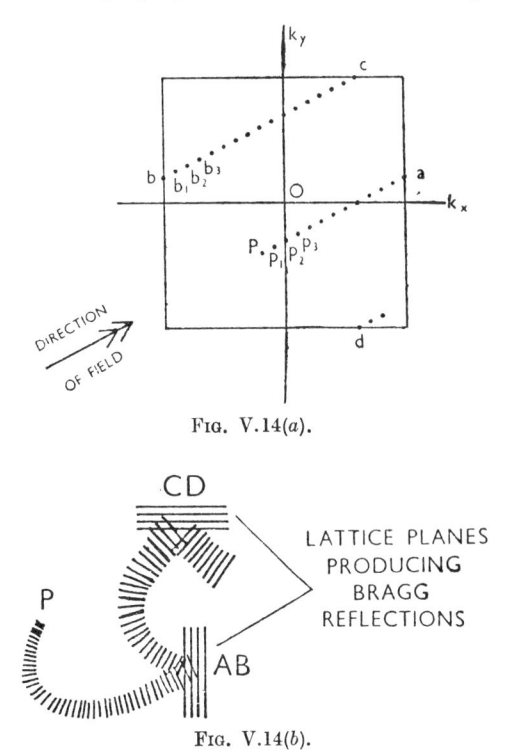

Fig. V.14(a).

Fig. V.14(b).

surface of the zone, when a Bragg reflection again occurs and the electron jumps from the state c to the state d, and a similar sequence of changes then takes place. The general effect of an electric field is thus to make the wave number k change continuously in the direction of the field until the wave number k reaches a state on the surface of the Brillouin zone where an energy gap occurs; at this stage there is an abrupt change in k, corresponding to a Bragg reflection in the lattice.

Fig. V.14(a) shows the way in which the wave number k changes under the influence of an electric field. The velocity, v, of the electron is proportional to dE/dk (see p. 209), and as long as the energy contours are spherical, v and k are in the same direction, since v is perpendicular to the energy contour (see p. 212). It follows, therefore, that so long as the energy contours are spherical, the effect of the electric field is to change v and k in the same way, and the electron is accelerated in the direction of the field. Where the energy contours are no longer spherical, the effect of the electric field is to change k in the direction of the field, as shown in Fig. V.14(a), and as k passes through the succession of points $p_1, p_2, p_3, \ldots$, the velocity, v, changes so that it is perpendicular to the energy contours at the succession of points p_1, p_2, p_3. ... It follows, therefore, that as the wave number k approaches the surface of the zone, where the energy contours are no longer spherical, and v and k are not in the same direction, the velocity of the electron, v, will no longer be changed in the direction of the field, and the acceleration of the wave packet which represents the electron will no longer be in the direction of the force. It is thus important to recognize that near to the surface of a zone, it is k and not v, which changes in the direction of the applied field. This is an example of the point previously emphasized (p. 197) that the individual waves moving in one direction may build up a wave group moving in another direction.

Fig. V.14(a) shows the way in which the points representing the states of an electron change under the influence of an electric field. If we are asked what path the electron follows, it will be appreciated that from Heisenberg's Principle we cannot follow the electron in detail along its path, but we may obtain a general idea of the process involved as follows. We may regard the velocity as resolved into components parallel and perpendicular to the field, and for simplicity we may assume that the energy contours are everywhere spherical, so that we ignore the complications near to the zone boundaries. We may then say that the general effect of the electric field is to accelerate the component of velocity parallel to the field, and to leave the perpendicular component unchanged. We may suppose that when the electron state is near the point P in Fig. V.14(a), the electron itself is somewhere near a point P in real space, as shown in Fig. V.14(b). Then by Heisenberg's Principle there will be a mutual uncertainty in our knowledge of the wave number and the position, and we may represent this by a shaded area in Fig. V.14(b). Since we are assuming the energy contours to be spherical, the velocity of the electron in the states $P, p_1, p_2, p_3, \ldots$ in Fig. V.14(a) will be in the directions $OP, Op_1, Op_2, Op_3 \ldots$ and the passage of the electron along the series of states $p_1, p_2, p_3, \ldots$ in Fig. V.14(a) will correspond to a curved trajectory in which the position becomes more and more uncertain as the wave packet spreads out. This curved trajectory is shown as a shaded region PAB in Fig. V.14(b) in order to emphasize the un-

The Brillouin-Zone Theory of Metals 219

certainty in the simultaneous position and wave number of the electron. At the point AB the wave number has reached the critical value for a Bragg reflection from the planes of atoms parallel to the side of the square lattice in Fig. V.14(b), and the reflection takes place as shown, so that the angle of incidence equals the angle of reflection, and the electron jumps from state a to b in Fig. V.14(a). The electron then sets out on a new curved trajectory as shown. This description is, of course, very greatly over-simplified, because in the neighbourhood of the zone boundaries the energy contours are really distorted, with the result that the force and the acceleration of the wave packet may no longer be in the same direction. The general process described above does, however, still take place, and the effect of an electric field on a single electron in a periodic field is to make the electron undergo a series of curved paths terminated by Bragg reflections. It should be emphasized that these Bragg reflections take place even at the absolute zero of temperature, and are quite distinct from the scattering of the electrons by " collisions " with the thermal vibrations of the lattice, which give rise to the resistance at temperatures above the absolute zero.

From the above description it will be appreciated that if there is an energy gap at the surface of a zone, an external electric field * will not enable an electron to pass from a state in one zone to a state in another. It follows, therefore, that at the absolute zero a substance must be an insulator if the number of valency electrons per atom completely fills a Brillouin zone with an energy gap across all of its surfaces (Fig. V.8(b)). In such a case, the application of an electric field can produce no resultant flow in any direction, because, since electrons are indistinguishable and all the states in the zone are occupied, no interchange of electrons among these states can affect the distribution of the assembly as a whole. In a completely filled zone, the application of an electric field may be regarded as resulting in a streaming of the states in the direction of the field, as in Fig. V.14(a) (p. 217), with continued Bragg reflections as the surface of the zone is reached, but as all the states are occupied, the distribution as a whole remains unaltered, and there can be no resultant flow in the direction of the field.

In the same way, the insulating properties of salts may be interpreted in terms of zone theory. For example, in the salt NaCl, the outermost electron states of the sodium ions are derived from the 2s-, 2p-, and 3s-states of the free ions. Each of these has broadened out as the atoms assemble to form the crystal, and the broadening results mainly from the interaction between a given sodium ion and its immediate

---

* In the whole of the present description we are ignoring the use of the very high voltages which cause the breakdown of an insulator.

sodium neighbours; the interaction between unlike ions produces a smaller effect. The 2s- and 2p-bands overlap to form a hybrid band which is fully occupied, and is separated by an energy gap from the empty 3s-band. The outermost electron states of the chlorine ions are derived from the 3s-, 3p-, and 4s-states of the free ions, and again the 3s- and 3p-states form a completely filled band which is separated by a finite gap from the unoccupied 4s-bands. The insulating properties of the crystals are then the result of the completely filled bands, and are not the result of the forces holding the electrons to the nuclei.

The above remarks refer to the effect of an electric field at the absolute zero of temperature and in the absence of other complicating factors, and they show that an insulator is essentially a structure in which the number of electrons is sufficient to fill one or more Brillouin zones, the outermost of which is completely separated from the next zone, as in Fig. V.8(b). If, however, the insulating crystal is illuminated by light of sufficiently short wave-length, electrons may be excited photoelectrically from the full zone to a higher zone, where they will be free to act as conduction electrons, and in this case the phenomenon of *photoconductivity* is shown. It is important to note that the excitation of the electron into the higher zone which produces the conductivity is due, not to the effect of the applied electric field, but to the absorption of the quantum of light. In the same way, if the energy gap at the surface of the filled zone is not too great, rise of temperature may result in the thermal excitation of electrons into the next zone, where they will be free to act as conduction electrons. This effect is discussed later (p. 228) in connection with semi-conductors.

It is, of course, well known that many insulating substances such as the diamond, can be regarded as co-valent structures in which each atom satisfies its normal valency by sharing electrons with (8—N) neighbours, where N is the number of the group to which the element belongs (p. 103). In the diamond structure, for example, each atom has four close neighbours, and the whole crystal may be regarded as a gigantic molecule in which the atoms satisfy their normal valencies. If it is asked how this concept affects the picture presented by the Brillouin-zone theories, the answer is that each represents part of the truth. The co-valency theory emphasizes that the electron-cloud density of the valency electron is greatest round the lines joining a given atom to its four neighbours. The Brillouin-zone theory emphasizes that the valency electrons belong to, and are free to move within, the crystal as a whole, but does not deal with the variation of the electron-cloud density in space. Each theory, therefore, represents part of the truth, and the two are complementary and not antagonistic to one another.

Strictly speaking (see p. 204), we should describe Brillouin-zone theory in terms of the number of electrons per unit cell of the crystal structure. So long as we are dealing with normal structures in which all the lattice sites are occupied, the number of electrons per unit cell of a given structure is a simple multiple of the number of electrons per atom (*i.e.*, of the electron concentration), and we shall therefore adopt the usual convention of describing the theory in terms of the number of electrons per atom.

Another example of the way in which a structure may be regarded from different points of view is that of the anti-isomorphous fluoride structure described on p. 116. Compounds such as Mg_2Si are quite properly regarded as normal valency compounds of electropositive magnesium with electronegative elements, and the relative stabilities of the series Mg_2Si, Mg_2Ge, Mg_2Sn, and Mg_2Pb are in agreement with the known electrochemical properties of the elements concerned. Mott and Jones [*] have, however, shown that the first Brillouin zone of the fluoride structure bounded by (220) faces can contain 8/3 electrons per atom, which is the exact number given by the above formulæ if a valency of 2 is allotted to magnesium, and a valency of 4 to silicon, germanium, tin, or lead. The same structure is also formed by Al_2Ca and Sn_2Pt, which cannot be regarded as ionic compounds, but which do correspond to an electron : atom ratio of 8 : 3 if the normal valencies (Ca = 2, Al = 3, Sn = 4, Pt = 0) used for electron compounds are employed. On the other hand, the same structure is formed by $AuAl_2$, and this can be regarded as an 8 : 3 electron compound only if gold is assumed to be divalent, whereas the stable valencies for gold are 1 and 3. The electrical properties suggest that Mg_2Sn is probably an intrinsic semiconductor (p. 228), whereas Mg_2Pb is more like a normal metal. This type of structure is thus not yet fully understood, and it is possible that the same structure can be formed by different processes.

We may now consider structures in which the number of valency electrons per atom is small compared with that required to fill a zone. In all simple translational lattices, each Brillouin zone contains $2N$ electron states, where N is the number of atoms in the crystal. It follows, therefore, that in the univalent alkali metals, and copper, silver, and gold, the occupied states only take up one-half of the first Brillouin zone. Fig. V.5(a) shows the first Brillouin zone for the face-centred cubic structure, and the Fermi surface is drawn so as to show the limits of the states occupied with one valency electron per atom if the surface is only slightly distorted from the spherical; there are unoccupied states

[*] N. F. Mott and H. Jones, "The Theory of the Properties of Metals and Alloys." Oxford : **1936** (Clarendon Press).

on all sides of those that are occupied. This is known to be the situation in the alkali metals, but in copper the Fermi surface touches some faces of the zone (Fig. VI.9, p. 304).

Under the conditions of Fig. V.5(a) the picture of electrical conductivity is very like that previously described for the simple free-electron model. In the absence of an electric field, the velocity distribution is symmetrical, so that for each electron with velocity $+v$, there is one with velocity $-v$, and the total flow of electricity in any one direction is zero. In the presence of an electric field, there is a slight shift in the velocity distribution, giving rise to a resultant current in the direction of the field, and a steady state is set up in which the effect of the field is neutralized by the collisions with the thermal vibrations of the lattice. The picture is complicated by the fact that if an electron escapes collision for a time sufficient to enable its wave number, under the influence of the electric field, to reach the surface of the Brillouin zone, then a Bragg reflection takes place just as was described in connection with Figs. V.14(a) and (b).

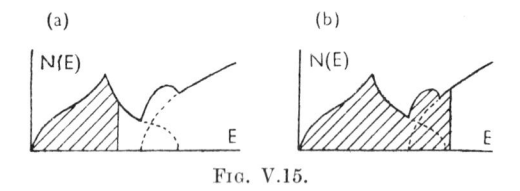

FIG. V.15.

We have already seen that although in some structures the different zones are completely separated (Figs. V.6(a) and V.8), there are others in which the overlapping of the zones takes place (Figs. V.6(b) and V.9). This latter case of overlapping zones is characteristic of most, if not all, structures with normal metallic properties. If the electron concentration is small compared with that required to fill the first zone completely, the condition of affairs will be as shown in Fig. V.15(a), where the shaded area represents the limit of the occupied states; this corresponds with Fig. V.5(a), discussed above. If, on the other hand, the electron concentration were greater, the condition of affairs might be as shown in Fig. V.15(b), where the whole of the states in the first zone, and some of those in the second zone, are occupied. In such a case, the electrons in the first zone do not contribute to the electrical conductivity. If the zone characteristics are such that in any given direction there is a break in the E/k curve, the external electric field will not enable an electron in the first zone to leave that zone, because when k reaches the surface of the zone, a Bragg reflection takes place.

The Brillouin-Zone Theory of Metals 223

The electrons in the first zone thus form a group of electrons, which, although free to move, do not produce a resultant flow in any direction. It must not, however, be imagined that a given electron which is in the first zone at one instant remains in this zone indefinitely. At finite temperatures there is always a certain probability of an electron passing from the first to the second zone as the result of thermal excitation, this probability increasing with rise of temperature. Mere acceleration under the action of an electric field only gives a very small probability of the electron passing from one zone to another if there is an energy gap at the surface. There are, however, some metallic structures with overlapping zones (*e.g.*, the close-packed hexagonal) in which, although there is an energy gap at the surface of the zone, this energy difference becomes zero along some of the edges of the zone (see p. 252), and in such cases the application of an electric field in certain directions relative to the crystal axes may enable an electron to pass through a series of states beginning in one zone and ending in another.

The above considerations serve to emphasize that although all the electrons take part in the process, the number effective in producing conductivity does not necessarily increase with the number of electrons per atom. Thus, in the case of Fig. V.15(*b*), the electrons which occupy the first Brillouin zone can contribute nothing to the conductivity and there are fewer electrons contributing to the conductivity than in the case of Fig. V.15(*a*), although the number of valency electrons per atom is greater for the case of Fig. V.15(*b*).

To describe the conductivity process further we may for simplicity use a two-dimensional model in which the first Brillouin zone is a square. Fig. V.16(*a*) shows a case in which there are relatively few electrons per atom, and the occupied states are represented by the shaded circle, the circumference of which corresponds to the surface of the Fermi sphere in the three-dimensional model. Under these conditions the picture of conductivity is very like that of the free-electron model (p. 184); all the electrons take part in the conductivity process, but only those at the edge of the circle can give a resultant flow in any one direction, or undergo collisions which affect the velocity distribution.

Fig. V.16(*b*) shows a case in which the first zone is completely filled and there is an overlap into the second zone. Here, in the general case,* only the electrons in the second zone produce a conductivity, and the position is analogous to that of Fig. V.15(*b*).

Fig. V.16(*c*) shows a case in which the overlap into the second zone has begun before the first zone is completely filled. The occupied states

---

* As explained above, in some structures the energy gaps may vanish at certain points on the zone surface.

Atomic Theory

are shown by single shading, and the corresponding $N(E)$ curve is given in Fig. V.17. We may now consider the effect of the application of an electric field from left to right as in Fig. V.16(d). In such a case the group of electrons whose states are within the rectangle *abcd*, which is shown

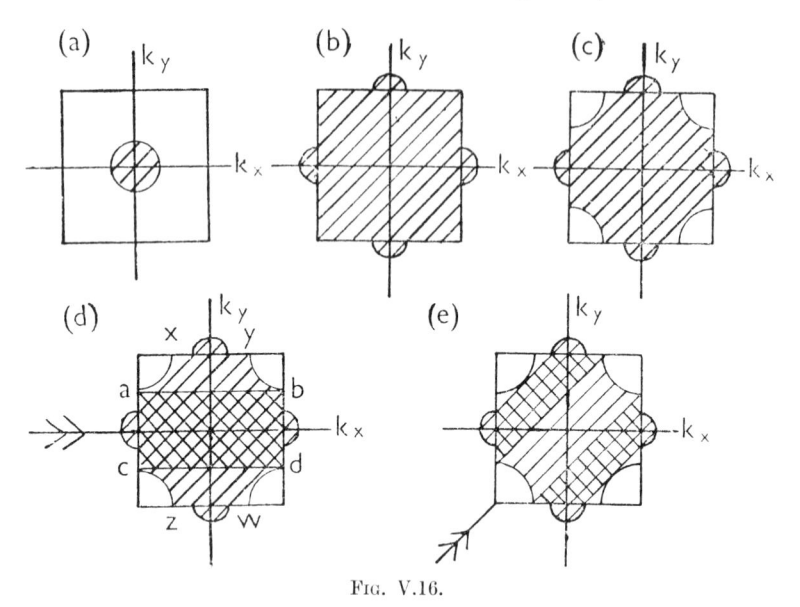

Fig. V.16.

in double shading, cannot contribute to a conductivity in the direction of the field, because for each state of $+k$ there is one of $-k$, and in the general case the energy gaps at the boundaries of the zone prevent the electrons from passing to a state lying outside the zone. The group of

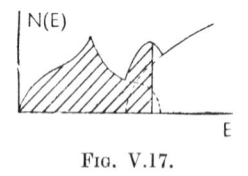

Fig. V.17.

electrons whose states lie in the areas *axyb* and *czwd* of the first zone, and those in the second zone can, however, undergo a displacement of the velocity distribution so as to produce a resultant flow in the direction of the field, although, as will be appreciated from p. 186, it is only those electrons in states near the Fermi surface which are really effective in determining the conductivity. If, for example, velocities are referred

The Brillouin-Zone Theory of Metals

to the direction of the field, then for each electron whose state is near the centre of the area $axyb$ of Fig. V.16(d) and which corresponds to a velocity $+v$, there is one with a velocity $-v$, and so it is only electrons near the Fermi surface which can contribute to the conductivity. On the other hand, the group of electrons with states inside the area $abcd$ of Fig. V.16(d) cannot give rise to a resultant flow at all. If the applied field has been in the direction of the diagonal of the square, as shown in Fig. V.16(e), then the group of electrons whose states are shown by the doubly shaded areas would be those which did not contribute to the conductivity.

From the preceding description it will have been realized that the electrons at the bottom of the first zone behave very like those of the free-electron theory. In one-electron theories,[*] the same applies to electrons at the bottom of higher zones, *provided that the energy is reckoned from the bottom of the zone concerned*. Thus, the $N(E)$ curves of the second zones in Figs. V.8(b) and V.9 are parabolas of the form $N(E) \propto (E-E')^{\frac{1}{2}}$, where E' is the energy at the bottom of the second zone. In general, for the few electrons in states at the bottom of zones, the relations between energy and other properties (*e.g.*, wave number, velocity v, $N(E)$) are similar to those of the free-electron model, provided that the energy is measured from the bottom of the zone concerned.

At the top of the zone the position is very different, and some of these complications may now be described. We have seen before (p. 209) that for the one-dimensional model, the group velocity v is

$$v = \frac{2\pi}{h} \cdot \frac{dE}{dk} \quad \ldots \ldots \quad \text{V (9)}$$

Under the influence of an electric field, F, the wave number, k, in the direction of the field increases uniformly with time, and the variation is given by the equation:

$$\frac{dk}{dt} = \frac{2\pi eF}{h} \quad \ldots \ldots \quad \text{V (10)}$$

It follows, therefore, that the rate of change of the group velocity is:

$$\frac{dv}{dt} = \frac{2\pi}{h} \cdot \frac{d}{dt}\left(\frac{dE}{dk}\right) = \frac{d^2E}{dk^2} \cdot \frac{4\pi^2}{h^2} \cdot eF. \quad \ldots \quad \text{V (11)}$$

If this is compared with the classical expression, eF/m, for the acceleration, it will be seen that the mass m is replaced by:

$$\frac{h^2}{4\pi^2 \dfrac{d^2E}{dk^2}}$$

and it is customary to call this last quantity the *effective mass* of the

[*] For a discussion of one-electron theories and their limitations, see p. 235.

electron, and the symbol m^* is sometimes used to denote this quantity.†
Examination of equation V (11) leads to interesting conclusions, for it will be seen that when d^2E/dk^2 is negative, the acceleration is negative. Examination of Figs.V.3(b) and V.6 will show that d^2E/dk^2 is negative for states in the upper portion of a band, and consequently for electrons in these states the effect of an electric field is to diminish the group velocity, v, and the effect is in some ways as though the electron had a negative mass. This is a characteristic of the behaviour of electron wave packets in a periodic field, and it is not possible to give a coherent picture of what happens in terms of the classical mechanics. It is probably justifiable to say that the electric field does not make the individual electron move backward, but changes the relative probabilities of finding the electron moving in opposite directions; but since the whole effect is within the length of the wave packet, no detailed picture can be given. For many purposes it is justifiable to regard the electrons at the top of a band in the one-dimensional model of Fig. V.3(b) as having a negative mass, or as behaving like particles with a positive charge. In the three-dimensional model, for electrons in states at the bottom of a zone, the effect of an electric field is to increase the value of k in the direction of the field, and the group velocity, v, is accelerated in the direction of F. As long as the energy E can be expressed in the form of $Ak_x^2 + Bk_y^2 + Ck_z^2$ (i.e., as long as the surfaces of constant energy are ellipsoids), equation V (11) is valid for the acceleration. As the surfaces of the zone are approached, the surfaces of constant energy become distorted, so that equation V (11) no longer holds, and in general a field F changes the group velocity, v, in directions other than that of F itself, so that force and acceleration are not in the same direction as regards the acceleration of the wave packet. For states near the top of a zone, the component of the velocity, v, perpendicular to the surface of the zone, decreases as the wave number k approaches the zone surface. The Brillouin-zone model of electrons in a periodic field has thus the characteristic that when there are comparatively few electrons in a zone, the group velocities, v, associated with the electron states behave very much like those of the free-electron model. On the other hand, if there is a completely filled or nearly filled zone, the electron states at the top of the zone are associated with the group velocities, v, which, under the influence of applied electric or magnetic fields, may vary as though the electron had a negative mass or a positive charge. It is in this way that the theory is able to account for the fact that in univalent metals the Hall coefficient (see p. 260) is negative, in agreement with the calculations of the

† The reader should be warned that departures from free-electron-like behaviour in many properties other than electrical conduction are often described in terms of an effective mass, the definition of which is not necessarily equivalent to that given above.

The Brillouin-Zone Theory of Metals

simple electron theory, whereas in crystals with nearly full zones the Hall coefficient may be positive.

The preceding remarks refer to the motion of electrons in a strictly periodic field, and the Bragg reflections referred to above are the result of the periodicity of the field, and the effects described take place even at the absolute zero. At this temperature, with small applied fields, crystals with perfectly regular structures will be insulators if they correspond with completely filled Brillouin zones of the type shown in Fig. V.8(b), or perfect conductors if they possess incompletely filled zones. It can be shown that the atomic vibrations corresponding with the zero-point energy * do not produce an electric resistance, but as the temperature is raised, the ordinary thermal oscillations interfere with the perfect periodicity of the lattice, and a resistance is created. The special phenomenon of superconductivity—the sudden disappearance of all resistivity at a given non-zero temperature—is discussed in Part VI, Section 3. The zero resistance of a perfectly pure metal at the absolute zero of temperature is the result of the wave-like characteristics of an electron and of the perfect periodicity of the lattice. In any real metal the perfect periodicity is disturbed by impurities or structural imperfections and, at any finite temperature, by lattice vibrations. The effects of these will be discussed in Part VII, Section 2. As the temperature is raised, the Fermi surface marking out the limit of the occupied states in k-space becomes more diffuse, but qualitatively the picture of electrical resistance is very like that previously described for the simple free-electron theory. It is again only electrons near to the Fermi surface of occupied states that can undergo collisions to affect the velocity distribution, and so create a resistance.

All detailed theories of the conductivities of pure metals use as their starting point a model of the electronic structure that involves only one zone of electrons, and usually assume a spherical Fermi surface. Approximate descriptions of a number of the transport properties of metals with more than one conduction electron per atom have been successfully produced in terms of two-band models. In these the electronic structure is taken as represented by the situation shown in Fig. V.16(c) with one set of parameters (n_1, m_1^*, v_1, &c.) for the states inside the zone and another set for those outside.

Semi-Conductors.

The zone theory has thus enabled a considerable insight to be gained into the processes involved in insulators and in normal metallic con-

---

* The zero-point energy is that due to atomic vibrations which persist at the absolute zero of temperature (see p. 281).

228   Atomic Theory

ductors. It remains to consider briefly the class of substance known as semi-conductors, which are in general characterized by the fact that their electrical conductivity is increased by rise of temperature and by the presence of impurities, in contrast to the behaviour of normal metals. We may consider first a substance with the zone-characteristics shown in Fig. V.18(a), where a fully occupied zone is separated from the next zone by a very small gap. This substance would be an insulator at the absolute zero, but at higher temperatures electrons would be thermally excited from the first to the second zone, and then, when in states at the bottom of the second zone, they would act as free electrons, and so give rise to a conductivity which would increase with the temperature, since

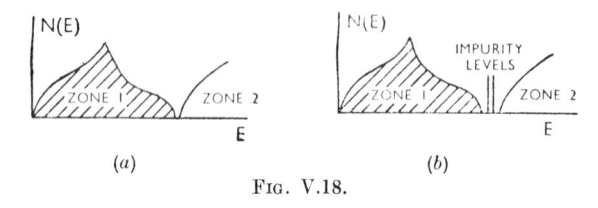

(a)   (b)
Fig. V.18.

the higher the temperature, the greater the number of electrons which would be excited into the second zone. It is important to note that it is the thermal excitation, and not the effect of the external electric field, which excites the electrons from one zone to another. In this type of semi-conductor, the electrons which are excited to the bottom of the upper zone behave like free electrons, and the Hall Coefficient (p. 260) has the normal negative sign. Substances of this kind which are semi-conductors in the pure state are known as *intrinsic semi-conductors*, and are comparatively rare. Graphite is of this type, and silicon and germanium are also intrinsic semi-conductors with respective energy gaps of 1·1 eV and 0·76 eV between the two zones of Fig. V.18(a). If the energy gap between the two zones is Q eV, the number of electrons per cm³ in the upper zone at a temperature θ is of the order $10^{19} \times e^{-Q/2k\theta}$ and thus increases exponentially with rise of temperature.* This increase is sufficient to outweigh the normal increase in resistance resulting from the greater amplitude of the thermal vibrations, and so the electrical conductivity rises with increasing temperature in contrast to the behaviour of a normal metal. At room temperature (p. 12) $k\theta$ is of the order of 3×10^{-2} eV, and consequently in silicon and germanium the number of conducting electrons is very much smaller than that in a normal metal (*ca.* 10^{22} per cm³).

* See D. A. Wright, " Semi-Conductors," p. 35. **1950:** London (Methuen and Co., Ltd.).

The current, i, carried by an intrinsic semi-conductor may be written in the form:

$$i = Ne\mu E \quad \ldots \ldots \quad \text{V (12)}$$

where N is the number of carriers, e the electronic charge, μ the mobility, and E the electric field strength. Measurements of the Hall coefficient enable N to be measured, and so μ can be determined.

For pure intrinsic semi-conductors, μ is often proportional to $T^{-3/2}$, where T is the absolute temperature, but in the presence of impurities the relations are more complicated. The $T^{-3/2}$ law involves the applicability of the Boltzmann statistics, and so does not hold if N is large.

In an intrinsic semi-conductor, the excitation of electrons from the first to the second zone produces unoccupied states at the top of the first zone. The electrons in the first zone can, therefore, also contribute to the conductivity, and as we have seen before, the effective mass of such electrons is usually negative, and they produce a Hall coefficient of abnormal, or positive sign. In most intrinsic semi-conductors, the predominant part is played by the electrons excited into the second zone but, as the mobilities of holes and electrons are different, the effect may become complicated.

The second class of semi-conductors owes its properties to the presence of impurities, and the condition of affairs is represented schematically in Fig. V.18(*b*). Here a completely filled zone * (marked zone 1) and an empty zone (marked zone 2) are separated by a comparatively narrow energy gap. The effect of introducing impurity atoms is, in general, to create new energy levels, and it may happen that these lie within the energy gap, as shown in Fig. V.18(*b*). At the absolute zero the structure will be an insulator, but as the temperature is raised, one or both of two processes may occur.

(1) Electrons may be excited thermally from the impurity levels into zone 2. These electrons then enter states at the bottom of zone 2, where they will act as free electrons, so that a limited conductivity will be shown, and the Hall coefficient will have the normal negative sign. Such semi-conductors are called *N-type semi-conductors*, because the electrons act as if they had a normal negative charge, and the impurity atoms are called *donor atoms* because they give electrons to the second zone. The conductivity of an N-type semi-conductor increases with the temperature, but if the concentration of donor atoms is very small a stage may be reached at

---

* The completely filled zone need not be the first zone. The full zone is commonly called the valence band and the empty one the conduction band, but these terms can be misleading.

which practically all of their available electrons have been given to the second zone, after which the conductivity may diminish owing to the usual effect of the thermal vibrations. Alternatively, if the gap between zones 1 and 2 is narrow, the substance may exhibit both intrinsic and N-type semi-conductivity, the latter predominating at low, and the former at high temperatures, so that the conductivity/temperature curve shows a change in direction at the temperature where the donor atoms have given up nearly all their electrons.

(2) In another class of semi-conductor electrons may be excited thermally from the top of zone 1 into the impurity levels, whose atoms are then called *acceptor atoms*. This process leaves some of the states at the top of zone 1 unoccupied, so that the application of an electric field can cause a slight shift in the velocity distribution in zone 1, and so create a conductivity. As the effective mass of electrons in states at the top of a zone is usually negative, these electrons act as though they had a positive charge, and where this process is concerned the substance is called a *P-type semi-conductor*, and the sign of the Hall effect is abnormal (i.e. positive). The increase in conductivity with rising temperature may again show saturation if the concentration of acceptor atoms is small, and if the energy gap between the zones is sufficiently small the processes of intrinsic and P-type semi-conductivity may be superimposed.

The properties of intrinsic semi-conductors are, of course, affected by the presence of impurity atoms, and in some cases clear valency effects are shown. Thus, silicon and germanium are normal 4-valent elements which crystallize in the diamond-type tetrahedral structure (Fig. III.11 p. 103), which gives rise to a completely filled Brillouin zone. The introduction of phosphorus or arsenic, whose atoms contain 5 valency electrons, causes electrons to enter the second zone with the production of an N-type semi-conductor. Conversely, the addition of boron, with only 3 electrons, results in unoccupied states at the top of the first zone with the production of a P-type semi-conductor.

From the above description it will be appreciated that according to the nature of the zone structure and the numbers of electrons and impurity centres present, semi-conductors may be produced whose conductivity/temperature relations vary greatly, and which may be extremely sensitive to the concentrations of the impurity atoms. It has also been shown that, in germanium, the presence of dislocations may affect the form of the band structure to such an extent that real metallic conductivity is exhibited. The semi-conductors discussed above

The Brillouin-Zone Theory of Metals

were elements, but semi-conductivity is also found in a wide variety of alloys and compounds. In some, like the III–V compounds (*e.g.*, InSb, GaAs), it seems appropriate to think of the full and empty bands as being produced by the same modification of the distribution of allowed states by the crystal-lattice potential as in Group IV elements. In others, like oxides and sulphides, the full band can be regarded as produced by a broadening of the atomic levels of the negative ion; in these, behaviour of the non-intrinsic semi-conductor type can often result from the departures from the exact stoichiometric ratio that can be obtained by suitable treatment. In this way the full band structure responsible for the insulating properties of a pure oxide is modified and examples of both N-type and P-type semi-conductors are known. The importance of electrochemical aspects is shown by the semi-conductivity of the compound CsAu. The change in the semi-conducting character of intermetallic compounds with increasing difference in electrochemical character of the components is well illustrated by the isomorphous series of compounds made by elements flanking tin in the Periodic Table. InSb is an intrinsic semi-conductor, CdTe a wide-gap semi-conductor and photo-conductor, and AgI an ionic conductor.

It is interesting to note that, while the conductivities of both metals and semi-conductors are dependent on purity and on temperature, these variables operate in different ways in the two groups of materials. In metals, impurities and change of temperature produce their main effects as means of scattering the current carriers, the number and nature of which are little affected, while in semi-conductors they modify the conductivity mainly by producing violent changes in the concentrations and character of the carriers.

SUGGESTIONS FOR FURTHER READING.

N. F. Mott and R. W. Gurney, " Electronic Processes in Ionic Crystals." Oxford: **1948** (Clarendon Press).
A. H. Wilson, " Semi-Conductors and Metals. An Introduction to the Electron Theory of Metals." Cambridge: **1939** (University Press). "The Theory of Metals," 2nd edn. Cambridge: **1953** (University Press).
D. A. Wright, " Semi-Conductors." London: **1950** (Methuen and Co., Ltd.).
W. Shockley, " Electrons and Holes in Semi-Conductors." New York: **1950** (D. Van Nostrand Co.).
W. Ehrenberg, " Electric Conduction in Semi-Conductors and Metals." Oxford: **1958** (Clarendon Press).

Note on Conduction of Positive Holes.

In all crystals the carriers of electricity are negatively charged electrons, but when a crystal has a nearly filled Brillouin zone, it is customary to speak of conduction by positive holes, for the following reasons. Consider a zone containing N electrons, and imagine these divided into one single electron, denoted 1_x, and the

Q

assembly of the remaining $(N - 1_x)$ electrons. The flow of electricity in the direction of an applied field may then be denoted:

$$C(N - 1_x) + Cl_x = 0$$

Here $C(N - 1_x)$ is the conductivity resulting from the $(N - 1_x)$ electrons, and Cl_x that from the single electron in the state x. The sum of these must equal zero, because it corresponds with a completely filled zone. We have, therefore:

$$C(N - 1_x) = -Cl_x$$

The conductivity resulting from a zone with a single electron missing from the state x, i.e., the conductivity of the assembly of $(N - 1_x)$ electrons, is equal to that of the state x with the sign reversed.

Suppose that in Fig. V.14(a) the zone is completely filled except for one state x, and that the direction of the field is such as to accelerate a normal electron in the direction of the arrows. Then all the states are to be regarded as drifting through the zone in the direction of the field, and the unoccupied state x also drifts in this direction. As the state x approaches the zone boundary, the effective mass becomes negative, and if the state x were occupied by an electron, the latter would be retarded instead of being accelerated by the field. The motion of the vacant state under the influence of the field is therefore the same as that of a positive particle with a positive mass. In a nearly filled band the " holes " or occupied states are, of course, always at the top of a band, and for these states, as the surface of the zone is approached, the total energy increases, but the kinetic energy decreases. It will be appreciated that in a real crystal the continuous drifting of the electron states in the direction of the field is opposed by the various scattering processes which contribute to the resistance.

3 ELECTRON THEORIES OF METALLIC CRYSTALS.

Introduction.

In Sections V.1 and 2 we have outlined the simple zone theories in which the electrons are treated as moving in a regular periodic field, the details of which have not been specified. We have regarded the Brillouin zones as arising from the reflection of electrons by the crystal lattice, so that the faces of the zones are directly related to the planes of the crystal with high structure factors in X-ray or electron diffraction. This is the easiest way of understanding the underlying ideas, but it is clearly only an approximation to the real state of affairs, and in the present chapter we shall outline some of the attempts that have been made to develop more complete theories.

It should be noted first that the relation between the Brillouin zone and the crystal structure is not really so simple as has been suggested in the preceding chapters. In particular, it is possible for a crystal plane with a zero structure factor to give rise to energy discontinuities in certain directions in k-space. This has led to confusion in the terminology as to exactly what constitutes a Brillouin zone. This point does not affect an elementary understanding of the subject, but for those who wish to go further an attempt has been made to explain the points at issue in Appendix I (p. 243), which may be ignored on a first reading.

The Brillouin-Zone Theory of Metals 233

A rigorous theory of the behaviour of electrons in a metallic crystal is a matter of extreme difficulty, and at present satisfactory theories exist only for the alkali metals, and for aluminium (p. 339). In most other cases the calculations involve such drastic approximations that the conclusions are of qualitative, rather then quantitative, value.

In Section III.1 we have explained how, when atoms are brought together to form the crystal of a metal, the general characteristic is for each electron state to broaden out into a band. We have seen (p. 91) that in crystalline sodium the bands derived from the $(3s)$ and $(3p)$ states of the free atoms have broadened out so much that they overlap, with the production of a hybrid band in which s-states predominate at the bottom, and p-states at the top of the first Brillouin zone. It is clear that when an electron is near to an atom, the wave function and general characteristics of the electron will be closely related to the general structure of the atom concerned. The more complete theories try to take this into account, and the general picture presented is one in which an electron, when relatively near to an atom, is treated as moving in the field of a free ion of the same structure as that existing in the metal. The electrons are thus described by wave functions which, when near to the nuclei, resemble those of free atoms in the appropriate electronic state, and which join up satisfactorily so as to produce acceptable solutions in the regions between the nuclei. Under these conditions it may be said that an electron spends a certain time near to a given nucleus, and for this period it behaves very much as though it were in a free atom. The electron then spends a further period in travelling to an adjacent nucleus, and during this passage the electron may, in some metals (*e.g.*, the alkalis), behave very much as in the simple *free-electron theory*, whilst in other metals the behaviour may be quite different.

The properties of the metallic crystal depend almost entirely on the outer groups of electrons, namely:

ns and np electrons for the elements of the First Two Short Periods, and of the B sub-groups,

ns, np, and $(n-1)d$ electrons for the transition elements, and probably also for the rare earths,

ns, np, $(n-1)d$, and $(n-2)f$ electrons for the elements of the actinide group.

In general, these electrons will exist in hybrid states, and the mathematical problem is to calculate accurate wave functions for electrons of these types moving in the periodic field of the crystal. For this purpose it is necessary to know the fields produced by the remaining electrons which build up the ions or cores of the atoms concerned. Usually, these fields cannot be calculated accurately, but reasonable * solutions can be obtained by the method of the self-consistent field referred to on p. 67. The details of these calculations lie outside the present book.†

In a purely theoretical approach the fields of the ions should be calculated, but an alternative method is sometimes used in which the field is assumed to be given by an empirical function whose constants are chosen so as to give the best possible agreement with properties such as the ionization potentials of the ion. Whichever method is adopted, it is necessary to allow for possible differences between the electronic states of atoms in the free state and in the metallic crystal. In the case of sodium, for example, it is reasonable to assume that, when near to a nucleus, a valency electron moves in the field of an Na^+ ion of electron configuration $(1s)^2 (2s)^2 (2p)^6$. On the other hand, in the case of nickel, allowance would have to be made for the fact that, although the outer electrons of the free atom have the configuration $(3d)^8 (4s)^2$, this configuration is not preserved in the solid.

Having determined or assumed the fields surrounding each ion, it is then in principle necessary to solve the Schrödinger equation for the motion of an electron in the field concerned. In practice, this problem is too difficult, and various approximate methods are employed, some of which are outlined below. The actual state of affairs is, however, far more complicated than that involved by the motion of one electron, since the different electrons influence each other's motions. These correlation effects result partly from the electrostatic repulsion between two electrons, and partly from effects of the Exclusion Principle analogous to those described on p. 140 in connection with the H_2 molecule. It is possible to estimate only the general magnitude of the correlation effects, and they become more and more serious as the valency of the metal increases. In the case of the alkali metals it can be shown qualitatively that the different correlation effects tend to cancel each other so that a satisfactory theory is obtained, but with metals of higher valency—aluminium is an exception—nearly all electron theories of the metallic crystal are but crude approximations.

Even if it were possible to obtain an accurate solution to the problem of the motion of the outer electrons in the field assumed, this by itself would be insufficient. For a rigorous solution it would be

---

* For criticisms of the method, see J. C. Slater, " Quantum Theory of Matter." **1951**: New York and London (McGraw-Hill).
† See D. R. Hartree, " The Calculation of Atomic Structures." **1957**: New York (John Wiley and Sons); London (Chapman and Hall).

The Brillouin-Zone Theory of Metals

necessary to determine how the calculated behaviour of the outer electrons affected that of the inner electrons, and how this modified the field in which the outer electrons had been assumed to move. In principle the calculation would then be refined until a completely consistent solution was obtained. In practice this is too difficult and it is only for a few metals that even approximate treatments of this type have been made.

Collective-Electron Theories of the Bloch Type.

In these theories (cf. p. 198) the electron states are described by Bloch functions of the general type:

$$\psi = e^{i\mathbf{k}\mathbf{r}}\phi_k(\mathbf{r})$$

where ϕ_k depends on the wave number k, and has the three-fold periodicity of the lattice. This represents a plane wave of wave-length $2\pi/k$, travelling in the direction of the vector $\mathbf{k}$, but the wave is modulated in the rhythm of the lattice. In the elementary developments, the ϕ function is given by a simple curve varying with the periodicity of the lattice. In the more complete theories, the curve is more complicated, and repeats the characteristics of s-states, p-states, &c., at regular intervals.

As explained in Section IV.2, for the free-electron theory, the models of wave mechanics lead to wave functions representing plane waves whose wave-lengths are related to the length of the edge of the cube assumed in the box or the cyclic models. In this case, if we consider a line through the centres of the atoms in an actual crystal, the smooth curve $ABCD$ of Fig. V.19 represents one-half a wave-length of a free-electron function, whilst the full lines show the real part of one form of Bloch function. There is a singularity at each atomic nucleus, and there are two nodes on each side of the nuclei 1, 2, 3.... If similar curves were obtained for all directions in which the internuclear spacing was the same as that in Fig. V.19, the function near to each nucleus would be of a $(3s)$ nature.

Theories based on Bloch functions are called *collective-electron theories*, and the Bloch function for each electron extends through the whole crystal. The electron states may be described in wave-number space, and the occupied states lie within the Fermi surface which, in contrast to that of the free-electron theory, is not spherical.

The term collective-electron theories is in some ways misleading, because Bloch wave functions are one-electron wave functions in which the electrons are considered singly, and in treating any one electron the remainder are regarded as smoothed out, so that the field which they produce has the same periodicity as the lattice. The Bloch functions imply that, in the three typical metallic structures, all the atoms of a pure metal are in the same state, and have the same spin. Certain Bloch functions (*e.g.*, those used in the tight-binding approximation described below) are analogous to the L.C.A.O. (Linear Combination of Atomic Orbitals) functions of molecular theory, but others (*e.g.*, those used in the cellular method of p. 239) are not.

Bloch functions, or at least the more simple types, do not take into account the correlation of positions of electrons and they would, for example, allow two electrons to be at the same point in space at the same time. Consequently, they become increasingly unsatisfactory as the volume per electron becomes smaller. If they are corrected to take correlation into account, the functions are no longer Bloch functions, but correspond with the configuration interactions of molecular theory.

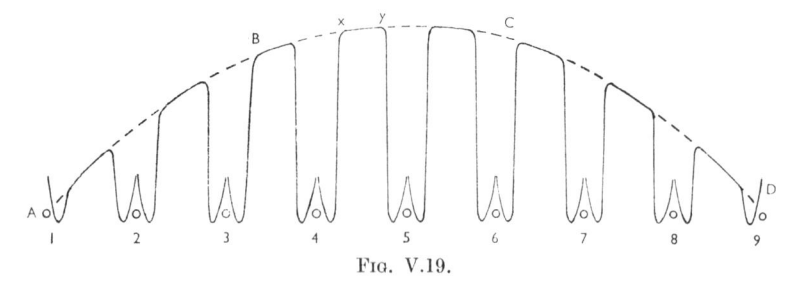

FIG. V.19.

In the development and application to particular metals of band-structure calculations of the collective-electron type several lines of approach have been used. Detailed accounts of these can be found in books and review articles.*

All these methods involve the expansion of the unknown wave function in terms of a set that is known or can be calculated; they differ in the choice of this set.

We shall not attempt a rigorous classification of these different lines of approach, but it is worth devoting a little space to ideas and methods that have proved especially useful in discussions of the measured properties of metals in terms of their electronic structure. For many of the theoretical physicists who make band-structure calculations the principal interest lies in finding solutions to formal problems, but in recent years

---

* H. Jones, " The Theory of Brillouin Zones." **1960**: Amsterdam (North-Holland Publishing Co.).
J. Callaway, " Energy-Band Theory." **1964**: New York (Academic Press).
L. H. Bennett and J. T. Waber, " Energy Bands in Metals and Alloys." **1968**: New York (Gordon and Breach).
L. Pincherle, *Rep. Progress Physics*, 1960, **23**, 355.

The Brillouin-Zone Theory of Metals

the extensive development of computer techniques for carrying out laborious computations has yielded results that can be used with some confidence by experimental physicists and physical metallurgists in the interpretation of their observations.

(a) *The Nearly-Free-Electron Approximation.*

This is not really a particular technique for carrying out band-structure calculations, but rather a first-order correction of the free-electron approach for the effects of the periodic potential. The wave functions $\psi_k(r)$ are assumed to be very similar to plane waves except when k satisfies, or comes close to satisfying, the diffraction condition. The wave function is then considered to be modified in such a way that a small discontinuity appears in the E/k relationship at the zone boundary. (This kind of treatment underlies most of the early developments of the theory of Brillouin zones, and has also been used extensively by Harrison* as an indication of the general shapes likely for the Fermi surfaces of polyvalent metals; such indications can help experimentalists who measure Fermi surfaces to interpret their results (see pp. 268–280).)

In terms of Fig. V.19, this means that the details of the curve in the regions of the atomic nuclei are ignored, and the departure from the smooth curve $ABCD$ is treated as a slight perturbation of the latter. The crystal structures are then used to deduce Brillouin zones for which the magnitudes of the energy gaps are made to agree with some observed physical property.

From Fig. V.19 it will be seen that, although the nearly-free-electron approximation may be valid for electron states of low wave number and long wave-length, it has no justification for states whose wave numbers are so high that the wave-lengths are of the order of the atomic diameter. For if the wave-length in Fig. V.19 is comparable with the distance 1–2, the singularities in the region of the nuclei occupy so great a proportion of the distance that the assumption of a small perturbation is no longer justified. However, the same philosophy underlies a simplified version of the next method to be discussed, which has been used with success on di- and trivalent metals.

(b) *The Orthogonalized Plane-Wave (O.P.W.) Approximation.*

The Schrödinger equation has the property that if ψ_0, ψ_1, ψ_2 are genuine solutions, they must be orthogonal to one another.† As explained above, the equation can never be solved for a real metal. It

---

* See his article in "The Fermi Surface" (edited by W. A. Harrison and M. B. Webb). **1960**: New York and London (John Wiley).

† For the meaning of orthogonality see Appendix II, p. 254.

is, therefore, necessary to assume functions of various types, and the assumptions should be such that the functions are orthogonal. In the nearly-free-electron approach, however, and in any approach where the unknown functions are expanded in terms of a set of plane waves, the wave functions produced for the Bloch electrons are not orthogonal to those of the inner-core electrons. The O.P.W. approximation is a method of correcting for this if the core wave functions are known or can be calculated, and the wave function is approximated as a linear combination of plane waves that have been made orthogonal (see p. 254) to the core functions. The resultant O.P.W. is a function (like that in Fig. V.19) which behaves as a plane wave at large distances from an atom but possesses the rapidly varying character of an atomic function near any nucleus. An important recent development has been* the use of a Schrödinger equation in which the true crystal potential is replaced by a "*pseudo-potential*" for which the solutions are smooth functions even near the nucleus, the orthogonality requirements that lead to the rapid variation in that region being taken care of by an appropriate pseudo-potential to which a reasonable approximation can often be made fairly easily.

Applications of the idea of a pseudo-potential have developed considerably in recent years and have proved useful in dealing with liquid metals as well as solid ones. A more extensive discussion of the concept than we have room for here is given in an article by Harrison.†

Its importance in the present context is that it justifies some of the earlier nearly-free-electron discussions of simple metals and provides ways of making them quantitative. Its basic idea is that one can represent by it the combined effects on an electron of the true potential due to an ion core and the effective repulsion produced by the Pauli principle (which forbids two electrons to occupy the same state) when the occupied states of the ion core are taken into account.

Spectroscopic information about the free atom can be used in the constructions of the pseudo-potential. Such information is also the basis of the *Quantum Defect Method* developed‡ to yield information about the energy states of electrons in metals without the construction of an explicit potential. This has led to good wave functions for the alkali metals and computed cohesive energies in very good agreement with experiment (see p. 293).

---

* J. C. Phillips and L. Kleinman, *Phys. Rev.*, 1959, **128**, 2098.
† W. A. Harrison, "Atomic and Electronic Structure of Metals," Paper 1. **1967**: Metals Park, Ohio (Amer. Soc. Metals).
‡ H. Brooks, *Phys. Rev.*, 1953, **91**, 1027; *Nuovo Cimento Suppl.*, 1958, **7**, 165.

The Brillouin-Zone Theory of Metals

(c) *The Cellular Method.*

In this method, each atom is regarded as surrounded by a polyhedron of such a shape that the different polyhedra pack together and fill the whole of space. For example, in the body-centred cubic structure, if a is the side of the unit cube, each atom is surrounded by eight atoms at a distance $a\sqrt{3}/2$, and by six second-closest neighbours at a distance a. If we draw planes to bisect the lines joining an atom to its closest and second-closest neighbours, these planes form a truncated octahedron, and if such polyhedra are stacked together, they fill the whole of space. Similar treatment can be applied to other crystal structures, and the resulting polyhedra are called " atomic polyhedra ", " s-polyhedra ", " atomic cells ", or Wigner–Seitz polyhedra.*

In the cellular method the wave function for the lowest state is periodic with the period of the lattice, and is symmetrical about any nucleus, and the boundary condition assumed is that at the boundary of any atomic polyhedron:

$$\frac{\partial \psi}{\partial n} = 0$$

where the differentiation is normal to the boundary. This boundary condition allows the wave functions to join up smoothly in the space between two atoms.

The potential within the atomic polyhedron is regarded as spherically symmetrical (ion-core field) and the Schrödinger equation is solved for an electron in this field, subject to the continuity of ψ and its normal derivative at the boundary. In the cellular method it is again necessary to assume or calculate the atomic field, and for this purpose use is generally made of the Hartree methods. For some states in simple crystals (especially those at the bottom of the conduction band in cubic metals) the boundary conditions are sufficiently simple for a solution to be obtained fairly easily. For many states of interest the complexity of the boundary condition causes serious mathematical difficulties, but Korringa and Kohn and Rostoker have shown that these can be avoided by a method (the variational method) that involves the examination of the effect of arbitrary variations in the wave function. In practice it is then necessary to use what is called a " muffin-tin " potential, which is constant outside a sphere containing the atomic nucleus.

* These polyhedra must in no way be confused with Brillouin zones. The latter are in k-space and show the surface in k-space at which the energy changes discontinuously. The atomic polyhedra are in real space, and show how real space may be divided up so that each atom is surrounded by a similar cell.

The above method is closely related to an approach (the *Augmented Plane-Wave* Method) which has gained greatly in importance since modern computers have been available to carry out the laborious work it involves, but which was first proposed by Slater in 1937.* It uses wave functions which are spherical waves inside the abovementioned sphere, joining on to plane waves outside it.

The details of all these methods and those in (b) above are too mathematical for the present book; suggestions for further reading will be found at the end of this Section and on p. 236.

(d) *The Tight-Binding Approximation.*

Before recent developments made the application of some of the above techniques to real band structures a practical possibility, a great deal of work had been carried out with a method which has much less claim to exactness but which provides a reasonably simple technique of approximating the energies of states at different points in the Brillouin zone. This is the Tight-Binding Approximation, the name of which is unfortunate because it does not mean that the atoms are tightly bound, but rather that they are so far apart that the electronic wave functions resemble those of the free atoms. In terms of Fig. V.19 this approximation takes into account the singularities round each atomic nucleus, and the wave functions can be built up from atomic wave functions with the perturbations which become greater on passing from the K to the L to the M ... electrons. The method satisfies the Bloch condition and it may be said that whereas the nearly-free-electron approximation takes the smooth curve $ABCD$ of Fig. V.19 as the fundamental function from which the behaviour at 1, 2, 3 ... is approximated by perturbation methods, the tight-binding approximation regards the singularities at 1, 2, 3 ... as the fundamental functions from which the behaviour in regions such as xy is a slight perturbation. The method is analogous to the molecular-orbital theories of the H_2 molecule, which regard the wave functions for the molecule as perturbations of those of the free atoms.

This approximation is in principle not suitable for cases where appreciable overlap of the wave functions on neighbouring atoms occurs. It has been suggested, for example, that it is unsuitable for the treatment of d electrons in transition metals, especially those of the $4d$ and $5d$ series, but the results of a number of applications to $3d$ elements accord reasonably well with those of more sophisticated methods, especially when it is combined with a nearly-free-electron treatment of the s electrons in a way that takes proper account of the hybridization between d and s states.

* J. C. Slater, *Phys. Rev.*, 1937, **51**, 846.

The Brillouin-Zone Theory of Metals 241

Heitler–London–Heisenberg Functions.

The method involving the use of Heitler–London–Heisenberg (H.L.H.) functions resembles that of the valence-bond theory of molecular structure. In these methods, the wave function for each electron is localized round an individual atom or pair of atoms, but exchange of electrons occurs, so that the electrons may move freely through the lattice as with Bloch functions. The electrons are considered two by two in pairs of opposite spin, and the correlation of the electrons in a pair is considered. There is also a partial correlation throughout the crystal, because, in considering any one pair of electrons, the other electrons are in different regions.

The H.L.H. functions may be regarded as derived from atoms in different states or with different spins, but since resonance is assumed to occur, and we cannot look within the resonance process, the atoms in the three normal metallic structures are usually regarded as being in the same state, as in the Bloch theory. In abnormal structures, such as that of α-manganese, the atoms are regarded as being in different states, and in principle there is no reason why superlattices of atoms in different states should not exist in the more normal metals.

In developments based on H.L.H. functions, there are no quantitative methods corresponding to the free-electron, tight-binding, or cellular methods of the Bloch theory, and the H.L.H. functions for a crystal cannot be calculated, even in approximate form. It is these functions that underlie the Pauling concept (p. 162) of the metallic bond as an unsaturated co-valent bond, in which the number of electrons per atom is insufficient to form co-valent bonds between an atom and all of its neighbours, so that resonance occurs between the different configurations which the available electrons might adopt. If by " theory " we mean the calculation of physical properties from fundamental principles, with the minimum number of assumptions, and without introducing numerical constants deliberately chosen to fit the facts, the H.L.H.–Pauling approach to metallic structure is less satisfactory than that based on Bloch functions. The advantage of the resonating-bond concept is that metallic behaviour of an element can be related to a considerable understanding of its chemical behaviour. The tetrahedral environment of each carbon atom in diamond is obviously related to the tetrahedral arrangement of hydrogen atoms around a carbon atom in the methane molecule. Furthermore, the ability of both copper and gold to show valencies in chemical combination greater than unity warns one that to regard all electrons in the elemental metals except the outermost s electrons as not involved in the bonding processes is likely to be misleading. This intuitive seizing on valency as a characteristic of an atom may often be the better way of trying to interpret, understand, and generalize the facts, but it is important that interpretations of this kind shall not be regarded as theories, when they have in fact made an array of assumptions, including numerical values, deliberately chosen to agree with the numerical values of the physical properties they are interpreting.

For insulators, there is general agreement that the collective-electron theory on the one hand, and the H.L.H. approach on the other, are merely different ways of looking at the same thing, and that each approach if carried sufficiently far would lead to the same picture. The crystal of the diamond, for example, may be regarded as a structure in which the electrons are described by Bloch functions, and the four electrons per atom completely fill a zone surrounded everywhere by an energy gap. It may also be regarded as a co-valent crystal structure in which the electrons are described by H.L.H. functions derived from the (sp^3) hybrid orbitals of the carbon atoms.

It can be shown that the use of H.L.H. functions derived from neutral atoms results in an insulator, and that for conductivity to occur it is necessary to introduce functions derived from ionized states. In the case of metallic sodium, for example, an insulator would result if the valency electrons were described by H.L.H. functions derived from neutral atoms in each of which the one valency electron resonated between the directions of the bonds to the eight nearest neighbours. In order to permit conductivity, functions must be derived also from ionized Na^+ and Na^- ions, and the electrons and ions must be widely separated on the atomic scale.

There is a difference of opinion as to whether both Bloch and H.L.H. functions may be used for conductors. According to Mott,* the separation of the electron and ion necessary to produce conductivity would result in an increase in energy, so that the process would not occur. Others consider that this increase in energy might be counterbalanced by a loss in energy resulting from resonance of configurations in which the electron and ion were not widely separated. For the moment it seems better to accept the possibility that for many conductors both approaches may be used for qualitative discussion. There are, however, some substances for which this view gives rise to difficulty. For example, nickel oxide, NiO, which crystallizes in the sodium-chloride structure (Fig. III.16, p. 115), is an insulator. If the substance is a salt of divalent nickel, the Ni^{++} ions will have the $(3d)^8$ configuration in which two electrons are missing from a completely filled sub-group,

* N. F. Mott, " Progress in Metal Physics," Vol. III, p. 76. **1952**: London (Pergamon Press).

$(3d)^{10}$. Mott,* therefore, argues that in this class of substance the $(3d)$ electrons must be described by non-conducting H.L.H. functions, and that the use of Bloch functions is not possible, since they would require conduction from the incompletely filled $(3d)$ band. This has led to a later suggestion by Mott and Stevens † that wave functions of electrons in solids should be divided into those which are conducting and those which are non-conducting, and that this is the real distinction, rather than the concept of bonding and non-bonding orbitals assumed by Pauling (see p. 360).

It should be noted that the concept of a Fermi surface is applicable only to theories of the Bloch type in which the wave function extends over large distances on the atomic scale. In this case the characteristics of the electrons in the Fermi surface are so closely related to many of the electrical and magnetic properties of the metal, that the Fermi surface can almost be regarded as a physical property of the metal.

More detailed discussions, at a fairly simple level, of some of the material of this Section can be found in:

S. Raimes, " The Wave Mechanics of Electrons in Metals." **1961**: Amsterdam (North-Holland Publishing Co.).
J. M. Ziman, " Electrons in Metals." **1963**: London (Taylor and Francis).

APPENDIX I.

CRYSTAL STRUCTURES AND BRILLOUIN ZONES.

1. *The Crystal Lattice.*

IN the previous chapters we have described the behaviour of electrons in a periodic field in the simplest possible way, related directly to the diffraction of electrons by the crystal lattice. This led to a description in terms of wave-number space or k-space, in which, in general, the increase in wave number of an electron corresponded with an increase in energy, the increase being continuous within the Brillouin zones, and discontinuous at the zone boundaries. This is sometimes referred to as a description in terms of *extended wave-number space*, and is the most useful way of visualizing the qualitative characteristics of electrons in metals where the number of valency electrons per atom is less than two. For more quantitative purposes, the theory is often expressed in other ways and, to appreciate these, the concept of a reciprocal lattice must be understood.

Every crystal structure may be regarded as built up by the three-dimensional periodic repetition of identical units. These units may be either single atoms, or groups of atoms, just as a wall-paper may involve the periodic repetition of single circles, or of more complicated units. The periodic repetition may be described in terms of a *space lattice,* which is an array of points such that, except at the surface of the crystal, each point has an identical arrangement of surrounding atoms. Every periodic arrangement may be described in terms of the repetition of a primitive cell which contains one unit. Fig. V.20 shows part of a lattice of this kind and in the simplest example there is one atom at each corner of the unit cells of

* N. F. Mott, " Progress in Metal Physics," Vol. III, p. 76. **1952**: London (Pergamon Press).
† N. F. Mott and K. W. H. Stevens, *Phil. Mag.*, 1958, **2**, 1364.

side a, b, c. In the general case, each corner is surrounded by an identical cluster of atoms, and the corners themselves need not be occupied by atoms. If O is taken as origin, the passage from O to P involves n_x steps of a along the x axis,

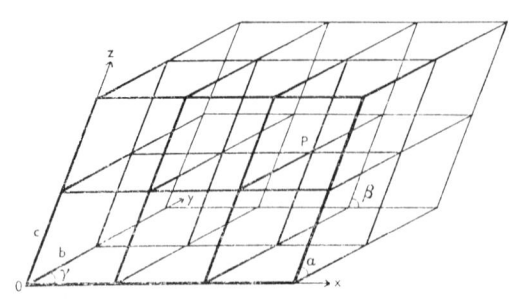

Fig. V.20.

n_y steps of b along the y axis, and n_z steps of c along the z axis. OP is thus a vector : *

$$n_x\boldsymbol{a} + n_y\boldsymbol{b} + n_z\boldsymbol{c}$$

The direction of the line OP may then be described by the set of numbers $n_x n_y n_z$ reduced to the smallest possible integers. In the example shown, OP has direction [211], where a square bracket is used to distinguish the indices from those used to specify particular planes, as described below.

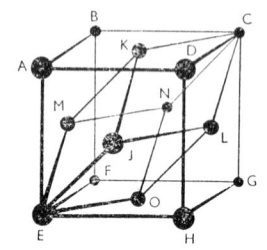

Fig. V.21. The face-centred cubic structure may be referred to the cubic cell $ABCDEFGH$ which contains 4 atoms, or to the rhombohedral cell $EMKJONCL$ which contains one atom.

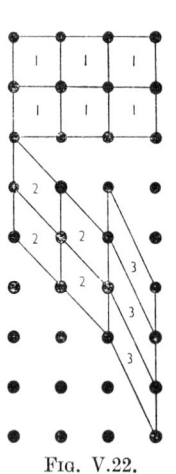

Fig. V.22.

The primitive cell is not necessarily the most convenient way of describing the periodic repetition. Thus, the face-centred cubic structure is generally described in terms of the unit of Fig. V.21 which contains 4 atoms. This emphasizes the cubic symmetry and if the edges of the cube are taken as axes, many of the equa-

---

* For an easy introduction to vector notation, the reader may consult " Theoretical Physics," by G. Joos. 3rd edn. **1958**: Glasgow and London (Blackie and Son, Ltd.).

The Brillouin-Zone Theory of Metals

tions for interplanar spacings and interatomic distances assume relatively simple forms.

Fig. V.21 shows how the face-centred cubic structure may be described in terms of a primitive rhombohedral cell with interplanar angle of 60°, and containing one atom per cell. This, however, is not a unique primitive cell, for as can be seen from the two-dimensional square array of Fig. V.22, the latter pattern can be regarded as resulting from the repetition of the square units 1, 1 . . . , or of the oblique units 2, 2 . . . , 3, 3. . . . In the same way, the three-dimensional lattice of Fig. V.20 may be described in terms of the cell which is drawn, or of others which are more oblique. It is usual to choose the primitive cell so that its shape reveals the highest symmetry of the lattice.

The direction of a plane is generally shown by its Miller indices, which are related to the lattice as shown in Fig. V.23. In this case a, b, and c are the lengths of the sides of the unit cell, and the plane shown cuts the axes which are parallel to the sides of the cell so that the intercepts on the x, y, and z axes are respectively 3 steps of a_1, 2 steps of b_1 and 1 step of c. The Miller indices of the plane are then the fractions:

$$\frac{1}{3}, \frac{1}{2}, \frac{1}{1}$$

reduced to the smallest integers, i.e., (236). The corresponding parallel plane on the opposite side of the origin would have negative intercepts on the axes, and be denoted $(\bar{2}\bar{3}\bar{6})$. In this notation, planes with indices $(hk0)$ cut the z axis at an infinite distance from the origin, and are thus parallel to the z axis.

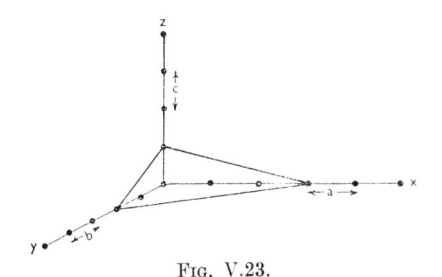

Fig. V.23.

Using the above systems, it follows that in cubic crystals a line with indices $[hkl]$ is perpendicular to a plane with indices (hkl), but this does not apply to crystals of other symmetries.

2. The Reciprocal Lattice.

In the above description the periodic repetition of a unit is in terms of the unit cells of a lattice in real space. For some purposes it is convenient to introduce the concept of reciprocal space or reciprocal lattices, and the underlying ideas may be summarized as follows.

We may suppose that Fig. V.24 shows a primitive unit cell, of sides a, b, and c. The *vector product* † of the vectors $\boldsymbol{a}$ and $\boldsymbol{b}$, which is written:

$$\boldsymbol{a} \times \boldsymbol{b}$$

is a vector directed perpendicular to the $\boldsymbol{ab}$ plane, and is of magnitude $ab \sin \gamma$, and this is equal to the area of the bottom face of the cell.

The volume of the unit cell is equal to the area of the base multiplied by the height, and is thus equal to $\boldsymbol{a} \times \boldsymbol{b}$, multiplied by the $\boldsymbol{c}$ interplanar spacing. If,

---

† See footnote to p. 244.

therefore, *ML* is drawn perpendicular to the *ab* plane, and meets the top face of the cell at *L*, the volume of the unit cell is equal to $a \times b$ multiplied by the length of *ML*, or in vector notation $a \times b \cdot c$.

We may now draw a new three-dimensional diagram in which a vector c^* is parallel to *ML*, and has a length equal to:

$$c^* = \frac{1}{ML} = \frac{1}{c \cos \theta}.$$

The length of the vector c^* in the new diagram is, thus, equal to the reciprocal of the *c* interplanar spacing of the crystal lattice. By definition, the *scalar product*

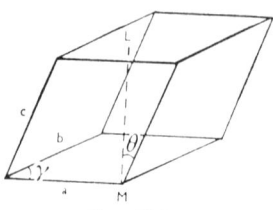

Fig. V.24.

of two vectors x and y is equal to the product of the magnitudes of the two vectors multiplied by the cosine of the angle between them, and is written $x \cdot y$. It follows therefore that c and c^* are defined by the equation:

$$c \cdot c^* = 1$$

and also by the equations:

$$c^* \cdot a = c^* \cdot b = 0$$

since *ML* in Fig. V.24 is perpendicular to a and b, and cos 90° = 0.

In the same way, we may define corresponding vectors b^* and a^*, whose directions are perpendicular to the *ac* and *bc* planes of the crystal, and whose lengths are the reciprocals of their interplanar spacings.

The unit cell of Fig. V.24 is only one of the large number of primitive cells to which the structure can be referred (Fig. V.22, p. 244), and a similar construction can be carried out for each of these. When this is done, each set of parallel planes of indices (*hkl*) in the crystal lattice is represented in the reciprocal lattice by a point whose distance from the origin is given by the vector translation ($ha^* + kb^* + lc^*$) and these points form a lattice known as the *reciprocal lattice*. This has the property that the vector $r^*(hkl)$ from the origin to the point (*hkl*) of the reciprocal lattice is normal to the (*hkl*) plane of the crystal, whilst the length of the vector $r^*(hkl)$ is equal to the reciprocal of the (*hkl*) interplanar spacing in the crystal.

As explained above, the length of the vector c^* in the reciprocal lattice is equal to the reciprocal of the *c* interplanar spacing of the unit cell. The length of the vector $2c^*$ in the reciprocal lattice corresponds to an interplanar spacing of $\frac{1}{2}c$, and the complete reciprocal lattice, therefore, shows not only the actual interplanar spacings in the crystal, but also spacings equal to $\frac{1}{2}$, $\frac{1}{3}$, $\frac{1}{4}$ . . . of these. This is of interest because, since the points in the reciprocal lattice give the directions of the planes, and the interplanar spacings, simple geometrical constructions exist †
which express the Bragg conditions for reflection in terms of the reciprocal lattice.

---

† For the geometrical aspect of reciprocal-lattice theory, the reader may consult " X-Ray Crystallography," by M. J. Buerger. **1942**: London (Chapman and Hall). The third edition of " Introduction to Solid-State Physics " by C. Kittel. **1966**: New York (John Wiley). London (Chapman and Hall), also contains much of interest.

The Brillouin-Zone Theory of Metals

When these are used, the reciprocal lattice point corresponding to an interplanar spacing of $\frac{1}{n}d$ is used to indicate the conditions for the nth order reflection from the actual crystal planes of spacing d.

The crystal lattice and the reciprocal lattice are mutually reciprocal with reference to each other. Just as the vector $r^*(hkl)$ from the origin of the reciprocal lattice is normal to the (hkl) planes of the crystal lattice, so the vector $r(hkl)$ from the origin of the crystal lattice is normal to the corresponding plane in the reciprocal lattice, and the length of the vector $r(hkl)$ in the real lattice is equal to the reciprocal of the interplanar spacing in the reciprocal lattice.

In cubic lattices, the vectors a^*, b^*, c^*, of the reciprocal lattice are parallel to the vectors a, b, c of the crystal lattice, and directions of the same indices [hkl] in the two lattices are parallel. In non-cubic crystals the relations are of the type explained above in connection with Fig. V.24.

In Fig. V.25 the upper part shows a simple two-dimensional square lattice. The directions of the (10), (31), (21), (32), (11), and (01) rows of points are marked and the interlinear spacings of these, and the directions of the normals are indicated. The lower part of the figure shows how the two-dimensional reciprocal lattice is related to these directions and spacings.

When this kind of construction is applied to the three-dimensional case of the face-centred cubic structure, the resulting reciprocal lattice is a body-centred cube. This can be understood from Fig. V.26, which shows that the interplanar spacing of the (111) octahedral planes is $\frac{a\sqrt{3}}{3} = 1 \bigg/ \frac{\sqrt{3}}{a}$, whilst the (200) spacing is $\frac{a}{2} = 1 \bigg/$

From Fig. V.27 it will be seen that if the reciprocal lattice is a body-centred cube of side $2/a$, the length of the vector from a corner to the centre is $\frac{\sqrt{3}}{2} \times \frac{2}{a} = \frac{\sqrt{3}}{a}$ and is thus the reciprocal of the interplanar spacing of the (111) planes in the face-centred cube. The distance between adjacent corners of the cube of Fig. V.27 is $2/a$, and is thus the reciprocal of the (200) interplanar spacing in the face-centred cube. Analogous methods can be used to show that, for a body-centred cubic structure, the reciprocal lattice is a face-centred cube.

3. Brillouin Zones and the Reciprocal Lattice.

In the preceding section we have described the relation between the reciprocal lattice and the crystal lattice, and have illustrated this by reference to the lattices of actual two-dimensional and three-dimensional structures. The underlying ideas are, however, independent of the existence of actual lattices, and we may imagine any representation in real space to be accompanied by a corresponding diagram in reciprocal space. If this is done, any vector drawn from the origin of reciprocal space will have a length equal to the reciprocal of the distance between a set of equidistant planes in real space, and the normal to these planes will be parallel to the corresponding vector drawn from the origin of reciprocal space.

In the description of the motion of electrons, the underlying reality is the probability wave function Ψ, and the associated wave-length λ is a vector in real space. The wave number $k = 2\pi/\lambda$ is a vector whose magnitude is equal to 2π multiplied by the reciprocal of the wave-length, and it may therefore be represented in reciprocal space. To illustrate the principles, we may consider the simple case of a two-dimensional square lattice. The reciprocal lattice of this is also a simple square lattice, and Fig. V.28 represents the assembly of points round the origin 0.

If a is the lattice spacing of the square lattice, the length of the line $0a_1$, in Fig. V.28, is equal to the reciprocal of the (10) interlinear spacing ($= 1/a$), so that if we regard Fig. V.28 as a wave-number diagram in reciprocal space, a_1 represents the point $k = 2\pi/a$. The point c which is mid-way between 0 and a_1 represents the wave number $k = \pi/a$, and the corresponding wave-length is:

$$\lambda = \frac{2\pi}{k} = 2a.$$

The condition for the first-order Bragg reflection with the incident ray at 90° from a set of lines of spacing d, is:

$$n\lambda = 2d \sin \theta$$

which for an interlinear spacing of a is the same as

$$\lambda = 2a.$$

The point c thus represents the electron state for which the wave-length satisfies the condition for the first-order reflection from the (100) lines of the square lattice,

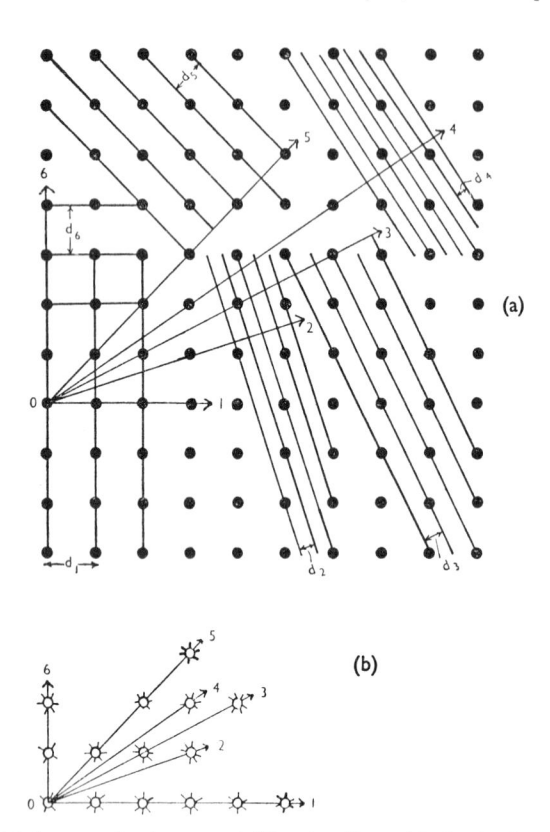

Fig. V.25 (a) shows a simple square lattice, together with 6 sets of parallel lines. The normals to these lines from the origin 0 are numbered 1, 2, 3, 4, 5, 6, and the interlinear spaces are marked d_1, d_2 . . .

Fig. V.25(b). The lines 01, 02 . . . are parallel to the corresponding lines 01, 02 . . . in (a). The distance from 0 to the first starred point on 01 in (b) is proportional to $1/d_1$ in (a), and the other starred points along 01 repeat this interval. The distance from 0 to the first starred point on 02 in (b) is proportional to $1/d_2$ in (a), and similar relations exist for the starred points along 03, 04, 05, and 06 in (b).

The starred points in (b) form the two-dimensional reciprocal lattice of the square lattice in (a).

and the point a_1 represents the electron state satisfying the condition for the second-order reflection.

We may now suppose that f is the mid-point of the diagonal $0b_1$. Then $0f = (0c)/\sin 45°$, and the corresponding wave-length is:

$$\lambda = \frac{2\pi}{k} = 2a \sin 45°.$$

This is the wave-length for the Bragg reflection of a ray inclined at 45° to the (10) lines of the lattice, and since this is a square array, the direction of $0f$ is that of a

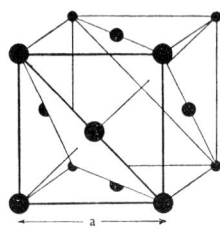

FIG. V.26. 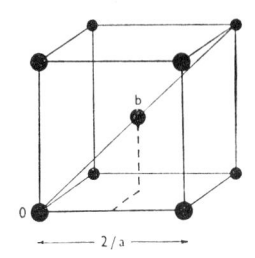 FIG. V.27.

ray at 45° to the (10) lines. In view of the symmetry of the square array, the wave-length corresponding to the point f also satisfies the Bragg equation for reflection from the (01) lines of the square lattice. Simple geometrical constructions of this kind show that the wave-lengths satisfying the Bragg equation for reflection from the (10), (01), ($\bar{1}$0), and (0$\bar{1}$) lines of the lattice lie on the dotted square $fghj$, and this forms the first Brillouin zone for the two-dimensional square lattice. This zone contains one point of the reciprocal lattice, and one electron state (*i.e.*, two electrons, one of each spin) per atom.

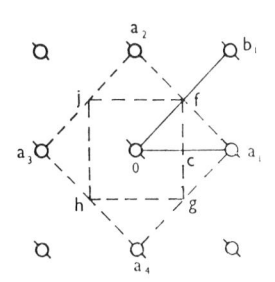

FIG. V.28.

Similar geometrical reasoning will show that the line a_1a_2 which bisects the vector from the origin of reciprocal space to the second-closest set of reciprocal lattice points, contains the states whose wave-lengths satisfy the Bragg equation for reflection from the (11) set of lattice lines. The second Brillouin zone comprises the regions of k-space lying between the first zone and the area $a_1a_2a_3a_4$. The four small sections of this zone can be pieced together to make an area of the same shape as the first zone, and the second zone, like the first, contains one state (*i.e.*, two electrons, one of each spin) per atom. This is a general characteristic of all Brillouin zones when properly constructed, *provided that the primitive unit cell of the structure contains only one atom.*

250 *Atomic Theory*

4. Three-Dimensional Brillouin Zones.

To obtain the Brillouin zones in three dimensions we repeat the construction of the previous section, using planes instead of lines. The crystal structure is first referred to its primitive lattice, and from this the reciprocal lattice is drawn in three dimensions. The origin of reciprocal space is then joined by straight lines to the nearest and next-nearest points of the reciprocal lattice, and planes are drawn to bisect these lines. The first Brillouin zone is then the smallest polyhedron which these planes form round the origin. In some structures this polyhedron results from the planes bisecting the lines to the nearest neighbours only; whilst for other structures (see p. 235 below) both nearest and next-nearest neighbours must be used.

When constructed in this way, geometrical arguments * show that the wave numbers on the faces of the zone correspond to wave-lengths for which the Bragg

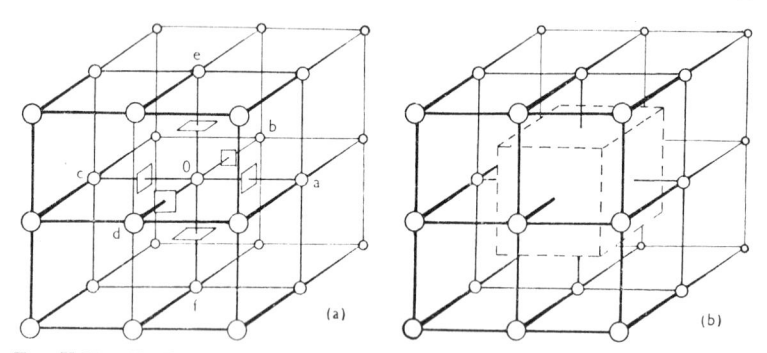

Fig. V.29. To illustrate the construction of a Brillouin zone in reciprocal space. (a) Shows the reciprocal lattice for a simple cubic structure. The reciprocal lattice is itself a cubic lattice and 0 is the origin. Planes are drawn bisecting the lines joining 0 to the nearest points (a, b, c, d, e, f) on the reciprocal lattice. As shown in (b), these planes intersect to enclose a cube surrounding the origin, and this cube is the first Brillouin zone.

equation is satisfied. If the primitive unit cell contains n atoms, the first Brillouin zone constructed as above contains $1/n$ electron states per atom which can accommodate $2/n$ electrons per atom. The volumes between the boundaries of the first and second zones can always be pieced together to form a volume of the same size and shape as the first zone, and so all zones contain the same number of electron states.

The Simple Cubic Lattice.

For the simple cubic lattice, the reciprocal lattice is also a simple cube, and Fig. V.29 shows how the above construction gives a first Brillouin zone which is itself a cube.

The Face-Centred Cubic Structure.

As explained above, the face-centred cubic structure gives a reciprocal lattice which is a body-centred cube. In this, the distances between the closest and second-closest neighbours do not differ greatly, and the above construction gives the Brillouin zone of Fig. V.5 (p. 202) which is a truncated octahedron. In this, the octahedral faces result from the planes bisecting the lines joining the origin of the

---

* See " Introduction to Solid-State Physics," by C. Kittel, 3rd edn. **1966**: New York and London (John Wiley).

The Brillouin-Zone Theory of Metals

reciprocal lattice to the 8 nearest points, whilst the small cube faces are parts of the planes bisecting the lines joining the origin to the 6 second-nearest points of the reciprocal lattice. This is the zone described previously (Fig. V.5), and it again contains one electron state (*i.e.*, two electrons of opposite spin) per atom.

A gradually expanding spherical Fermi surface first touches the zone boundary at the centres of the octahedral faces at an electron concentration of 1·36. If the Fermi surface continues to expand spherically inside the first zone, with no overlap into the second zone, it will touch the (200) cube face of the zone at an electron concentration of 1·88.

The Body-Centred Cubic Structure.

For the body-centred cubic structure, the reciprocal lattice is a face-centred cube, and the origin is surrounded by 12 equidistant points. The first Brillouin zone obtained by the above construction is thus the rhombic dodecahedron of

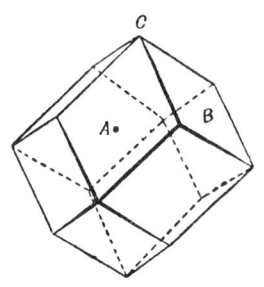

[*Courtesy Clarendon Press*]

FIG. V.30. The First Brillouin Zone of the body-centred-cubic lattice. The zone is a dodecahedron; *A* is the centre of a face, and is one of the points on the surface which is nearest to the origin; this is therefore one of the points where a spherical Fermi surface first touches the zone boundary. It should be noted that the corners of the zone are not all equivalent; the corner *C* lies at a greater distance from the centre of the zone than the corner *B*.

Fig. V.30. The zone again contains one electron state (*i.e.*, two electrons, one of each spin) per atom, and an expanding spherical Fermi surface first touches the face of the zone at an electron concentration of 1·48.

The Close-Packed Hexagonal Structure.

Unlike the face-centred and body-centred cubic structures, the close-packed hexagonal arrangement cannot be described in terms of a unit cell containing one atom. In such cases, if the smallest unit cell contains n atoms, the construction given above (*i.e.*, the mid-way planes in reciprocal space) gives a first Brillouin zone containing $1/n$ states per atom, which can accommodate $2/n$ electrons per atom.

For the close-packed hexagonal structure, $n = 2$, and the first Brillouin zone is of the form of Fig. V.31(*a*), and contains ½ state per atom, which can accommodate one electron per atom. The boundaries of the second zone are shown in Fig. V.31 (*b*). The second zone is the volume contained between these two polyhedra, and again contains ½ state per atom, so that the total volume enclosed within the planes of Fig. V.31(*b*) contains one state which can accommodate two electrons per atom.

For the first zone of Fig. V.31(*a*), the energy gap across the top and bottom faces

252   *Atomic Theory*

is zero, and this led Jones to draw a "first zone" of the form of Fig. V.31(c). This zone has certain curious characteristics which are described in connection with the ζ-phases of copper and silver alloys (p. 318).

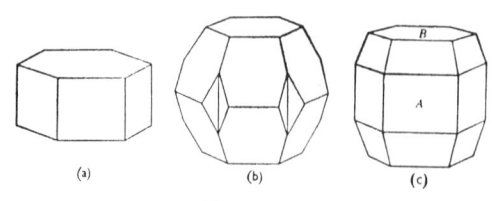

FIG. V.31.

The terminology in the literature is very confusing, and the term "first Brillouin zone" has been used to describe the polyhedra of all three types (Fig. V.31(a), (b), and (c)). It is becoming the practice to define the first Brillouin zone as that constructed as shown above, and in this case the zone of Fig. V.31(a) is the first zone for the close-packed hexagonal structure.

5. *Extended and Reduced Zones.*

In the above description, the construction described on p. 250 has been used to build up the first, second, third . . . Brillouin zones. With increasing wave numbers these zones extend outwards in k-space, and in the nearly-free-electron approximation the E/k relations are of the form shown in Fig. V.32(a), and involve a continual extension along the k-axis. An alternative method of showing this relation is given in Fig. V.32(b), in which the curve BOA of Fig. V.32(a) is unchanged and is marked $n = 1$ to show that it refers to the first zone or band. The curves

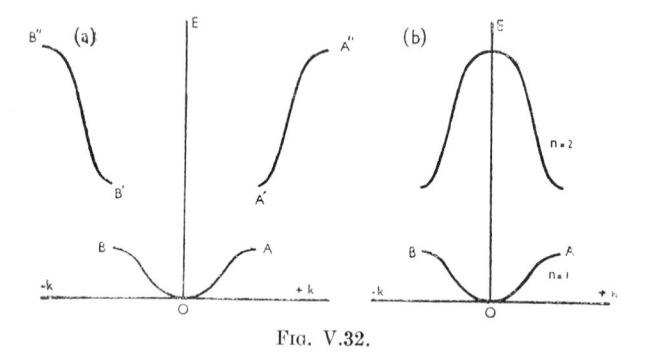

FIG. V.32.

$B'B''$ and $A'A''$ are then displaced as shown and join together to form the E/k curve for the second band marked * $n = 2$. We have already seen that points such as A'' and B'', where the E/k curve is horizontal, represent states for which the mean velocity of the electron is zero (p. 209), and the representation of Fig. V.32(b) serves to emphasize that the points A'' and B'' of Fig. V.32(a) represent the same electron state.

---

* The integers $n = 1$ and $n = 2$ enter as quantum numbers in the solution of the wave equation.

The Brillouin-Zone Theory of Metals

In the same way, instead of drawing successive Brillouin zones extending farther and farther outwards into k-space, it is possible to use a reduced zone notation in which the outer zones are transferred to the space within the first zone. Suppose, for example, that Fig. V.33(a) shows the first three zones for a simple two-dimensional square lattice in ordinary or extended k-space, with the energy contours as shown, and that the different regions of the second zone are numbered 1, 2, 3. . . . These regions may then be transferred, as in Fig. V.33(b), so as to lie within the square of the first zone, in which case the energy contours will take the form shown. Fig. V.33(c) illustrates how the portions $abcdefgh$ of the next part of the extended zone may also be placed within the square of the reduced zone.

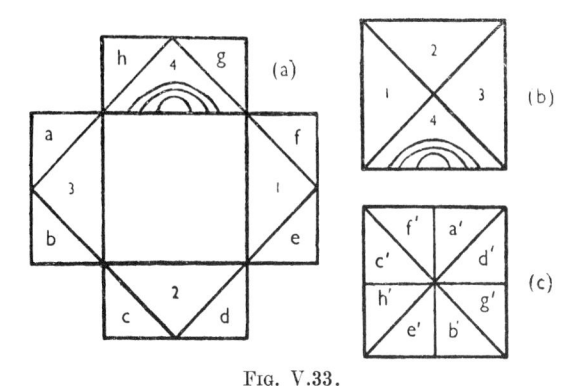

Fig. V.33.

If this method is adopted, the fundamental zone is referred to by some writers as *the* Brillouin zone, and, when properly constructed, it is a characteristic of all the outer Brillouin zones that they can be pieced together so as to fit into the fundamental Brillouin zone. The methods of extended zones, and of reduced zones are completely equivalent, and any description in terms of the reduced-zone scheme can be expressed in terms of extended zones, and vice versa. The reduced-zone scheme is very convenient for some purposes. When large numbers of electrons per atom are involved, the Fermi surface may extend into a number of zones, and the reduced-zone scheme is often most valuable in showing how the different small volumes can be pieced together, and regarded as representing one group of electrons.

6. Brillouin Zones and Jones Zones.

The Brillouin zones set up by the method of p. 250 are clearly related to the interplanar spacings of the crystal, and in the early days of the theory it was noted by Jones that the zone faces corresponded to crystal planes with high structure factors. For example, in the truncated octahedron (Fig. V.5, p. 202) which is the first Brillouin zone for the face-centred cubic structure, the octahedral and square faces of the zone correspond with the satisfying of the Bragg equation for reflection from the (111) and (200) planes of the crystal, and these are of high structure factor, and give rise to prominent diffraction lines on X-ray or electron-diffraction films. It can be shown that, in general, a high structure factor (S) corresponds with a large energy gap at the zone boundary, and in the nearly-free-electron approximation the energy gap is proportional to S. Jones, therefore, adopted the policy of constructing zones so that all planes with $S = 0$ were omitted, and the zones were built up as far as possible from planes with large values of S. Such zones may be called Jones zones.

For many simple structures, the Jones zones of lower number are identical with the Brillouin zones of the extended-zone scheme, but differences and difficulties may arise if the structure contains screw axes or glide planes of symmetry, because a zero structure factor does not necessarily imply a zero energy gap.

In contrast to the true Brillouin zones, the Jones zones do not always contain one electron state (*i.e.*, two electrons, one of each spin) per atom. The Jones zones are, however, of great value in studying many qualitative principles underlying the properties and structures of metals and alloys, because the general correspondence between large energy gaps and large structure factors means that the boundaries of the Jones zones correspond with large breaks in the E/k curve.

APPENDIX II.

ORTHOGONAL FUNCTIONS.*

Two vectors are orthogonal when they are at right angles to one another, and their scalar product (p. 246) is zero. In this case, since $\cos 90° = 0$, neither vector has a component in the direction of the other, and we may say that neither vector can contribute to the other. A three-dimensional vector a can be expressed in terms of its components a_x, a_y, a_z parallel to the x, y, and z axes, where $a_x = a \cos \theta_x$, if θ_x is the angle made by a with the x axis, and similarly for a_y and a_z. In this case the condition for two vectors being orthogonal is that:

$$a_x b_x + a_y b_y + a_z b_z = 0.$$

If we imagine a space of N dimensions, two vectors are orthogonal if:

$$a_1 b_1 + a_2 b_2 + \ldots a_n b_n = \sum_{i=1}^{n} a_i b_i = 0.$$

If this condition is satisfied, we may again say that neither vector can contribute to the other.

A function can be regarded as being built up from an array of vectors, and if the function is continuous there is an infinite number of components. In this case if $f(x)$ and $g(x)$ are two functions of x, the functions are said to be orthogonal if:

$$\int f(x) g(x) dx = 0.$$

In this case we can say that the infinite number of vector components which build up the one function can make no contribution to, and cannot be expressed in terms of, those which build up the other functions.

4. THE EXPERIMENTAL EXAMINATION OF THE ELECTRONIC STRUCTURE OF METALS.

In the preceding sections we have described some of the developments of the electron theory of metals, and in Part VI we deal with the application of the theory to individual metals and alloys. Except for the alkali metals, a purely theoretical approach is generally too difficult to yield an unambiguous solution for a particular metal. It is, therefore, necessary to gain as much information as possible from the measurement of physical properties which are directly related to the electronic

---

* For further details the reader may consult:

H. Margenau and G. M. Murphy, " The Mathematics of Physics and Chemistry." **1956**: New York (D. Van Nostrand Co. Inc.).

F. Mandl, " Quantum Mechanics," 2nd edn., Chapter I. **1957**: London (Butterworth).

structure. In the present section, we shall review briefly some of the methods which have been developed for this purpose.

The different types of measurements fall into two distinct groups:

(a) those which provide only rather general information about the character of the energy bands and the occupied energy states, but which can be applied, at least for cubic materials, to polycrystalline samples and to alloys as well as to metals of very high purity:

(b) those which provide detailed information about the geometry of the Fermi surface and the character of the states comprising it, but which require very high-purity single-crystal samples to be examined in the liquid-helium temperature range.

For convenience we shall refer to these two groups as General Measurements and Fermi-Surface Measurements, respectively.

A. General Measurements.

(a) *X-ray Spectroscopy and Optical Properties: Band Widths and Positions.*

The general ideas of soft X-ray spectroscopy have been outlined in Section III.1. In principle, the method should permit the determination of the width of the band, and the form of the $N(E)$ curve, and should also distinguish between the proportion of the p fraction of a hybrid band on the one hand, and the s and d fractions on the other. In practice, the method has been less useful than had been expected, partly owing to the difficulty of preparing uncontaminated surfaces, and also because of the existence of secondary effects, some of which are referred to below. Helpful discussions have been published recently.*

The Form and Width of $N(E)$ Curves.

Using the notation of Skinner (p. 88), we denote by E_0 the energy of the lowest state of a metallic crystal, and by E_m the highest occupied state at the absolute zero. (See Fig. V.34.) When a vacancy has been

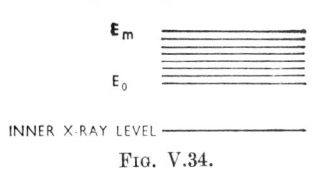

Fig. V.34.

created in an inner level, let $v^3 f(E)$ be the probability per unit time that an electron of energy E falls into an inner level. Let $I(E)dE$ be the

* " Soft X-Ray Band Spectra " (edited by D. J. Fabian). **1968**: New York and London (Academic Press).

intensity (measured as numbers of quanta per second) of the radiation in the emitted band in the range of quantum energies from E to $E + dE$. Then, if the energy of the inner X-ray level is taken as zero:

$$I(E) \propto \nu^3 f(E) \, N(E) \text{ if } E \leqslant E_m$$
and
$$I(E) = 0 \text{ if } E > E_m$$

Here $h\nu$ is the energy of the transition, and the factor ν^3 enters into all transition probabilities.

In the corresponding absorption process, the energy absorbed must raise the electron from an inner level to a state above E_m, since all states below E_m are occupied; we refer here to the absolute zero, and ignore the slight broadening of the edge. If μ_E is the partial absorption coefficient of a thin layer of the metal as a function of the quantum energy of the absorbed radiation, then:

$$\mu_E = 0, \text{ if } E < E_m$$
$$\mu_E \propto f(E) \, N(E), \text{ if } E > E_m$$

There should thus be a sharp edge on the high-energy (short-wavelength) side of the emission band, and on the low-energy (long-wave-

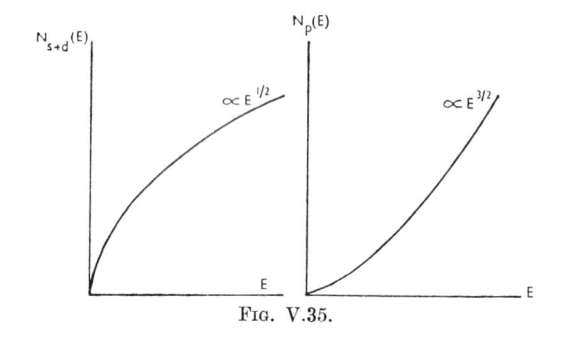

FIG. V.35.

length) side of the absorption band, and these sharp edges should be at the same wave-length. This is confirmed experimentally, and is unaffected by the presence of exciton states (see p. 259).

Fig. V.34 shows clearly that, since emission spectra result from electrons falling from the occupied portion of the valency band to the inner level, an emission spectrum can give information only about occupied states. Conversely, absorption spectra give information about unoccupied states. Both types of spectra are needed to give the complete $N(E)$ curve for the valency band. Further, since the electrons in the valency bands of real metals are in hybrid (sp) or (spd) states, both K and L spectra must be examined in order to obtain the proportions of p or of (s and d) functions in the hybrid.

The Brillouin-Zone Theory of Metals 257

The first difficulty in determining the band width lies in the accurate estimation of the low-energy limit of the band. In one-electron theories of the Bloch type, the $N(E)$ curve is such that, if E is measured from the bottom of a band, the initial portions of the $N(E)$ curves are of the form:

$$N_{(s+d)}(E) \propto E^{1/2}$$
$$N_p(E) \propto E^{3/2}$$

and are of the shapes shown in Fig. V.35 above. In practice, there is always some tailing off of the band at the low-energy end, and as the point $N(E) = 0$ cannot be established accurately, the policy adopted in much early work is to fit the main portion of the curve to one involving $E^{1/2}$ or $E^{3/2}$, and to extrapolate back to $E = 0$.

It is now recognized that the above relations may be quite invalid when electron correlation is taken into account, and for some metals the band widths are known only approximately.

It must be emphasized that, owing to the selection rules (p. 86), each

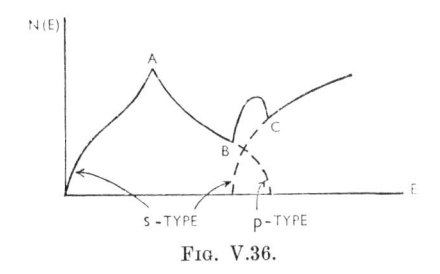

Fig. V.36.

emission spectrum can give information only about the appropriate part of a hybrid function. In a K-emission spectrum, an electron from the valency band falls back to a $(1s)$ state. The s part of the hybrid is, therefore, not revealed in a K-emission spectrum, which does, however, reveal the p part of the hybrid. If a particular $N(E)$ curve is of the form of Fig. V.36, with s states at the bottom of each zone, and p states at the top, the sudden rise at B results from the introduction of new s states, and will not be revealed by a K-emission spectrum, although the latter will reveal the kink at C, since this corresponds to the dying away of p states. It is, thus, necessary to use both K and L spectra in order to reveal the full $N(E)$ curve, and a single emission curve has, in general, a shape different from that of the total $N(E)$ curve.

Solid Solutions.

In some cases soft X-ray spectroscopy can give information about the electronic structure of solid solutions. In a binary solid solution, each kind of atom will have its own K level, and emission spectra may result from the outer electrons falling into both of these. In this way it may be possible to distinguish between solid solutions in which the two kinds of atom form a common valency band whose width increases with the number of valency electrons per atom, and those solid solutions which give rise to bound states whose electrons are associated only with solute atoms. Even where a collective-band scheme exists, however, the observed spectra will be modified because the character of the screening cloud of electrons around the two types of atom will differ.

Auger Transitions and Probabilities.

As explained on p. 86, the possible electronic transitions are controlled by selection rules, of which the most important is that the probability of a transition occurring is small, unless the quantum number l changes by 1. It is sometimes possible for an electron to fall from an energy level E_1 to a level E_2, so that the resulting energy is given to another electron instead of being emitted as radiation. In such a case the probability of the transition $E_1 \longrightarrow E_2$ may be large, even though it is a forbidden transition for the emission of radiant energy.

This process may lead to a broadening of energy states because if, for example, an electron is expelled from the bottom of the valency band, the resultant hole may be filled so quickly that the energy becomes uncertain (*i.e.*, is broadened) owing to the Heisenberg principle.

Satellite Lines and Perturbation Effects.

The ordinary X-ray emission spectra result from the return of an electron to the inner levels of singly ionized atoms. If electron bombardment results in the ejection of an electron from a particular atom, then for this atom the screening effect of the K-shell is reduced, and its

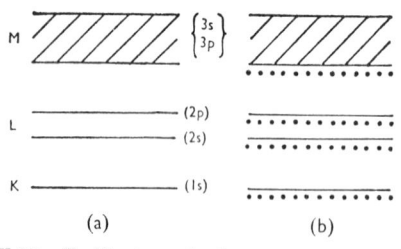

Fig. V.37. To illustrate the formation of exciton states.

outer electrons are held more firmly than in a normal atom. We have, thus, what is essentially a solid solution of an impurity atom of greater atomic number, and this produces discrete one-electron or *exciton* states below the normal states (Fig. V.37).

It has, therefore, been argued that, by their very nature, soft X-ray spectra always relate to a perturbed part of the lattice, and not to the normal lattice. Recent work by Catterall and Trotter † suggests that the above effect is unimportant, and cannot be responsible for some of the discrepancies between experimental results and theoretical calculation. Catterall and Trotter studied emission spectra from lithium and beryllium under conditions in which *satellite spectra* were produced from atoms in which the electronic bombardment resulted in the ejection of two electrons (instead of the normal one electron) from the K-shell. Had the error referred to above been serious, the satellite and normal spectra should have differed appreciably, but actually the satellite bands had the same shape, and almost the same width, as the main bands.

Except in special circumstances the ordinary optical properties of metals and alloys are even less easy to interpret, since the response of the conduction electrons to electromagnetic fields is affected by many factors, but in recent years efforts have been made to identify features of the absorption spectra with various types of interband transition and to trace the changes in these features with alloying. As with X-ray spectra, the excitations of an electron are governed by selection rules and an absorption may often be not to the lowest-energy empty state but to the lowest-energy state of a certain symmetry. However, our knowledge of the positions of the d bands relative to the Fermi surface in copper, silver, and gold is based on such measurements, and recent optical work‡ has enabled some definite statements to be made about the positions and widths of virtual bound states (see Section VI.6) on transition metals dissolved in silver.

(b) *Electronic Specific Heats, Paramagnetic Susceptibilities, and the Density of States.*

In Section IV.1 we have seen that, for the free-electron theory, the energy of the electron gas at the lower temperatures is given by the equation:

$$E = N[\tfrac{3}{5}\varepsilon^* + C\theta^2] \quad . \quad . \quad . \quad \text{V (13)}$$

where $\tfrac{3}{5}N\varepsilon^*$ is the energy of the assembly of N electrons at the absolute zero, and ε^* is the energy of the electrons in the Fermi surface. The specific heat of the electron gas c_v, being the derivative of this energy

† See J. A. Catterall, "The Physical Chemistry of Metallic Solutions and Intermetallic Compounds," Vol. I, paper 1J. 1959: London (H.M. Stationery Office).
‡ A. Karlsson, H. P. Myers, and I. Wallden, *Solid-State Commun.*, 1967, **5**, 971.

with respect to the absolute temperature θ, is thus proportional to θ, and it can be shown* that the relation can be expressed in the form:

$$c_v = \tfrac{2}{3}\pi^2 k^2 N(\varepsilon) \cdot \theta = \gamma\theta. \qquad \text{V (14)}$$

where $N(\varepsilon)$ is the density of states at the Fermi surface. Results are normally expressed in terms of the coefficient γ.

In the more general case, where the electrons move in the periodic field of the lattice, the energy of an assembly of N electrons is again the sum of their energy at the absolute zero, and that of the excitation of the electrons in the immediate vicinity of the Fermi surface, and equation V (14) still applies.

At room temperatures the electronic specific heat is only a small fraction of the specific heat, which all solids possess, arising from the lattice vibrations. At low temperatures, however, the latter varies as θ^3 so that the former heat which varies as θ is a larger fraction of the total. The electronic specific heat is therefore usually determined from measurements at low temperatures ($1 - 20°$ K), where the total specific heat divided by the absolute temperature has the form $A\theta^2 + \gamma$, and γ is related to $N(\varepsilon)$ by equation V (14). In principle, γ could be determined from high-temperature measurements where the lattice term is nearly constant and the electronic term still proportional to θ, but in practice this is difficult since the lattice term contains a small correction term linear in θ. A brief account of lattice specific heats is given in an appendix to this Section. (More detailed considerations show that a complete separation of electronic and lattice terms is not quite correct; a coupling between the two causes the values of $N(\varepsilon)$ derived from experimental γ values to be somewhat larger than the true ones.)

In Section IV.3 it was shown that free electrons contribute a paramagnetic susceptibility to metals which will be further discussed in Section VII.1. Simple theories show this to be proportional to $N(\varepsilon)$, but values of $N(\varepsilon)$ derived from it are, for reasons we shall discuss later, not generally in good agreement with those derived from γ; and the latter are to be preferred.

(c) *The Hall Coefficient and the Current Carriers.*

The passage of a current down a rectangular conductor in the presence of a transverse magnetic field causes a difference in potential

---

* See N. F. Mott and H. Jones, "Theory of the Properties of Metals and Alloys," p. 176; and J. C. Slater, "Introduction to Chemical Physics," p. 76.

The Brillouin-Zone Theory of Metals 261

to be generated between the sides of the conducting plate. The Hall coefficient r is defined by the relation:

$$r = \frac{Ed}{IH} \quad \cdots \cdots \quad \text{V (15)}$$

where E is the transverse potential difference, d the thickness of the plate parallel to the field of strength H, and I the total current. By convention, the coefficient is positive if the electric potential is raised on that side of the plate on which the Amperean current generating the magnetic field has the same direction as the current I (Fig. V.38).

The Hall effect is due to the drifting of the electric carriers in a direction perpendicular to the field, and to their own direction of motion

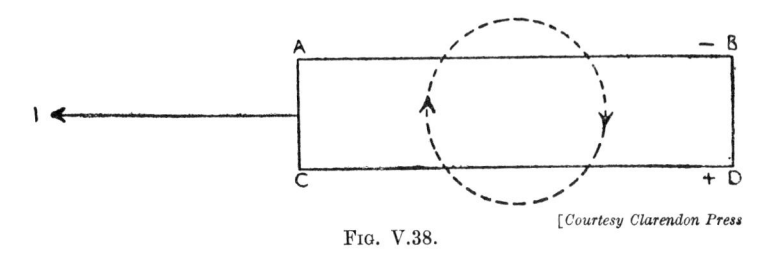

Fig. V.38.

[Courtesy Clarendon Press]

(Fig. V.38). If the electric carriers are free, and carry a charge e, the Hall coefficient * r is given by:

$$r = \frac{1}{eNc} \quad \cdots \cdots \quad \text{V (16)}$$

where c is the velocity of light and N is the number of carriers per unit volume.

For free electrons, the electronic charge e is negative, and the Hall coefficient has a negative sign. For univalent metals, the calculated and observed values are in good agreement, and this confirms that these metals conduct electricity by the motion of free electrons present in proportion of about one per atom.

For divalent metals, the Hall coefficient is positive, and this is taken to indicate that conduction by positive holes is the predominant effect. In such a case, where there is an overlap from the first into the second zone, the observed coefficient is the net result of the negative contribution of the relatively free electrons at the bottom of the second zone, and of the positive contribution from the holes at the top of the first zone. The expression for the Hall coefficient then becomes more complicated,

---
* This is sometimes called the Hall constant.

and involves the numbers and mobilities of the negative and positive carriers.*

For bismuth, the observed value of the Hall coefficient is negative and is about 250 times as great as the free-electron value; this implies that the effective number of charge carriers is only a very small fraction of the number of atoms, *i.e.*, that there is a very small Brillouin-zone overlap.

It is by considerations such as the above that measurements of the Hall coefficients are able to throw light on the electronic structure of metals. With metals of higher valency, where conduction occurs by means of both holes and relatively free electrons, the results are naturally complicated, and the relative mobilities can seldom be calculated. Even in such cases the study of alloys may reveal valuable information. If, for example, a metal A is alloyed with a metal B of higher valency with the formation of a solid solution, the onset of a Brillouin-zone overlap will cause the Hall coefficient/composition curve to bend in the direction of a more negative coefficient.

(d) *Nuclear Magnetic Resonance.*

General.

The finer details of the hydrogen spectrum are due to the electron possessing a spin angular momentum (p. 63) and an associated magnetic moment. The magnitude of the spin angular momentum is $\frac{h}{2\pi}\sqrt{s(s+1)}$, where $s = \frac{1}{2}$. The maximum component of this in the direction of an applied magnetic field is $\frac{h}{2\pi} m_s$, where $m_s = \pm \frac{1}{2}$, and the component of the magnetic moment in the direction of the field, which is the observed quantity, is equal to:

$$g \cdot \frac{e}{2mc} \cdot \frac{h}{2\pi} \cdot m_s$$

where, for an electron, the splitting factor $g = 2$, so that the magnetic moment, the Bohr magneton μ_B, is equal to $\frac{eh}{4\pi mc}$.

The further examination of the spectra of hydrogen, and of other elements, led to the discovery of hyperfine structures of certain spectral lines in the presence of a magnetic field. These arise from the fact that

---

* *For further reading see:*
N. F. Mott and J. Jones, " The Theory of the Properties of Metals and Alloys."
D. K. C. MacDonald and K. Sarginson, *Rep. Progress Physics*, 1952, **15**, 249.

The Brillouin-Zone Theory of Metals

the atomic nuclei possess spin angular momenta, and corresponding nuclear magnetic moments. The intrinsic angular momentum or spin of a nucleus is described by the nuclear spin number I, which may be an integer or a half-integer. Half-integral values are found for nuclei of odd mass number, whilst $I = 0$ for elements whose mass numbers and atomic numbers are both even.

The magnitude of the intrinsic nuclear angular momentum is equal to $\frac{h}{2\pi}\sqrt{I(I+1)}$, and its component in the direction of an applied magnetic field is equal to $\frac{h}{2\pi}$ m, where m can take values $I, I-1 \ldots -I$. Thus, for $I = \frac{3}{2}$ the possible values of m are $\frac{3}{2}, \frac{1}{2}, -\frac{1}{2}$, and $-\frac{3}{2}$. The relations between I and m are thus the same as those between s and m_s.

There is no way of calculating the magnetic moment of a nucleus, but by analogy with the expression for an electron the nuclear magnetic moment is regarded as having magnitude:

$$\frac{ge}{2Mc}\frac{h}{2\pi}\sqrt{I(I+1)}$$

and the component of the magnetic moment in the direction of an applied magnetic field, which is the observable quantity, is:

$$\mu = \frac{ge}{2Mc} \cdot \frac{h}{2\pi} \text{ m}$$
$$= g \cdot \frac{eh}{4\pi Mc} \text{ m } [\text{m} = I, I-1 \ldots -I].$$

Here M is the mass of a proton, m is a quantum number, and g is chosen so as to agree with the observed values. The quantity $\frac{eh}{4\pi Mc}$ is known as the *nuclear magneton*, and is a convenient unit of nuclear magnetic moments. The values of g are of the order of unity, and nuclear magnetic moments are thus of the order of $\frac{1}{2000}$ of a Bohr magneton, since M is 1840 times greater than the mass of an electron. It is important to note that the nuclear magneton is not the magnetic moment of the proton itself—the definition is only by analogy with the expression $\frac{eh}{4\pi mc}$ for the electron.

The Resonance Condition.

Under the influence of a magnetic field of strength H, the degenerate nuclear energy level is split into the $(2I+1)$ levels corresponding to

s

the values $I, I-1, \ldots -I$. The separation between these levels is equal to $g\mu_0 H$, where μ_0 is the nuclear magneton. If, therefore, transitions occur between adjacent levels and are accompanied by the emission or absorption of radiation, the frequency of the latter will be given by the Einstein relation:

$$h\nu = g\mu_0 H \quad \ldots \ldots \quad \text{V (17)}$$

We may imagine a beam of atoms to be passed through a magnetic field of strength H which is increased slowly. If the beam is submitted to electromagnetic radiation circularly polarized in a plane perpendicular to the magnetic field, transitions between the different sub-levels will occur when the frequency, ν, satisfies relation V (17). The magnitudes are such that with field strengths of the order of 10,000 gauss, the critical frequencies are in the radio-frequency range.

Analogous effects are produced if a beam of atoms is passed through a field of strength H, perpendicular to which a small oscillating field is superimposed. In this case strong resonance effects are observed when the frequency satisfies equation V (17), and the experiments may be made at constant H and variable frequency, or vice versa.

These nuclear-resonance effects are found not only for beams of free atoms, but also for atoms in solids or liquids. In such cases a given nuclear magnet is subject not only to the applied magnetic field H, but also to local fluctuating magnetic fields produced by neighbouring nuclei. In this case the resonance condition may be written:

$$h\nu = g\mu_0(H + H_{loc.}) \quad \ldots \ldots \quad \text{V (18)}$$

$H_{loc.}$ is of the order of 5–10 gauss, and as it is a variable quantity the resonance frequency is no longer sharp.

Spin-Relaxation Time.

For simplicity we may consider a system of identical nuclear magnets for which $I = \frac{1}{2}$. In the presence of a magnetic field H, there will be two spin states $m = +\frac{1}{2}$, and $m = -\frac{1}{2}$, and the magnets will tend to align themselves parallel to the field, and to occupy the state for which $m = +\frac{1}{2}$. The nuclear magnetic moments are, however, very small, and at room temperatures the thermal energy of a nucleus greatly exceeds the difference in energy between the two spin states, and the magnetic order is almost completely destroyed. The excess number of nuclei in the lower state can be calculated from the Boltzmann distribution and, according to Pake,* for protons in a field of 20,000 gauss at room temperature, the ratio of the number of nuclei in the lower state

* G. E. Pake, *Amer. J. Physics*, 1950, **18**, 438, 473.

The Brillouin-Zone Theory of Metals 265

to that in the upper is as 1·000,014 : 1 when thermal equilibrium has been reached.

We may now suppose that the field of 20,000 gauss is suddenly replaced by a very small field. The energy difference between the two states will then become vanishingly small, and the equilibrium condition will be that of equal population of the two states. The establishment of this equilibrium requires interchange of energy between the nuclear magnets and the surrounding atoms in the liquid or solid. The *time of relaxation*, T_1, is the time required for all but $1/e$ of the excess electrons in the upper state to reach the lower state. For solids, the values of T_1 are of the order of 10^{-2}–10^{-4} sec.

*Effects in Metals.**

The experimental investigation of nuclear magnetic resonance in metals is complicated by the fact that the radio-frequency magnetic field penetrates only a thin layer ($\sim 5 \times 10^{-3}$ cm) of the surface of the specimen, and finely divided materials have to be used. The following effects may be noted briefly :

Spin-Lattice Relaxation Times. Many of the mechanisms responsible for spin-lattice relaxation in non-metallic substances are also effective in metals. In particular, the presence of paramagnetic ion impurities may greatly reduce the value of T_1. In metals, however, a new process is encountered, which has been interpreted as due to an interaction between the magnetic energy sub-levels of the nuclei and the electron states of the conduction electrons.

When an electron passes near to a nucleus, the latter is subject to a relatively large magnetic field. A transition from one nuclear sub-level to another may be possible if the electron can change into a new state. This is possible only for electrons near the Fermi surface, and it can be shown that the probability W of such a transition is of the order of :

$$W = \frac{1}{2T_1} \sim \frac{E_1^2 kT}{E^2_{max.} h}$$

where $E_{max.}$ is the energy at the top of the Fermi level, and E_1 is of the order of the hyperfine splitting of the ground state of the metal atom.† This relation has received some experimental confirmation as regards the inverse proportionality between T_1 and T, and that between T_1 and E_1^2, but there are discrepancies between some of the data.

Electrons responsible for relaxation effects are mainly of *s*-type, because these have the highest density near the nucleus. From the

---
* For a very thorough review of nuclear magnetic resonance see L. E. Drain, *Met. Rev.*, 1967, **12**, 195.
† In this notation T is the absolute temperature and T_1 is a time.

relaxation times, one may calculate the density of states of the electrons responsible for the relaxation, and compare this with the total density of states as determined from specific-heat measurements (p. 260), and so determine the proportion of electrons at the Fermi surface which are in s-type states. In this way it has been shown by Butterworth * that in vanadium and niobium only about 0·17 and 0·13 of the electronic states at the Fermi surface are concerned in relaxation, in agreement with the view (p. 353) that in these metals a high proportion of d-type bonding is involved.

The Knight Shift. It was first shown by W. D. Knight (1949) that, for a fixed external field, the nuclear magnetic resonance effect for a given nucleus in a metal occurs at a slightly higher frequency than for the same nucleus in a non-metallic substance—chlorides are generally taken as the standard. At constant frequency the *Knight Shift*, ΔH, is in the direction of lower field strength, and is proportional to the field strength H; it is generally expressed as the value of $\Delta H/H$.

In pure metals, for a constant field, the Knight shift is always in the direction of higher frequency, although shifts in the opposite direction are known for intermetallic compounds. The effect is progressively larger for heavier elements, and for pure metals it is greater when the conduction electrons are predominantly of the s-type.

In the theory proposed by Townes, Herring, and Knight,† the shift is regarded as a result of the magnetic field aligning the spins of electrons at the top of the Fermi surface, which then interact with the nuclei. If these electrons are in s-states, they have a high probability density near the nucleus, and so the local susceptibility may be much greater than the average in the metal as a whole.

It can be shown that the extra field ΔH due to electron spins is $8\pi/3$ times the mean density of spin moment at the nucleus for structures of cubic or higher symmetry. The magnitude of the effect can be related to the magnitude of the hyperfine splitting for an s electron in the free atom, and to the ratio of the probability densities at the nucleus in the free atom and in the metal. The expression obtained is:

$$\frac{\Delta H}{H} = \frac{hc\Delta\nu I\chi_p M}{\mu_I\mu_B(2I+1)} \cdot \frac{<|\psi_F(0)|^2>_{\text{AV}}}{|\psi_a(0)|^2}$$

where $\psi_a(0)$ is the wave function at the nucleus for the s electron in the free atom, and $\psi_F(0)$ is the average function for all electronic states on the Fermi surface; $hc\Delta\nu$ is the energy of the hyperfine splitting in

* J. Butterworth, *Phys. Rev. Letters*, 1960, 5, 305; *Bull. Ampère*, Special No., 1960, p. 417.
† C. H. Townes, C. Herring, and W. D. Knight, *Phys. Rev.*, 1950, 77, 852.

the free atom and is known to increase with the atomic number; M is the mass of the nucleus of spin I and moment μ_I, whilst μ_B is the Bohr magneton for the moment of an electron; χ_p is the spin contribution to the macroscopic susceptibility per unit mass.

For sodium the value of $\psi_F(0)$ had been calculated accurately, and the calculated value of $\Delta H/H$ was in good agreement with the experimental results. It was shown by Kohn and Bloembergen * that the method did not give such good agreement for lithium as had at first been imagined. It was later shown by Jones and Schiff † that this difference was due to the electrons at the Fermi surface in lithium being of a predominantly p-like nature.

The Knight shift may thus be used to indicate whether electrons in the Fermi surface are of a predominantly s-like type (see p. 300 for lithium), and in favourable cases it can be used to determine the probability density at the nucleus for electrons in the Fermi surface. Measurements of the Knight shift, supplemented by independent determinations of electronic specific heat, may lead to more detailed information on the electronic structure of alloys. Within a terminal solid solution the position of an absorption line is usually not a rapid function of the composition. Local concentrations of one kind of atom, and changes from disordered to ordered structures, affect the environment of a given kind of atom and so alter the positions of the absorption lines. The details of the process depend upon whether the electrons in the Fermi surface are in predominantly s-states or not. Thus, no marked change in the Knight shift occurs near the solubility limit of an α solid solution in copper or silver, because the wave functions of the electrons on the zone faces are mainly p-like.

Diffusion Effects. For the alkali metals lithium, sodium, rubidium (and probably also caesium), the width of the resonance line decreases rapidly with increasing temperature. The same effect has been observed for aluminium, and has been explained as being due to self-diffusion of the metallic ions, and in this way nuclear magnetic resonance throws light on diffusion processes.

Magnetic Effects. Measurements of nuclear magnetic resonance effects may also throw some light on the magnetic state of atoms in alloys. If, in a solid solution of B in A, the resonance line width and intensity for A are little affected it may be concluded that the outer electrons of the B atoms enter into a common conduction band. If, on the other hand, the B atoms enter into solid solution with a localized magnetic moment (resulting from an incomplete d or f shell), the resonance

* W. Kohn and N. Bloembergen, *Phys. Rev.*, 1950, **80**, 913; 1951, **82**, 283.
† H. Jones and B. Schiff, *Proc. Phys. Soc.*, 1954 [A], **67**, 217.

line width is greatly broadened. For dilute solutions of manganese in copper the effect has been studied in detail, and the ^{63}Cu resonance is no longer observable when a few per cent. of manganese has been added.

Quadrupole effects. Nuclei with $I > \frac{1}{2}$ possess also a nuclear quadrupole moment that is sensitive to departures from, say, local cubic symmetry around the nucleus. As a consequence, loss of intensity of NMR signals from host cubic metals is observed in many solid-solution alloys, even when the solute does not possess a magnetic moment. Thus the resonance of ^{63}Cu is not observable in α-brasses with more than $\sim 20\%$ zinc, although in Ag–Cd alloys a similar loss of signal is not observed for ^{109}Ag which, having $I = \frac{1}{2}$, has no quadrupole moment.

B. Fermi-Surface Studies.

A large number of techniques has been applied in recent years to the direct measurement of the geometry of Fermi surfaces in pure metals, and recently some extensions have been made to alloys. The subject is too wide to cover in detail here but the reader is referred to some review articles and conference proceedings.* As illustrations we shall discuss one of the most widely applied methods, de Haas–van Alphen-effect studies, and that which first led to an experimental Fermi surface, the measurement of the anomalous skin effect.

(a) *The de Haas–van Alphen Effect.*
Introductory.

In 1930 de Haas and van Alphen found that at very low temperatures the magnetic susceptibility of single crystals of bismuth was not constant, but showed an oscillatory variation with the field. The effect was later found to exist with many multivalent metals, also with copper, and recently even for the alkali metals. In all cases, the variations of the magnetic susceptibility (χ) with the field strength (H) are such that, if χ is plotted against $1/H$, the periodicity is constant for a given orientation of the crystal relative to the field. In some cases there is one single periodic variation of χ with $1/H$, whilst in others two or more periodic effects may be superimposed, with the production of more complicated periodic curves. The periodicity is independent of the temperature, but the amplitude diminishes rapidly with rising temperature, and the effect disappears at 30°–40° K, so that the experiments are usually made at liquid-helium temperatures.

* " The Fermi Surface " (edited by W. A. Harrison and W. Webb). **1962**: New York and London (John Wiley). A. B. Pippard, *Rep. Progress Physics*, 1960, 23, 176.

The Brillouin-Zone Theory of Metals

Simple Two-Dimensional Models.

The origin of the de Haas–van Alphen effect may be understood most simply by considering a two-dimensional crystal of a metal, as was done by Pippard * and by Shoenberg.† It can be shown that the effect is due to the orbital motion rather than to the spin of the electrons, and we may first ignore the effects of spin, and consider a two-dimensional metal in which the electrons behave as free electrons. If we draw a two-dimensional momentum diagram, an assembly of N electrons will occupy the N lowest energy states which are uniformly distributed within the Fermi circle of Fig. V.39(a). The radius of this circle, $p_{\text{max.}}$,

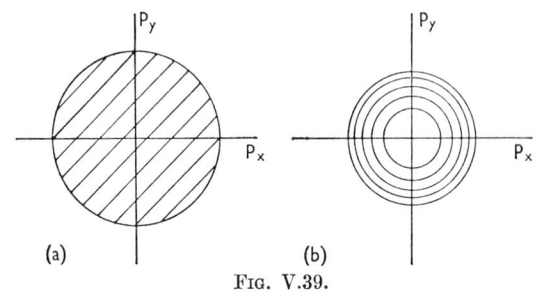

Fig. V.39.
(a) Without magnetic field.   (b) With magnetic field.
The diagrams are schematic and are not to the same scale.

corresponds to the energy $E_{\text{max.}}$ or ε of the electrons in the Fermi surface, the relation being:

$$E_{\text{max.}} = \varepsilon = p^2_{\text{max.}}/2m \quad \ldots \ldots \quad \text{V (19)}$$

In the absence of a magnetic field, the electrons move in straight lines, but if a field of strength H is applied perpendicular to the plane of the metal, each trajectory is bent into a circular orbit of radius p/eH. At the same time the quantum restrictions imply that only certain orbits are possible, namely those which satisfy the conditions that the orbit in real space must enclose $(n + \tfrac{1}{2})$ quanta of magnetic flux, one quantum of flux being h/e.

In the momentum diagram, states of constant energy $E = p^2/2m$, lie on a circle of radius p, which is thus eH times the radius (p/eH) of the circular orbit in real space. The effect of the magnetic field is thus to change the motion of the electrons in real space from straight lines to circles, and to restrict the energies to a definite series of values, so that

---

* A. B. Pippard, " Les Electrons dans les Métaux." (Tenth Solvay Congress, Brussels, 1955.) This report is elementary, but contains unfortunate misprints.
† D. Shoenberg, *Progress in Low-Temperature Physics*, 1957, 2, 226.

the uniformly distributed points in the momentum diagram of Fig. V.39(a) assemble on circles, as in Fig. V.39(b), the radii of these circles being such as to give the energies required by the quantum condition. The area of the circle of radius p in momentum space is thus $(eH)^2$ times the area of the corresponding circular orbit in real space. From this it can be shown that the quantizing condition becomes:

$$a(E) = (n + \tfrac{1}{2})eHh \quad \ldots \ldots \quad \text{V (20)}$$

where $a(E)$ is the area of the circle of constant energy in momentum space.

Each of the occupied circles in momentum space contains a high concentration of representative points, and this may be understood by remembering that the same kind of permitted circular orbit may be traversed by electrons in different parts of the crystal.

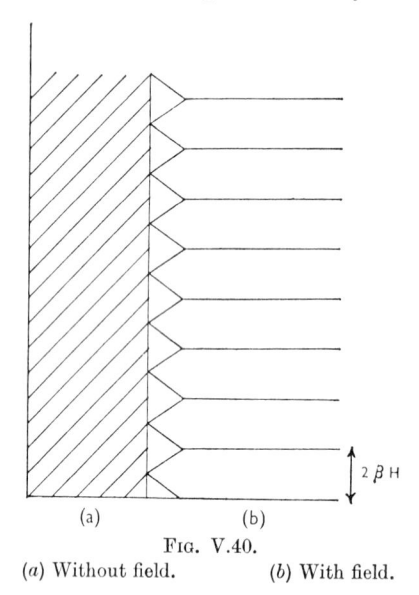

Fig. V.40.
(a) Without field.   (b) With field.

The permitted momenta are given by:

$$p^2 = (n + \tfrac{1}{2})eHH/\pi \quad \ldots \ldots \quad \text{V (21)}$$

and the corresponding permitted energies are given by:

$$E = p^2/2m = (n + \tfrac{1}{2})eHH/2m\pi$$
$$= (2n + 1)\frac{eh}{4\pi m}H = (2n + 1)\mu_B H \quad \ldots \ldots \quad \text{V (22)}$$

where μ_B is the Bohr magneton. The effect of the magnetic field is thus to split up the continuum of energy levels into a set of levels separated by uniform gaps * of $2\beta H$, as shown in Fig. V.40. On applying a magnetic field, the average energy of most of the electrons is unaltered by the change from the continuum to the fixed levels, but electrons with energies $<\mu_B H$ can enter the lowest state, and this produces the ordinary paramagnetic effect.

For electrons of the highest energy (*i.e.*, those in the Fermi surface) the application of a magnetic field will in general raise the energy, but if the field is such that the highest level of Fig. V.40 is a distance $\mu_B H$ below the Fermi surface, then the number of electrons whose energy is increased will be equal to that whose energy is decreased, and the average energy will remain unchanged by the sorting out process. If the field strength H is gradually increased, the energy steps between the permitted levels increase, and each time the condition $E_F = (2n + 1)\mu_B H$ is satisfied the energy of the electrons near the Fermi surface will be unaffected. There is thus a periodic variation of the energy E with the field strength H, and hence a corresponding variation in the magnetic moment—$\partial E/\partial H$, and the effect may be expected to show a regular periodicity in terms of $1/H$. From equation V (22), it will be seen that the periodic variation is such that equivalent points are reached when:

$$\frac{1}{H} = \frac{(2n+1)eh}{2a(E)} \quad \ldots \quad \text{V (23)}$$

In this way, the area of the Fermi circle is related to the period of the variations in the curve obtained by plotting χ against $1/H$. In this simplified free-electron model of a two-dimensional metal, the measurement of χ as a function of $1/H$ would thus enable the area of the Fermi circle to be obtained.

The Effect of a Periodic Field in Two Dimensions.

The above two-dimensional model over-emphasizes the amplitude of the oscillations observed under experimental conditions. At any temperature above the absolute zero, the Fermi surface of occupied states is not perfectly sharp, the probability of occupation dropping rapidly from unity to zero over an energy range of the order of kT. This reduces the amplitude of the oscillations in the χ versus $1/H$ curves, and a further reduction results from the fact that the energy levels are broadened by " collisions " of the electrons with the thermal vibrations of the lattice. The main reduction in amplitude

* If we consider the effect of electron spin, the application of a field H raises or lowers the levels by an amount βH according to whether the spin is antiparallel or parallel to the field. The permitted levels thus again form a series separated by steps of $2\beta H$.

results from the neglect of the third dimension in the two-dimensional model.

For the two-dimensional free-electron model considered above, in the absence of a magnetic field an electron moves in a straight line, and its representative point in the momentum diagram is stationary until the electron reaches the boundary of the crystal. In the presence of a magnetic field the electron moves in a circular orbit in real space, and the representative point moves round one of the permitted circles of Fig. V.39(b). As explained on p. 213, the direction of motion of the electron considered as a wave-packet is perpendicular to the curve of constant energy at the occupied state concerned, so that as the electron moves round the circular orbit, the representative point moves round the permitted circle in momentum space with an angle of 90° between the two directions, as shown in Fig. V.41(a) and (b).

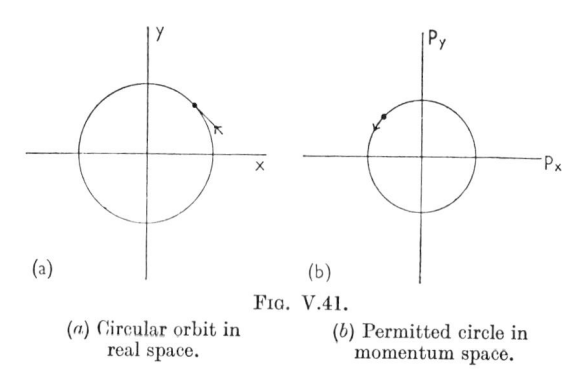

Fig. V.41.

(a) Circular orbit in real space.   (b) Permitted circle in momentum space.

In a two-dimensional crystal in which the electrons move in a periodic field, the electrons no longer behave as though they were free, but are subject to the Brillouin-zone conditions described in Section V.1. The arguments given above can still be applied if the wave-number k is used instead of the momentum p; for free electrons $k = p/\hbar$. The energy contours, which are circles for the two-dimensional free-electron model, become distorted, particularly for wave-numbers near the edges of the Brillouin zone. Since the velocity associated with a state is always in a direction at right angles to the energy contour, and since the force produced by a magnetic field is at right angles to the motion of an electron and does no work on it, the effect of the magnetic field is to move the representative point along the energy contour. In the presence of a magnetic field, the path of an electron in real space is thus of the same shape as that of the energy contour, but twisted through 90°, and reduced in scale by a factor eH.

The Brillouin-Zone Theory of Metals

As with the free-electron model described above, the application of a magnetic field at right angles to the plane of a two-dimensional crystal with a periodic field produces a series of quantized orbits. The continuum of representative points in the wave-number diagram is replaced by a series of permitted curves, and as the field is increased the permitted energy levels become more and more widely separated. The curve connecting the magnetic susceptibility χ and $1/H$ is again of a periodic nature, and the period again depends upon the area of the Fermi curve in the wave-number diagram. The exact relations are more complicated, and if the effective mass of the electrons is no longer equal to the normal electronic mass, the Bohr magneton μ_B of p. 270 has to be replaced by an effective Bohr magneton number $\mu_{\text{eff}.B}$.

Three-Dimensional Crystals.

To extend the above argument to a three-dimensional crystal, we may imagine the magnetic field to be applied along the z-axis of the crystal. The Fermi surface will then be in three dimensions and for

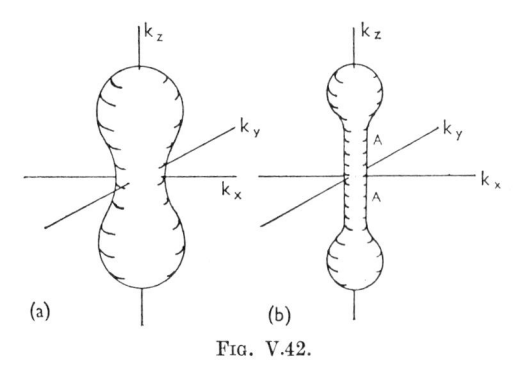

Fig. V.42.

any one state the component in the direction of k_z will be unaffected by the application of the magnetic field. We may, therefore, imagine the volume of occupied states divided up into a number of two-dimensional slices parallel to the $k_x k_y$ plane, and each of these slices may be treated as for a two-dimensional metal, with the appropriate area $a(E)$ of the Fermi surface. The resulting variation of χ with $1/H$ will thus involve a range of periodicities which will tend to produce a smooth curve. A periodic effect will not, however, be completely destroyed, because there will in general be one or more regions of the Fermi surface for which $a(E)$ varies relatively slowly with k_z, so that contributions from these regions are still evident, although the amplitude of the effect is much less than that for the two-dimensional model discussed above.

These dominant regions will in general be those for which $a(E)$ is a maximum or a minimum. Qualitatively, the effect will be greater, the slower the variation of $a(E)$ with k_z. It would thus be less for a Fermi surface with the shape of Fig. V.42(a) than for the Fermi surface of Fig. V.42(b), where the portion AA has the same cross-sectional area over a wide range of values of k_z.

It is by considerations such as these that a study of the de Haas-van Alphen effect for different orientations of a crystal relative to the magnetic field is able to give information about the shape of the Fermi surface. The details of the analysis are too complicated for the present book, and reference should be made to the papers listed on p. 269.

General Results.

An estimate of the general magnitudes to be expected can be obtained by first using the free-electron model to calculate the size of the Fermi sphere for the known atomic volume and number of conducting electrons per atom for the metal concerned. According to Pippard,[*] the radius of the Fermi sphere for copper in k-space is $1 \cdot 3 \times 10^8$ cm^{-1}. The corresponding period of the de Haas–van Alphen oscillations plotted against $1/H$ is $1 \cdot 7 \times 10^{-9}$ gauss^{-1}, so that even in a field of 10^5 gauss the spacing of the oscillations is only 17 gauss. Such a close spacing cannot be measured by ordinary methods using steady fields, because these cannot be maintained sufficiently constant. A method devised by Shoenberg [†] enables rapidly varying fields of the order of 10^5 gauss to be obtained by the discharge of a battery of condensers through a solenoid in which the specimen is placed. The whole process of the rise and fall of the magnetic field takes only a few thousandths of a second, during which the magnetization of the specimen is measured, and the de Haas–van Alphen oscillation is revealed as a high-frequency ripple on the magnetization curve.

For some of the metals which have been examined, the period of the de Haas–van Alphen oscillations is much larger. This means that they cannot be due to the main body of electrons, but must be caused by very small numbers of electrons overlapping the faces of Brillouin zones, or by very small pockets of positive holes in the corners of zones. If the number of electrons (or holes) concerned is greater than about 10^{-3} per atom, the period of the oscillations becomes so small that resolution is extremely difficult in fields of the order of 20 k.gauss. The de Haas–van Alphen effect, thus, usually gives information about very small overlap regions, or very small numbers of holes in the Fermi surface. The work

[*] A. B. Pippard, " Les Électrons dans les Métaux " (Tenth Solvay Congress, Brussels, 1955), p. 141.
[†] D. Shoenberg, *Phil. Trans. Roy. Soc.*, 1952–53, [A], **245**, 1.

The Brillouin-Zone Theory of Metals 275

of Gold * on lead has, however, revealed what is probably the largest part of the Fermi surface present.

(b) *The Anomalous Skin Effect.*

Introductory.

When a high-frequency current is carried by a plane metallic surface, the fields and the currents are confined to a thin surface layer. This may be called the *skin effect*. Under normal conditions the thickness of the conducting layer can be calculated, but at very low temperatures complications arise because the mean free path of an electron between two collisions with the thermal vibrations of the lattice becomes greater than the thickness of the conducting surface layer; this gives rise to what is called the *anomalous skin effect*.

At room temperatures, under normal conditions the current J, the field E, and the direct-current conductivity σ are related by Ohm's Law:

$$J = \sigma E \quad \ldots \ldots \quad \text{V (24)}$$

For a semi-infinite plane metal surface, the field E and current J are related by Maxwell's equation for an oscillation † of angular frequency ω:

$$\frac{d^2 E}{dz^2} = 4\pi i \omega J \quad \ldots \ldots \quad \text{V (25)}$$

where the surface is in the xy plane. The solution of this for a semi-infinite plane leads to the relation:

$$R = \sqrt{\frac{2\pi\omega}{\sigma}} \quad \ldots \ldots \quad \text{V (26)}$$

where R is the surface resistance for alternating currents. The field and the current decay exponentially into the metal, and the skin depth δ is defined by:

$$\delta = \frac{R}{2\pi\omega} = (2\pi\omega\sigma)^{-1/2} \quad \ldots \ldots \quad \text{V (27)}$$

The skin depth gives the thickness of the layer in which most of the current is carried, and except at very low temperatures the above relations are satisfied, and R is proportional to $\sqrt{(1/\sigma)}$. At temperatures below about 20° K, σ increases very greatly, and at microwave frequencies of the order of 10^9–10^{11} cycles/sec, R does not decrease sufficiently rapidly to satisfy the proportionality to $\sqrt{(1/\sigma)}$. This is the

* A. V. Gold, *Phil. Trans. Roy. Soc.*, 1958, [A], **251**, 85.
† See C. A. Coulson, "Electricity," p. 229.

anomalous skin effect, and with a really pure metal at a sufficiently low temperature, R becomes independent of σ at what is known as the extreme anomalous limit.

The anomalous skin effect is due to the fact that at these low temperatures, the mean free path of an electron between two collisions with the thermal vibrations of the lattice is much greater than the skin depth δ. This means that, unless an electron is moving nearly parallel to the surface, it is spending only a very small fraction of its time in the region where it is subject to the field, and so the conditions which lead to Ohm's Law are no longer satisfied.

Effective and Ineffective Electrons.

In the treatment adopted by Pippard,[*] the term "ineffectiveness concept" is used to describe the hypothesis that all except the electrons

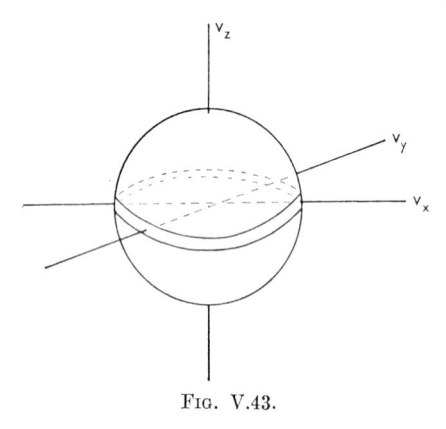

Fig. V.43.

moving nearly parallel to the surface may be neglected in considering the anomalous skin effect. It is further assumed that all the electrons which may be considered as effective in carrying the current can be treated as though they move parallel to the surface, so that they remain in a constant field during the execution of a free path. There is of course no sharp distinction between effective and ineffective electrons, and as an approximation it is assumed that the electrons are divided into two groups. Those with components of velocity normal to the surface greater than $\beta\delta/\tau$ are regarded as ineffective, whilst those with components less than $\beta\delta/\tau$ are considered effective. Here τ is a relaxation

[*] A. B. Pippard, *Proc. Roy. Soc.*, 1954, [A], **224**, 273. A summary is given by the same author in the Report of the Tenth Solvay Congress, "Les Électrons dans les Métaux," but suffers from misprints and the omission of a critical diagram.

time, defined by $l = V_F\tau$, l being the free path, V_F the velocity of the electrons in the Fermi surface, and β a numerical constant.

If we consider a free-electron model for which the Fermi surface is spherical, we may draw a velocity diagram in which the occupied states lie within a sphere of radius equal to the Fermi velocity V_F. If the surface of the metal is in the xy plane, the effective electrons will lie in a thin slice of the sphere parallel to the $v_x v_y$ plane, as shown in Fig. V.43. The thickness of this slice is $2\beta\delta/\tau$, and the fraction of the total number of electrons lying within the slice is $f = \frac{3}{2}\beta\delta/l$ where $l = V_F\tau$. Since all the electrons in the slice are assumed to move in a constant field, the effective conductivity $\sigma_{\text{eff.}}$ may be written:

$$\sigma_{\text{eff}} = f\sigma = \tfrac{3}{2}\beta\delta\sigma/l \quad \ldots \ldots \text{V (28)}$$

and hence, if in V (27) $\sigma_{\text{eff.}}$ is written for σ,

$$\delta = (3\pi\omega\beta\sigma/l)^{-1/3} \quad \ldots \ldots \text{V (29)}$$

and

$$R = \left(\frac{8\pi^2\omega^2 l}{3\beta\sigma}\right)^{1/3} \quad \ldots \ldots \text{V (30)}$$

and also

$$R = (4\pi^2\omega^2\delta/\sigma_{\text{eff.}})^{1/3} \quad \ldots \ldots \text{V (31)}$$

In equation V (30), σ appears in the form l/σ, which is a constant of the metal, so that R is independent of σ, in agreement with observation at very low temperatures.

A more complete theory was developed by Reuter and Sondheimer,[*] and comparison with this shows that, if electrons are diffusely scattered at the surface of the metal, the constant β must be given by:

$$\beta = \frac{8\pi}{3^{3/2}} = 4\cdot 837 \quad \ldots \ldots \text{V (32)}$$

A comparison with the more detailed theory shows that it is justifiable to consider β as constant for all shapes of the Fermi surface.

Two-Dimensional Models.

Experiment shows that R depends on the orientation of the crystal relative to the direction of the electric field, and an analysis of this effect enables information to be obtained about the shape of the Fermi surface. The principles can be understood by considering a two-dimensional metal with an oscillating field applied parallel to an edge, whose direction may be taken as that of the x-axis. The anomalous skin effect then concerns a thin layer parallel to the edge of the two-dimensional specimen.

[*] G. E. H. Reuter and E. H. Sondheimer, *Proc. Roy. Soc.*, 1948, [A], **195**, 336.

In the corresponding momentum diagram the energy contours are closed curves, and the direction of motion of the electrons in real space is perpendicular to the energy contour at the representative point concerned. If, for example, the Fermi surface were of the shape of Fig. V.44, and a field were applied to accelerate the electrons in the direction A, the effective electrons would be those in states such as a and b, where the energy contour is perpendicular to A.

In a crystal where the anomalous skin effect is anisotropic, there will in general be some axis of the crystal parallel to a characteristic direction of the Fermi surface. If, for example, the Fermi surface had the shape of the closed curve of Fig. V.44, the direction AB in the momentum

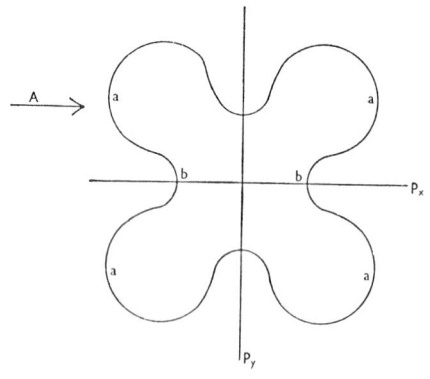

Fig. V.44.

diagram would correspond to a characteristic axis in the crystal, and we may denote by θ the angle between this axis and the edge of the metal, which is taken to be the x-direction.

If the field is such as to accelerate electrons in the direction F (Fig. V.45), the effective electrons will be those in states such as pp' where the energy contour is perpendicular to F. The shape of the Fermi surface at any point may be described in terms of the radius of curvature ρ, and the angle ϕ between the normal to the energy contour and the fixed axis AB. For the effective electrons, therefore, $\theta = \phi$.

If a field is applied so as to accelerate the electrons in the direction F, the energy contour at p will be displaced to a position such as that shown by the dotted line in Fig. V.45, and a new steady state will be set up as in the free-electron model described on p. 184. At the same time the energy contour at p' will undergo a corresponding displacement.

If $\bar{v}$ and τ are the electronic velocity and relaxation time at P, the

The Brillouin-Zone Theory of Metals

effective electrons move within an angle $2\beta\delta/\bar{v}\tau$, and occupy a length of arc $2\beta\delta|\rho|/(\bar{v}\tau)$. Under the influence of a constant field E, the displacement of the Fermi surface relative to itself will be $eE\tau$, and the general effect is therefore to move electrons occupying an area $2\beta\delta e|\rho|E/\bar{v}$ from

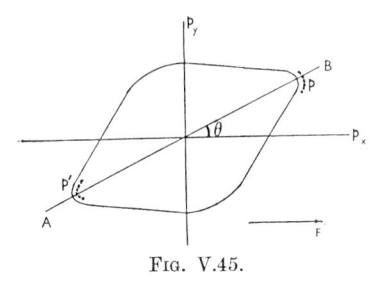

FIG. V.45.

p' to p with a resultant flow from left to right. The density of states within the Fermi surface is $2/h^2$ per unit area of momentum space and of metal, and hence the current density is:

$$J = \frac{8\beta\delta|\rho|e^2}{h^2} \cdot E \qquad \ldots \ldots \quad \text{V (33)}$$

so that
$$\sigma_{\text{eff.}} = 8\beta\delta|\rho|e^2/h^2 \qquad \ldots \ldots \quad \text{V (34)}$$
and hence from V (31):

$$R = \left(\frac{\pi^2\omega^2 h^2}{2\beta|\rho|e^2}\right)^{1/3} = \frac{\sqrt{3}}{2}\left(\frac{\pi^2\omega^2 h^2}{2|\rho|e^2}\right)^{1/3} \quad \ldots \quad \text{V (35)}$$

This implies that, in the extreme anomalous limit, the surface resistance R depends only on the radius of curvature $|\rho|$ of the Fermi surface. A study of R as a function of θ thus permits a deduction of the relation between $|\rho|$ and ϕ for the Fermi surface, because $|\rho|$ is always determined for the point at which $\phi = 0$, and:

$$|\rho| = \frac{3^{3/2}\pi\omega^2 h^2}{16 e^2 R^3} \qquad \ldots \ldots \quad \text{V (36)}$$

so that a plot of $1/R^3$ against θ is equivalent to a plot of $|\rho|$ against ϕ, except for the constant scaling factor.

Three-Dimensional Metals.

It can be shown that a three-dimensional Fermi surface is equivalent to a stack of two-dimensional slices formed by sectioning the surface by planes parallel to the principal axis and the normal to the surface of the metal. On these sections the points where the normal to the Fermi

T

surface lies parallel to the surface of the metal are the points representing the effective electrons. In this way, by measuring R as a function of the orientation, information may be gained about the curvature of the Fermi surface, subject to the limitations mentioned below, but the details are too complicated for consideration here.

Scope and Limitations of the Method.

The anomalous skin effect is a surface effect, and is therefore very sensitive to imperfections in the surface, such as the straining that may occur during mechanical polishing. It is suitable for use only with very pure metals, because small quantities of impurities may create a considerable residual resistance at low temperatures, with a corresponding shortening of the free path.

The analysis of the results is difficult for metals in which the free Fermi surface (*i.e.*, the part of the surface not in contact with zone faces) lies in more than one Brillouin zone. In this case the value of R will be determined by the sum of the effects of the free surfaces in the different zones, and the method will not give an unambiguous solution. Even where only one zone is concerned, the method is not unambiguous if the Fermi surface is inflected. The method is, therefore, best suited for the study of univalent metals, and has been used with great success to determine the shape of the Fermi surface in copper.

APPENDIX: SPECIFIC HEAT OF METALS.

The total specific heat of a metal may be regarded as the sum of the normal specific heat resulting from the thermal vibrations of the atoms, and that due to the valency electrons. We may consider first the normal specific heat due to the atomic vibrations. If U is the internal energy associated with the atomic vibrations, the resulting normal specific heat at constant volume is defined by $c_v = dU/d\theta$, where θ is the temperature on the absolute scale. If a crystal consists of N atoms held together by elastic forces, the lattice possesses $3N$ vibrational degrees * of freedom, and we may say that there are $3N$ independent normal modes of vibration. These normal modes may be regarded as having frequencies $v_1, v_2, v_3 \ldots$ and the actual condition of the crystal results from the superposition of these $3N$ frequencies, each having an amplitude which depends on the temperature. As the temperature falls, the amplitude and hence the energies of the vibrations

* More strictly $(3N - 6)$, since the solid as a whole has 3 translational and 3 rotational degrees of freedom. In any crystal of normal size, the number 6 is negligible compared with $3N$.

The Brillouin-Zone Theory of Metals

diminish, and according to the quantum theory, the energy associated with a vibration of frequency ν is:

$$\tfrac{1}{2} h\nu + \frac{h\nu}{e^{\frac{h\nu}{k\theta}} - 1},$$

where h is Planck's constant, k is Boltzmann's constant, and θ is the absolute temperature.

In this expression the term $\tfrac{1}{2}h\nu$ represents the zero-point energy, i.e., the energy which an oscillator still retains at the absolute zero. The second term represents the temperature-dependent part of the energy resulting from a vibration of frequency ν. This term is of such a form that as the temperature becomes very low, and $\frac{h\nu}{k\theta}$ becomes very large, the temperature-dependent part of the energy and its specific heat become very small, and vanish at the absolute zero. The higher the frequency ν, the higher the temperature below which its energy is negligible. It follows, therefore, that as a metal is cooled, the energies associated with the highest frequencies die away most rapidly, and the lower the temperature the more the energy is associated with vibrations of low frequency and hence of long wave-length. The general picture presented is one in which we have at all temperatures a constant term (the zero-point energy) which does not contribute to the specific heat, and an energy term dependent on the temperature. The latter term and the associated specific heat are zero at the absolute zero, and on raising the temperature the vibrational energy increases in such a way that the energy is first associated with the lowest frequencies.

The details of the process depend on the relative numbers of the different frequencies, or in other words on the frequency-distribution curve, or the vibrational spectrum of the crystal structure. Just as we have drawn $N(E)$ curves to show how the electron states are distributed over different ranges of energy, so we may draw frequency-distribution curves showing how the frequencies are distributed over different ranges. Each type of crystal structure gives rise to a characteristic vibrational spectrum just as it gives rise to a characteristic type of $N(E)$ curve. The form of the vibrational spectrum has been calculated for the body-centred cubic structure, but not for the face-centred cubic structure or for more complicated structures, although various simplified two-dimensional models have been dealt with. The details of this work are complicated and lie outside the present book, although reference may be made to a report by Blackman.* For most

---
* M. Blackman, *Rep. Progress Physics*, 1941, **8**, 11.

purposes the problem is drastically simplified, and the first step was taken by Einstein, who assumed that all the atoms vibrated with the same frequency, so that the vibrational spectrum consisted of one single frequency. This was clearly too simple a model, and the next step was taken by Debye, who assumed that the solid crystal could be treated as an elastic continuum. In this case the theory of elasticity indicates that a specimen of definite size and shape can undergo only certain definite vibrations whose frequencies are related to the elastic properties of the material, and depend also on the size and shape,* although the shape has an appreciable effect only on the lowest frequencies. In this case if $N(\nu)$ is the number of frequencies in the range ν and $\nu + d\nu$, it is easy to show that $N(\nu) \propto \nu^2$; the constant of proportionality involves the velocities of transmission of waves in the solid, and so forms a connecting link between the theory of elasticity and the theory of specific heat. In the case of a true continuum there is no upper limit to the frequency, but in a crystal the atomic structure imposes an upper limit to the frequency, since there is a minimum wave-length of the order of the distance between two atoms. The assumption made by Debye was then that the equations for an elastic continuum could be applied to a crystal, provided that the spectral distribution curve was cut off sharply at a limiting frequency ν_D such that the total number of frequencies is equal to $3N$. This assumption is purely arbitrary, and the theory is a rather illogical mixture of what would be expected for a continuum, and what is required for an actual lattice. It has become increasingly clear that for some problems the Debye theory is too simple, but in many cases it is a justifiable approximation at low temperatures, and it leads to the following conclusions:

(1) If the limiting frequency is ν_D, we may define a characteristic temperature $\bar{\theta}$ by the relation $\bar{\theta} = \dfrac{h\nu_D}{k}$. In this case the specific heat of all metals should be the same function of $\bar{\theta}$, so that different metals should have the same specific heats if they are compared at temperatures which are the same fraction of the characteristic temperature. This is found to be the case for many cubic metals. For non-cubic metals it is necessary to introduce more than one characteristic temperature because the elastic properties are different in different directions, and similar difficulties are encountered in some cubic metals if these are elastically very anisotropic.

(2) At very low temperatures the theoretical specific heat is proportional to θ^3, and this is in reasonable agreement with the facts for

* This is the three-dimensional analogue to the tones and overtones of a stretched string.

The Brillouin-Zone Theory of Metals

some metals, but for others the θ^3 law is not satisfied. Really critical tests are not easy to make because it is sometimes difficult to allow for the electronic specific heat, but the general conclusion is that the theory is not confirmed so well as was originally thought.

(3) At high temperatures the atomic heat should approach asymptotically to the value $C_v = 3R = 6$ cal. units, in agreement with Dulong and Petit's law. The fact that at very high temperatures the atomic heats exceed this value is partly because the amplitudes of the vibrations become too large to be treated as simple harmonic vibrations, and also because of the increase of the electronic specific heats.

Considered empirically the Debye theory has achieved great success, but there is increasing evidence that there are discrepancies between the thermal and elastic data, and it is generally recognized that the use of the continuum hypothesis, although often satisfactory for some low-temperature properties, is too great an approximation.

In some interesting recent neutron-scattering studies it has proved possible to establish by experiment the form of the vibrational spectrum (p. 281) for some crystals.

Suggestions for Further Reading.

G. E. Pake, *Amer. J. Physics*, 1950, **18**, 438, 473. Two general review papers.
E. R. Andrew, "Nuclear Magnetic Resonance." **1955** : Cambridge (University Press).
G. E. Pake, " Solid State Physics " (edited by F. Seitz and D. Turnbull), Vol. II, p. 1. **1956** : New York (Academic Press); London (Academic Books).
W. D. Knight, *ibid.*, p. 93. These two papers are reviews of the subject at an advanced level.
T. J. Rowland, " Progress in Materials Science ", Vol. 9, No. 1, p. 1. **1961**: Oxford (Pergamon Press).

PART VI. ELECTRONS, ATOMS, METALS, AND ALLOYS.

The object of this Part is to examine the extent to which the ideas introduced in earlier sections can be usefully applied to the properties of individual metals and alloys; and, especially, to their alloying behaviour. It should be recognized that the objectives of the electron theory of metals are much broader than the interpretation and prediction of crystal structures and phase diagrams of alloys, and its overall success should not be evaluated purely in terms of this area. These aspects, however, inevitably are those of greatest immediate interest to the student of metallurgy, and he must be on his guard against rejecting a particular description of a particular group of metals simply because it seems to account less well for their alloying behaviour than some *ad hoc* scheme aimed specifically at that aspect. He must not, on the other hand, be blind to the clues afforded by, for example, the chemical behaviour of metals even when these are difficult to incorporate formally in a mathematical theory of these metals.

1. The Alkali Metals.

General.

The alkali metals lithium, sodium, potassium, rubidium, and caesium all crystallize in the body-centred cubic structure, the unit cell of which is shown in Fig. III.12 (p. 104). These elements are the nearest approach to the "ideal metals" of the electron theories, and they are characterized by the fact that the electron clouds of their ions are small compared with the distances between the atoms in the crystals of the metals. In Fig. VI.1 we show the $\rho(r)$ curves for the outer parts of the *free ions* of the alkali metals as a function of the distance from the nucleus, and we indicate also one-half the nearest distance of approach of the *atoms in the crystals of the metallic elements*. It will be seen that in lithium and sodium the electron cloud of the ion has almost vanished at the point midway between the atoms in the metal, whilst in potassium, rubidium, and caesium the density of the cloud is very small. As the centre of the ion is approached the electron-cloud density becomes very large.

Strictly speaking, the electron-cloud patterns of the ions are slightly different in the metals and the free atoms, but the difference is small,

Electrons, Atoms, Metals, and Alloys

and Fig. VI.1 is a good approximation for the electron clouds of the ions in the solid metals. When it is remembered that the volume varies as the cube of the dimensions, it will be seen that the volumes in which the electron-cloud densities of the ions are appreciable are only small fractions of the total volumes of the metallic crystals. Metals of this

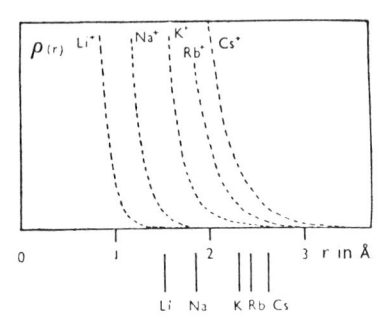

FIG. VI.1. The $\rho(r)$ curves for the outer part of the ions of the five alkali metals as a function of the distance (r) from the nucleus. The five vertical lines are at distances equal to one-half the closest distance of approach of the atoms in the crystals of the pure metals. The vertical scale is arbitrary, but is the same for all five curves, and is the same as that of Fig. VI.7.

kind may conveniently be called "open metals." Metals for which the electron clouds of the ions overlap considerably may be called "full metals." The alkali metals form the most open group of metals, and they may be regarded as consisting of relatively small ions maintained in regular array by the valency electrons. The overlap of the ions is thus very small, and the resulting forces have a negligible effect in determining the binding energies or the lattice spacings. (The forces resulting from the slight overlap of the ions vary rapidly with the distance, and these forces play a slight part in determining the compressibilities, and a significant part in determining the elastic constants of the alkali metals.) The properties of the alkali metals are determined mainly by the valency electrons, and it is for this reason that the alkali metals are of such interest in developing theories of metals, even though they are too reactive to be of any practical value as the main constituents of alloys.

As explained on p. 251, the first Brillouin zone of the body-centred cubic structure is the rhombic dodecahedron of Fig. V.30 and can accommodate two electrons per atom. Since the alkali metals contain one valency electron per atom, the zone is only half-filled, and it was the natural assumption of all the earlier theoretical work that the Fermi surface of occupied states is approximately spherical and does not touch the surface of the zone.

A significant amount of direct experimental evidence about the Fermi surfaces of the alkali metals is now available and the following data are from a useful theoretical paper by Lee,* which gives references to the original work. In Table IX k_0 is the free-electron-sphere Fermi-surface wave-vector and k_{hkl} the observed value of the Fermi wave-vector in the [hkl] direction in k-space. The distortions are percentages and are most accurate for sodium and potassium.

TABLE IX.

	Li	Na	K	Rb	Cs
$(k_{100} - k_0)/k_0$	− 1·0	− 0·08	+ 0·15	− 0·2	− 0·9
$(k_{110} - k_0)/k_0$	+ 4·0	+ 0·1	+ 0·11	+ 0·9⁵	+ 3·3
$(k_{111} - k_0)/k_0$	− 1·0	− 0·01⁵	− 0·11	− 0·5	− 1·4

The distortions are smallest for sodium, increasing along the series to caesium, and also being large for lithium, the abnormal properties of which are discussed at the end of this section. It should be noted that a [110] distortion of about $+ 12\%$ would be needed for the Fermi surface to touch the closest zone face.

Alloys of the Alkali Metals

In Part III.4 we have seen how the tendency for two metals to form primary solid solutions is controlled first by the " size-factor." (Appendix B, p. 422) shows the interatomic distances in the crystals of the elements; the well-marked periodicity is apparent, and in each Period the alkali metal has the largest atomic diameter. In Table X are shown the lattice spacings and interatomic distances in the crystals of the alkali metals, and also the percentage increase in atomic diameter

TABLE X.—*Lattice Spacings and Interatomic Distances of Alkali Metals.*

Metal.	Lattice Spacing, kX.	Interatomic Distance, kX.	Increase in Atomic Diameter, %.
Li	3·5023 at 20° C.	3·0331	22
Na	4·2820 at 25° C.	3·708	25
K	5·333 ,,	4·618	
Rb	5·62 at −173° C.	4·87	5·5
Cs	6·05 ,,	5·24	7·6

* M. F. G. Lee, *Phys. Rev.*, 1969, **178**, 953.

(defined by the closest distance of approach of the atoms) on passing from one element to the next. Fig. VI.2 shows diagrammatically the equilibrium diagrams of the alloys formed by different pairs of the alkali metals, so far as they are known. It will be seen that no wide range of solid solutions is formed in the systems where the atomic diameters differ by more than the critical value (14-15%) which was referred to in Part III.4. On the other hand, continuous solid solutions are formed in the systems K-Rb, K-Cs, and Rb-Cs, where the size-factors are favourable. This tendency for elements of the same sub-group, and of the transitional elements, to form solid solutions with one another was pointed out by van Liempt as long ago as 1926, and later work has confirmed that, apart from the size-factor effect, different kinds of atom fit together more easily on a common lattice if their valencies are the same, *i.e.*, if they contribute the same number of valency electrons per atom.

Fig. VI.2 shows that, whereas sodium, potassium, rubidium, and

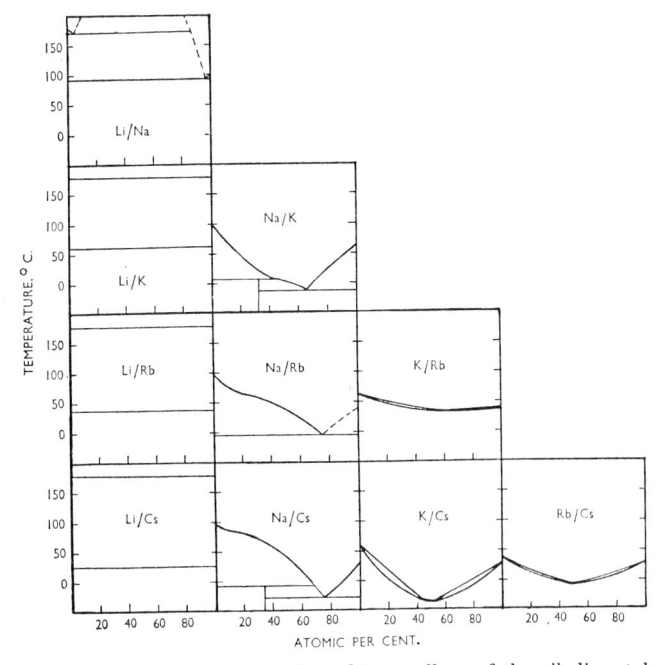

FIG. VI.2. Equilibrium diagrams of the binary alloys of the alkali metals with one another. Immiscible liquids are formed in the systems Li–Na, Li–K, Li–Rb, and Li–Cs. Continuous solid solutions are formed in the systems K–Rb, K–Cs, and Rb–Cs.

caesium are totally miscible in the liquid state, lithium behaves abnormally, so that immiscible liquids are formed in the systems Li–Na, Li–K, Li–Rb, and Li–Cs.

We have already seen in Part III.4 that when we deal with pairs of metals which differ markedly in the electrochemical series, there is a tendency to form stable intermediate phases at the expense of the primary solid solutions. In general, the alkali metals are so electropositive compared with other metals that solid solutions in the alkali metals are comparatively rare, even where the size-factors are favourable.

Wave Functions and Binding Energies in the Alkali Metals.

We have seen that actual calculations of electronic structure (p. 235) involve wave functions that resemble those of a free atom near each nucleus but join up satisfactorily in the regions between the atoms. Fig. VI.3 shows the value of ψ as a function of distance along the line joining two neighbouring atoms in a sodium crystal. (This is taken from the classic work of Wigner and Seitz.) It will be seen that the ψ curve has two nodal points on each side of a given nucleus, in agreement with the general principle that the electron cloud of a $3s$ state has $3 - 1 = 2$ nodes. It will also be noted that the curve is almost horizontal for a relatively large part of the distance between two adjacent atoms, and it is this fact which makes the free-electron model a justifiable approximation for the alkali metals. When it is remembered that the volume is proportional to the cube of the distance, it will be appreciated that for the greater part of the volume of the crystal of sodium, the valency electron may be treated as though it were an electron in the free-electron model.

The energy associated with the ψ curve of Fig. VI.3 can then be used to calculate the binding energy of the crystal, which is defined as the work required to dissociate the crystal into free neutral atoms, and is thus the same as the heat of sublimation. It is very necessary to make quite clear the meaning of the symbols used in this connection, since the descriptions in some papers and books are unfortunately confusing. We may denote by E_I the ionization potential of a free atom of sodium, so that E_I is the work required to separate a neutral atom into a free positive ion and a free electron. We may now imagine that we have an assembly of atoms at very large distances from one another. Then, since the atoms are widely separated, their interaction may be ignored, and the work required to remove separately one electron and one positive ion will be equal to E_I. If the positive ion and the electron recombine to form a neutral atom, the energy evolved will again be E_I, and the binding energy will then be zero. We may now suppose that the atoms assemble together to form a crystal of sodium. The energy of the valency electrons is then spread over a range or band, and the work required to remove the electron in the lowest state, *together with one positive ion*,* may be denoted $-E_0$, and is a function of the distance between the atoms.

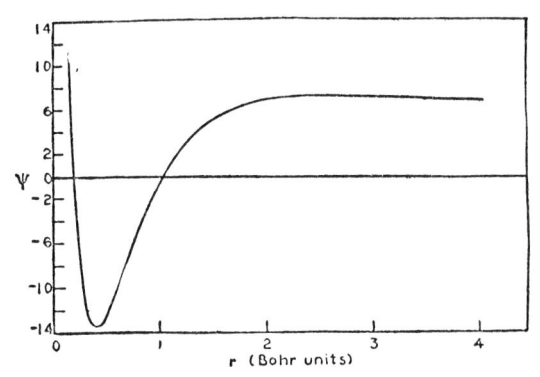

Fig. VI.3. The wave function of the lowest state of the valency electrons in a crystal of pure sodium. The electron cloud density of the valency electron at a distance r from the nucleus is proportional to the value of ψ^2 at the distance concerned. The distance between two adjacent atoms in the crystal is $3\cdot71$ Å $= 7\cdot01$ Bohr units.

The convention of a negative sign is used so that the values of E_0 will indicate the energy of the crystal in the sense that the firmer the binding the lower the energy and the more stable the structure.† If the electron and the positive ion then recombine to form a neutral atom, the energy evolved will be E_I, and consequently $(-E_0 - E_I)$ represents the net work required to remove from the crystal one positive ion and one electron in the lowest state, and then to allow them to recombine to form a neutral atom. If all the electrons in the crystal were in the lowest state, then $-(E_0 + E_I)$ would give the binding energy or heat of sublimation per atom.

* It is here that some of the descriptions are so very confusing, since they readily give the impression that E_0 is the work required to remove *an electron* in the lowest state from the crystal. The heat of sublimation, however, refers to the removal of a complete atom (electron plus ion), and the term E_0 must therefore be defined so as to include the work required to remove the positive ion.

† Thus if the work required to remove an electron and an ion from one structure were 10 cal, and from another structure 1000 cal, we should have $-E_1 = 10$ and $-E_2 = 1000$, and hence $E_1 = -10$ and $E_2 = -1000$, and the energy E_2 would be lower than the energy E_1.

In the crystal of a metal, the electronic energies extend over the whole range of the Fermi distribution, and an electron in a high-energy state naturally requires less work for its removal than does an electron in the lowest energy state. If the symbol E_F is used to denote the mean Fermi energy of the electrons, with the lowest electronic state taken as zero, then the average binding energy of the crystal per atom will clearly be $-(E_0 + E_F + E_I)$ as compared * with the value $-(E_0 + E_I)$ which would hold if all the electrons were in the lowest state. Since $-(E_0 + E_F + E_I)$ is the average energy per atom required to evaporate the electrons and ions from the crystal and convert them into neutral atoms, the energy of the crystal relative to the free atom is $+(E_0 + E_F + E_I)$ per atom. Large values of E_F or E_I thus tend to make the energy of the crystal greater than that of free atoms, and hence to make the crystal relatively less stable. For any given metal, E_I is a constant of the free atom, whilst both E_0 and E_F are functions of the distance between the atoms. E_F is the mean Fermi energy, and in the free-electron model E_F is simply the kinetic energy of the translatory motion of the electrons, while in the zone theories E_F is that part of the energy associated with the motion of the electrons in the periodic field of the lattice; as explained before (p. 210) this cannot always be interpreted as the energy of long-range translatory motion. As defined above, E_0 is the work required per atom to break up or dissociate into free electrons and positive ions a hypothetical crystal in which all the electrons are in the lowest energy state. In a simple model which ignored the atomic characteristics, E_0 would involve only the *potential energy of the hypothetical crystal* in which all the electrons occupied the lowest energy state.† In the more complete theories, E_0 involves both potential and kinetic energy, because, as can be seen from Fig. VI.3, even in an alkali metal the wave function near an atom is no longer constant, and this leads to a kinetic energy.

For the alkali metals it is possible to calculate both E_0 and E_F for different values of r_0, and, as shown in Fig. VI.4, the curve for $(E_0 + E_F)$ passes through a minimum which is not very far removed from the minimum on the curve for E_0 alone. The equilibrium condition will clearly be that corresponding to the minimum on the curve for $(E_0 + E_F)$, and in this way it is possible to calculate the value of r_0, and hence also the atomic volume and lattice spacing of the metal at the absolute zero of temperature. The binding energy is then equal to

---

* It will be appreciated that with the above convention of signs, E_0 is negative and E_F and E_I are positive. When the atoms are infinitely far apart E_F is zero and E_0 and E_1 are numerically equal but of opposite sign.

† It is essential to note that the term E_0 refers to the energy of the crystal, and is not the work required to remove one electron from the lowest state.

$-(E_0 + E_F + E_I)$, and since the curves of Fig. VI.4 express the energy as a function of the atomic volume,* it is possible to calculate the compressibility K, by means of the relation:

$$\frac{1}{K} = -v\left(\frac{dp}{dv}\right) = v\left(\frac{d^2E}{dv^2}\right) \quad \ldots \quad \text{VI (2)}$$

where E is the total energy.

In general it may be said that for the alkali metals the agreement between theory and experiment is satisfactory. The various methods

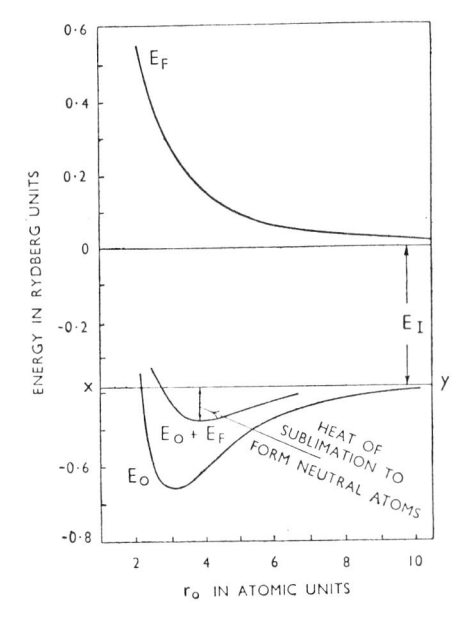

FIG. VI.4. The lowest curve shows the variation of E_0 with r_0 (for the definition of these quantities see p. 289). If all the electrons were in the lowest state, the position of the minimum on this curve would indicate the equilibrium value of r_0 at the absolute zero of temperature. Actually only two electrons can occupy the lowest state, and it is necessary to add the mean Fermi energy E_F, which is itself a function of r_0. The uppermost curve shows the variation of E_F with r_0, and the curve marked $E_0 + E_F$ is obtained by the summation of the curves for E_0 and E_F. The minimum on the curve marked $E_0 + E_F$ indicates the equilibrium value of r_0, and the curvature in the region of the minimum permits a calculation of the compressibility. The line xy lies below the zero of energy at a distance equal to E_I, the ionization potential of a free atom of sodium. The distance between xy and the minimum of the curve $E_0 + E_F$ indicates the heat of sublimation.

---

* It should be pointed out that r_0 is defined by the relation $\frac{4}{3}\pi r_0^3 = \Omega$, where Ω is the atomic volume.

used differ in the degree to which simplifying assumptions are introduced. The simplest treatment of all is due to Fröhlich and may be called that of *the ideal metal*. In this the essential assumptions are that:

(1) The electron clouds of the ions are small compared with the distances between the atoms, so that their overlapping can be ignored.

(2) The effective electronic mass is unity and the Fermi distribution is spherical, so that the mean Fermi energy E_F can be calculated from the simple free-electron theory. In Part IV we have seen that E_F is proportional to $\left(\dfrac{N}{V}\right)^{2/3}$, where N is the number of electrons in volume V, and hence E_F is proportional to $\dfrac{1}{r_0^2}$; the constant of proportionality is approximately 2·2 if energy is measured in Rydberg units, and the unit of length is the radius of the first Bohr orbit (see p. 73).

(3) All correlation effects arising from the fact that two electrons cannot be at the same place at once can be ignored.

Subject to these simplifying assumptions, Fröhlich shows that the energy of the crystal per atom which we have previously written $(E_0 + E_F + E_I)$ can be expressed in the form:

$$-\frac{3}{2}\frac{e^2}{r_0} + \frac{2\cdot 2}{r_0^2} + \frac{1}{2}\frac{e^2 R^2}{r_0^3} + E_I \ . \quad . \quad . \quad . \quad \text{VI (3)}$$

Here the first term represents the potential energy, and is inversely proportional to r_0. The second term is the Fermi energy E_F referred to above, and is inversely proportional to r_0^2. The third term represents the kinetic energy of the lowest electron state (see p. 290), and the term assumes its comparatively simple form—inversely proportional to r_0^3—because the parts of the wave function of Fig. VI.3 which give rise to the kinetic energy are near the nucleus and are not affected by small variations in r_0. The constant R differs from one alkali metal to another, and can be calculated from the ionization energy, or from the experimental value of r_0. In this way it is possible to calculate the binding energy and the compressibility of the alkali metals at the absolute zero, and, except for lithium, reasonable agreement is obtained with the values found by experiment at other temperatures. The failure of the theory in the case of lithium is thought to be due to the effective electronic mass m^* being different from m.

The equations for the ideal metal of Fröhlich have been generalized by Bardeen, who writes the energy of the crystal of an alkali metal in the form:

$$-\frac{A}{r_0} + \frac{B}{r_0^2} + \frac{C}{r_0^3} + E_F \quad \ldots \quad \text{VI (4)}$$

and then determines A, B, and C from the experimental values of the lattice constant, the heat of sublimation, and the compressibility at zero pressure. The equation thus obtained is then used to calculate the variation of the compressibility with the pressure, and reasonable agreement with experimental data is obtained.† The method is sufficiently general to include the case where the effective electronic mass m^* (p. 225) differs from the normal mass m, and the energy and wave number are related, at least approximately, by an equation of the type:

$$E = \frac{h^2 k^2}{8 m^* \pi^2} \quad \ldots \quad \text{VI (5)}$$

For lithium m^*/m is in the range 1·2–1·5, and equation VI (5) is not strictly true.

The most accurate values of the binding energies‡ of the alkali metals are given by the Quantum Defect method (p. 238) and are shown below in kcal/mole:

	Li	Na	K	Rb	Cs
Observed binding energy	36·5	26·0	22·6	18·9	18·8
Calculated values	43·8	25·8	20·3	20·3	18·2

The Alkali Metals and Electron Correlation.

The above theories of the alkali metals involve the assumption that at any instant each atomic cell contains only one electron, and that further correlation effects, and also the effects of the electron clouds of the ions on one another, can be ignored. Attempts have been made to extend the theory, and the following points have been established for the alkali metals.§

The assumption that the atomic polyhedron may be replaced by an atomic sphere produces a negligible error. The mutual interaction of the electron clouds of the ions is very small in the alkali metals, and has a negligible effect on the binding energy, but the forces involved vary rapidly with the distances. Fuchs extended the theory to include the

† The data must, of course, be corrected to refer to the absolute zero.
‡ *See* J. Callaway, "Energy Band Theory." **1964**: New York and London (Academic Press).
§ For references see list on p. 301.

calculation of the elastic constants by a method in which the energy changes were calculated for deformations which produced a change of shape without a change of volume, and it was found that the interaction of the electron clouds of the ions produced a considerable effect in determining the elastic constants, and a slight effect on the compressibilities.

The most important corrections to the theory of the alkali metals concern the assumption that there is at a given instant only one electron in an atomic polyhedron, and that further correlation effects concerning the mutual positions of the electrons can be ignored. In the simplest Wigner–Seitz theory, it is assumed that there is at any instant only one valency electron in a given atomic cell, and that the Fermi energy of these electrons is given by the free-electron theory. In this case, if the energy is expressed in Rydberg units, and the unit of length is the first Bohr orbit:

$$E_F = +\frac{2\cdot 21}{r_0^2} \quad \ldots \ldots \quad \text{VI (6)}$$

and, using the previous notation, the energy of the crystal per atom may be written:

$$E_0 + E_I + \frac{2\cdot 21}{r_0^2} \quad \ldots \ldots \quad \text{VI (7)}$$

where E_0 is calculated (Fig. VI.4), and E_I is a constant of the free atom.

This expression requires correcting because, since each electron in the free-electron theory is regarded as distributed through the whole crystal, the electron in a given atomic sphere moves not only in the field of the free ion, but also in that of the negative charge of the valency electrons as a whole. It can be shown * that this produces an increase in the energy of the crystal, equal to $\frac{1\cdot 2}{r_0}$, so that the energy of the crystal per atom may be written:

$$E_0 + E_I + \frac{2\cdot 21}{r_0^2} + \frac{1\cdot 2}{r_0} \quad \ldots \ldots \quad \text{VI (8)}$$

This expression gives too large an energy, and would in fact make the crystal unstable. This is because VI (8) ignores completely the correlation of the positions of the electrons. These effects are generally considered under two headings, namely, effects between electrons of parallel spin, and effects between those of anti-parallel spin.

Electrons of parallel spin tend to avoid each other on account of the Exclusion Principle which makes it impossible for two electrons

* See N. F. Mott and H. Jones, " Theory of the Properties of Metals and Alloys," p. 140.

with parallel spins to be at the same place at the same time (p. 140). By thus keeping the electrons apart, it lowers their mutual electrostatic energy of repulsion by $\dfrac{0\cdot 92}{r_0}$ Rydberg units. This treatment, which again gives a set of plane waves for the electronic wave functions, is known as the *Hartree-Fock approximation*.

Electrons of opposite spin also tend to avoid each other, and according to Wigner this lowers the energy of the crystal by an amount equal to $\dfrac{b}{r_0}$, where b is a slowly varying function of r_0, and equals about 0·2 Rydberg units for the alkali metals. The total lowering of the energy of the crystal is thus $\dfrac{0\cdot 92 + 0\cdot 2}{r_0} = \dfrac{1\cdot 12}{r_0}$ Rydberg units, and when this is subtracted from VI (8), the energy of the crystal per atom becomes:

$$E_0 + E_I + \frac{2\cdot 21}{r_0{}^2} + \frac{0\cdot 08}{r_0}. \quad \ldots \quad \text{VI (9)}$$

so that the different correction terms have nearly cancelled out, and VI (7) is almost unaltered. In spite of the approximate nature of the calculations, the binding energies for the alkali metals calculated by means of VI (7) are in good agreement with the facts, as may be seen from the following table given by Pines.*

TABLE XI.—*Binding Energies of Alkali Metals.*

Metal.	Effective Mass of Valency Electrons.	Binding Energy, kcal/mole.	
		Wigner (calculated).	Observed.
Li	1·45	− 41	− 36·5
Na	0·98	− 30	− 26
K	0·93	− 26	− 23
Rb	0·89	− 24	− 19
Cs	0·83	− 23	− 19

Electron Correlation and Plasma Vibrations.

The whole problem of electron correlation has been discussed at length by Bohm and Pines,† who have shown that, owing to the relatively long range of the Coulomb interactions between the electrons in a metal, the latter may give rise to plasma oscillations analogous to

* D. Pines, " Solid-State Physics," Vol. I, p. 373. **1955**: New York (Academic Press); London (Academic Books).
† D. Bohm and D. Pines, *Phys. Rev.*, 1951, [ii], **82**, 625; 1953, **92**, 609; D. Pines and D. Bohm, 1952, **85**, 338; D. Pines, 1953, **92**, 626. For a general review see D. Pines, ' Solid-State Physics " (edited by F. Seitz and D. Turnbull), Vol. 1, p. 367. **1955**: New York (Academic Press); London (Academic Books).

those observed in electrical discharges. We may suppose that in a particular region of the metal there is at a given instant a deficit of electrons and hence a region of positive charge. The surrounding electrons will then rush towards this region in order to restore the charge balance, and in so doing they will acquire velocities which carry them past the positions for electrical neutrality, and a region of negative charge will be built up. The electrons will then be repelled from this region and will again overshoot the positions for electrical neutrality, so that a zone of positive charge is once more built up. The plasma oscillations result from this kind of process, and in metals the plasma frequencies are of the order of 10^{16} per sec, and the quantum of energy associated with this frequency is greater than the energy of an electron at the top of the Fermi distribution. Consequently no electron in a metal has sufficient energy to excite a plasma oscillation into a higher state, and these oscillations exist in metals only in their lowest or ground state. They may, however, be excited by the passage of fast charged particles through metals. These effects have been noted when high-speed electrons are passed through metallic films, and the transmitted beam is found to contain electrons of different velocities corresponding to those which have excited one or more quanta of the plasma vibrations.

The plasma oscillations result from the mobility of the electrons, and this mobility enables the electrons to exert a screening action on each other's charges. The Coulomb repulsion between electrons means that each electron is surrounded by a cloud of charge which screens the field of the electron within a distance of the order of the Debye length, λ_D, defined by the relation:

$$\lambda_D = \left(\frac{kT}{4\pi n_0 e^2}\right)^{1/2} = \left(\frac{m\langle v^2\rangle_{AV}}{12\pi n_0 e^2}\right)^{1/2}$$

where n_0 is the number of electrons per unit volume, and $\langle v^2\rangle_{AV}$ is the mean square velocity of the electrons. For a metal $\lambda_D \sim 0.5$ Å. For phenomena involving distances $> \lambda_D$, the system behaves collectively and is characterized by harmonic oscillators representing the plasma vibrations. For effects involving distances $< \lambda_D$ the system is best described as a collection of independent electrons interacting weakly via a screened Coulomb force, and it is this which provides the justification for treating the valency electrons as a gas of free particles. The electrons in the plasma are thus capable of both collective and independent particle behaviour, and the Debye length provides a rough measure of the distances within and outside which the different effects predominate.

The theory of the electron-plasma effects lies quite outside the

Electrons, Atoms, Metals, and Alloys

present book. It has been examined in detail by Bohm and Pines for the case in which the periodic field of the ions is replaced by a uniform background of positive charge, but even in this simplified model the calculations are only approximate. So far as the binding energy is concerned, the results agree with those of Wigner to within a few per cent. The division of the correlation energy between the electrons of parallel and anti-parallel spin is, however, considerably altered, and the fact that the electrons of parallel spin avoid each other owing to Coulomb interaction, produces a reduction in the term $\dfrac{0.92}{r_0}$ referred to on p. 295, and the correlations arising from the Pauli Principle are important only when two electrons are closer together than the screening radius λ_D.

Using the simplified model referred to above, Pines * has shown that

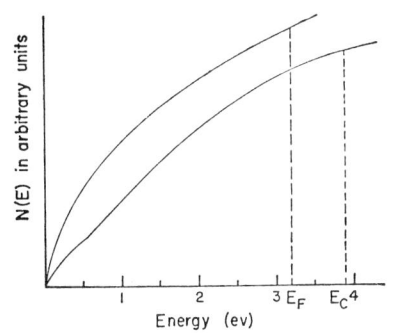

FIG. VI.5. $N(E)$ curve for sodium. The upper curve is that given by Raimes (*Phil. Mag.*, 1954, [vii], **45**, 727) on the basis of the Sommerfeld free-electron theory. The lower curve is that corrected for electron correlation, subject to the approximations described in the text. (*Pines*.)

the $N(E)$ curve for sodium takes the form of Fig. VI.5, where the kink in the low-energy region results from the sharp cut-off in the plasma oscillations. The new band width is a little greater than the old, but it is doubtful whether the difference is significant, and detailed comparison with the experimental data is difficult because of the uncertainties in the low-energy region.

For the electronic specific heats, the electrical conductivities σ, and the thermo-electric powers s, the calculated values allowing for correlation are so near to those given by the free-electron theory that it is not possible to say whether the calculations of correlation effects are correct. For the magnetic properties, the new calculations have, however,

* D. Pines, " Solid-State Physics," Vol. 1, p. 367.

removed one of the older difficulties. Since the expressions VI (6)–VI (8) above involve terms of the nature of $\dfrac{x}{r_0^2}$ and $\dfrac{x}{r_0}$, they imply that, at a sufficiently large value of r_0, the increase in Fermi energy (p. 191) would be less than the decrease in energy resulting from the reversal of spin. Dilute electron gases would, therefore, be ferromagnetic, and the Hartree–Fock approximation had in fact the effect of requiring caesium to be ferromagnetic. According to the new collective theories, this difficulty is removed, although the calculated susceptibilities are not yet in good agreement with the observed values for the alkali metals.

In general, it may be said that the calculations of correlation effects are still too approximate to be of great value to metallurgical theory. The examination of the simplified model has, however, explained why the effects of the different complicating factors cancel each other in such a way that the free-electron theory remains a good approximation for the alkali metals.

Lithium and Its Abnormal Properties.

From Fig. VI.2 it will be seen that lithium differs from the remaining alkali metals in that it gives rise to immiscible liquids in the systems Li–Na, Li–K, Li–Rb, and Li–Cs, whereas the remaining alkali metals are all miscible in the liquid state. In much of its alloy chemistry, lithium is abnormal and appears to behave as though it had a valency greater than unity. There is, for example, an Ag–Li phase, of composition approximately Ag_3Li_{10}, with a γ-brass type of structure. No combination of univalent elements can give the characteristic electron/atom ratio of 21/13 although, if lithium were divalent, the above composition would be nearly that required for the 21/13 ratio. In the system Li–Mg, there is a very wide solid solution of magnesium in lithium, and a considerable solid solution of lithium in magnesium, and as can be seen from Fig. VI.6 the two elements behave as though they were of the same valency. This is in contrast to systems such as Ag–Mg and Na–Mg, where the size-factors are favourable, but the solid solutions in the divalent magnesium are restricted; in the system Na–Mg, the metals are almost immiscible in the liquid state.

This characteristic of lithium resembles the well-known "diagonal principle" of Inorganic Chemistry

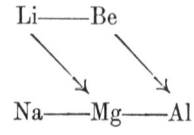

```
Li────Be
   ╲    ╲
    ↘    ↘
Na────Mg────Al
```

according to which the compounds of the elements lithium and beryllium in the First Short Row of the Periodic Table tend to resemble those of magnesium and aluminium, respectively, in the Second Row.

Some of the physical properties of lithium are also abnormal. Its thermo-electric power s is positive, whereas for the remaining alkali metals s is negative. The electrical conductivity of lithium is decreased by hydrostatic pressure, whereas the general effect of pressure is to increase the conductivity by reducing the amplitude of the lattice vibrations.

According to Cohen and Heine,* the differences in physical properties

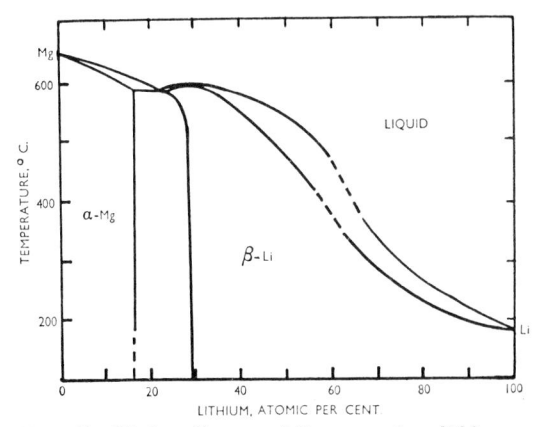

Fig. VI.6. Equilibrium diagram of the magnesium–lithium system.

are connected with the distribution of s and p states in the Brillouin zone of lithium. In the free atoms of all the alkali metals, the valency electrons occupy the ns levels, which are lower than the unoccupied np levels. The difference between the two levels may be denoted Δ_{sp}, and as shown below, these values increase steadily on passing from Cs ⟶ Rb ⟶ K ⟶ Na, but on proceeding to lithium this tendency

	Li	Na	K	Rb	Cs
Δ_{sp}	1·85	2·10	1·61	1·58	1·44

is reversed. In lithium, therefore, the $2p$ level is not so far above the $2s$ level as would be expected from the behaviour of the remaining alkali metals.

When the free atoms assemble to form a crystal of the metal, the energy levels of the ns and np states broaden out to such an extent that they overlap (see p. 91). The occupied states then lie within the

* M. H. Cohen and V. Heine, *Advances in Physics*, 1958, **7**, 395.

Fermi surface, and if this touches the surface of the first Brillouin zone it will do so first at points such as A (Fig. V.30) at the centres of the zone faces. The theory then works in such a way that the states at A just inside the first zone are either purely s-like or purely p-like, but are not a mixture of the two.

If the energy states at A, just inside the zone, are p-like, the states in the second zone, just outside A, are s-like, and vice versa. Cohen and Heine suggested that in lithium the states at A inside the zone are p-like, and that there is a large energy gap (ca. 2–3 eV) across the zone face. This produces a distortion of the Fermi surface from the spherical, which is so marked that, although the zone is only half-filled (1 electron/atom), the Fermi surface touches the centres of the dodecahedral faces at points such as A (Fig. V.30, p. 251). This touching of the zone faces results in a fall in the $N(E)$ curve, and since the states in the region of A are of a p-like nature this fall is revealed in the K-emission spectrum (Fig. III.5, p. 95).

For sodium the energy gap across the face of the zone is much smaller, and the Fermi surface is very nearly spherical and does not reach the zone boundary. It is uncertain whether the s-like or p-like state has the lower energy at the centre of the zone face, but this is immaterial because the Fermi surface is relatively far from the zone face, so that in either case the $N(E)$ curve rises continuously until the Fermi limit is reached, and this is shown in the K-emission spectrum of Fig. III.5.

For potassium, the Fermi surface is again nearly spherical, and the calculations suggest that the states at A inside the first zone are s-like. This characteristic is retained in rubidium and caesium, for which the Fermi surfaces become somewhat distorted from the spherical, but do not actually reach the zone faces.

More recent theoretical work[*] has yielded results in good accord with the distorted (but *not* zone-contacting) Fermi surface found for lithium, the almost spherical surface of sodium, and the gradual increase (p. 286) in distortion in the series K → Rb → Cs. The details are not easily described in simple terms but the relative amounts of s- and p-character are still a central feature, with considerations of d-character for the later elements. The absence of p levels in the ion core of lithium means that p-character plays a larger role in its conduction electrons than in sodium (with a $2s^2 2p^6$ shell in its ion core), while an increasing role of the $3d$, $4d$, and $5d$ excited states of the potassium, rubidium, and caesium atoms can be correlated with the increasingly distorted Fermi surfaces in the metals.

[*] *See* V. Heine and I. Abarenkov, *Phil. Mag.*, 1964, **9**, 451. M. F. G. Lee, *Phys. Rev.*, 1969, **178**, 953.

REFERENCES AND SUGGESTIONS FOR FURTHER READING.

Electronic Structure of Alkali Metals.
E. Wigner and F. Seitz, *Phys. Rev.*, 1933, [ii], **43**, 804; 1934, [ii], **46**, 509.
E. Wigner, *ibid.*, 1934, [ii], **46**, 1002.
J. Bardeen, *J. Chem. Physics*, 1938, **6**, 367.
D. J. Howarth and H. Jones, *Proc. Phys. Soc.*, 1952, [A], **65**, 355.
M. H. Cohen and V. Heine, *Advances in Physics*, 1958, **7**, 395.
B. Schiff, *Proc. Phys. Soc.*, 1954, [A], **67**, 2.
F. Ham, *Phys. Rev.*, 1962, **128**, 82, 2524.
V. Heine and I. Abarenkov, *Phil. Mag.*, 1964, **9**, 451.
M. F. G. Lee, *Phys. Rev.*, 1969, **178**, 953.

Theory of Compressibility and Elastic Constants.
K. Fuchs, *Proc. Roy. Soc.*, 1936, [A], **153**, 622; **157**, 444. (The second paper corrects mistakes in the first.)

General Theory.
H. Fröhlich, *Proc. Roy. Soc.*, 1937, [A], **158**, 97.

Much of this work is summarized in "The Modern Theory of Solids," by F. Seitz. **1940**: New York (McGraw-Hill Book Co., Inc.); and in "The Theory of the Properties of Metals and Alloys," by N. F. Mott and H. Jones. **1936**: Oxford (Clarendon Press).

2. COPPER, SILVER, AND GOLD.

General.

COPPER, silver, and gold in Group IB are univalent metals whose atoms contain one valency electron outside a closed group of 18. They all crystallize in the face-centred-cubic structure (Fig. III.12, p. 104). Fig. VI.7 shows the electron-cloud density $\rho(r)$ curve for the free *ion*, Cu^+, and the lower vertical line shows one-half the closest distances of approach of the atoms in the *crystal of pure copper*. When this figure is compared with the corresponding diagram (Fig. VI.1) for the alkali metals, it will be seen that in the crystal of copper there is a much greater overlap of the electron clouds of the ions. The fact that the electron clouds of the ions overlap so much means, of course, that the curve of Fig. VI.6 is not an accurate representation of the actual state of affairs in the metal, although it does show clearly that the ionic overlap is considerable. The repulsive forces resulting from this ionic overlap vary rapidly with the interatomic distance, and it is for this reason that the compressibility of copper is so much smaller than those of the alkali metals. There is a general tendency for the compressibility of a metal to increase with the atomic volume, but as can be seen from Fig. VI.8, the compressibilities of the alkali metals are so much greater than that of copper that the effect cannot be due solely to the greater atomic volumes of the alkali

302   *Atomic Theory*

metals. Similar considerations apply to silver and gold, and the pronounced ionic overlap in these metals means that, as pointed out by Mott and Jones, the crystals of copper, silver, and gold may be regarded as built up of hard spheres (ions) held in contact by the valency electrons, in contrast to the alkali metals, where the ionic overlap is so small that it accounts for only a small fraction of the repulsive force.

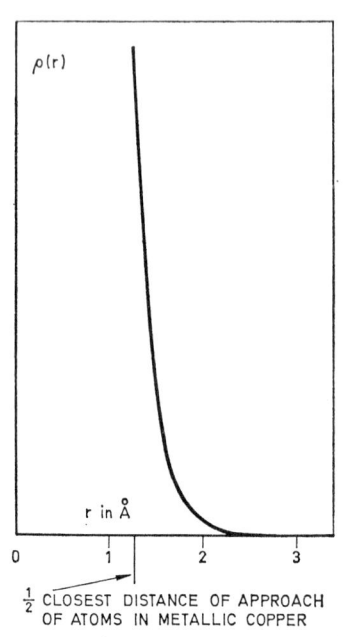

Fig. VI.7. The $\rho(r)$ curve for the outermost part of the Cu$^+$ ion. The vertical scale is arbitrary, but is the same as that of Fig. VI.1.

It has been shown by Fuchs that if the compressibility of copper is calculated by the methods of Wigner and Seitz without allowing for the effect of the ionic overlap, the calculated value is from 4 to 8 times that observed experimentally; whilst if the ionic overlap is allowed for, a good agreement is obtained between the calculated and observed values. The resistance of copper to compression is thus largely the result of the ionic overlap, and the same factor plays a predominant effect in determining the elastic constants.

The work of Fuchs has also shown that for copper, when the energy of the ionic overlap is taken into account, the face-centred cubic structure is more stable than that of a body-centred cube, and we can thus understand why copper crystallizes in the face-centred cubic structure. In the case of the alkali metals, however, there is little difference between the energies calculated for the two structures, and the fact that all these elements crystallize in the body-centred cubic structure has not yet been

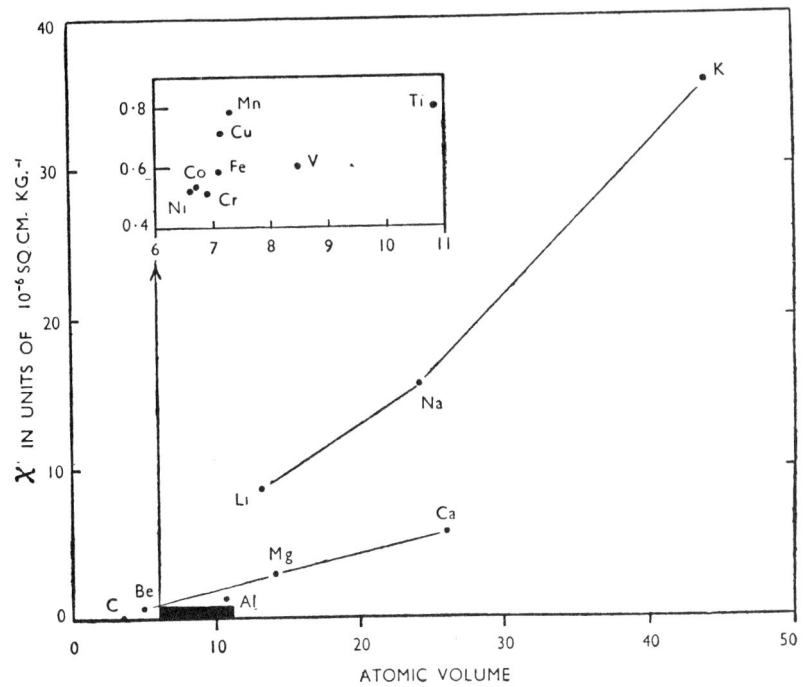

FIG. VI.8. Compressibilities of the metals of lower atomic number plotted against their atomic volumes. The points for Li, Na, and K, and for Be, Mg and Ca show the general tendency for the compressibilities to increase with atomic volume. The compressibilities of the elements Ti–Cu are very much smaller, and this is due to the overlap of the electron clouds of the ions.

explained. It is, however, possible to understand (p. 322) why, in lithium and sodium, where close-packed modifications are stable at low temperatures, the close-packed forms exist at the low, rather than at higher temperatures.

first Brillouin zone of the face-centred cubic structure has been illustrated in Fig. V.5 (p. 202). The zone is a truncated octahedron, and the faces of the zone nearest the origin are the (111) planes, and those next nearest are the (200) planes. This corresponds with the fact that the first two strong X-ray (or electron-wave) reflections from face-centred cubic crystals have indices (111) and (200).

The zone of Fig.V.5 can contain two electrons per atom, and as there is only one electron per atom in copper, silver, and gold, the zone is only half-filled. In the early theories it was thought that, at an e.c.* of 1·0, the occupied states would lie inside an almost spherical Fermi surface which had not touched the surface of the zone (Fig. V.5 (a)). Later calculations by H. Jones (p. 311) suggested that the Fermi surface was highly distorted and touched the octahedral face of the zone at an e.c. of about 1·0. Measurements of the anomalous skin effect by Pippard † and of the de Haas–van Alphen effect by Shoenberg ‡ showed that in pure copper the Fermi surface has touched the zone, and Fig. VI.9 gives the Fermi sur-

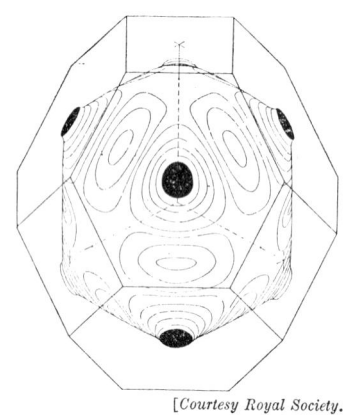

[Courtesy Royal Society.]

FIG. VI.9. Suggested Fermi surface of copper. (*Pippard*).

face deduced by Pippard for copper. Later work by Shoenberg § showed that, in silver and gold, the Fermi surfaces also touched the zone. These conclusions are confirmed by other physical properties, and many arguments based on spherical Fermi surfaces must now be discarded.

Recent calculations of the band structure of copper are in reasonable agreement with experimental indications of a full $3d$ band with a width of about 2 eV centred about an energy 3 eV below the Fermi level.

The calculations of several authors indicate that the binding energy of the copper crystal is distinctly greater than can be accounted for by the valency electrons alone, and there is now general agreement that the

---

* For brevity we denote electron concentration by e.c.
† A. B. Pippard, *Phil. Trans. Roy. Soc.*, 1958, [A], **250**, 325.
‡ D. Shoenberg, *Nature*, 1959, **183**, 171.
§ D. Shoenberg, *Phil. Trans. Roy. Soc.*, 1962, [A], **255**, 85.

interaction between the outermost electrons of adjacent ions gives rise to an additional cohesive force. The fact that copper is diamagnetic suggests strongly that the $(3d)^{10}$ band is filled, and so prevents the otherwise natural assumption of the existence of some divalent copper atoms (analogous to the cupric compounds) in the solid element. As explained later (p. 363), the resonance-bond theories of Pauling assumed that on the average as many as 5·4 electrons per atom took part in metallic resonance-bond cohesion in copper, but such a high valency for copper is regarded as improbable by most other authors, and there is no chemical evidence to justify the assumption of such high valency states.

According to Mott,* the wave function of the copper atom in metallic copper is strongly perturbed by the influence of the surrounding atoms, and this is the case for the $3d$ electrons as well as for the $4s$ or valency electrons. In this sense the $3d$ electrons may be said to take part in the cohesion of the copper crystal, giving rise to forces which may be of a van der Waals or correlation nature. This view was originally criticized because Fuchs' calculations (p. 327) indicated a small van der Waals force, but the later work of Friedel † showed that the forces due to the interaction between the electrons of the ions were considerable, and might account for about 30% of the binding energy. Some writers consider that, with increasing strength of the perturbation, there may be a continuous transition from van der Waals forces to those of a co-valent-resonance nature, and that intermediate forms may exist which are not yet understood, but may correspond to the condition suggested by Mott.

A further possibility for copper, also suggested by Mott, is that the additional cohesion results, not from the perturbation of the $3d$ electrons of one atom by those of its neighbours, but from the perturbation of the $3d$ electrons of an atom by its own $4s$ or valency electron.

It is thus certain that the normal univalent cohesion in the copper crystal is reinforced by forces in which the $(3d)$ electrons are concerned, even though the exact nature of the process is unknown. Many facts show that the $(4d)^{10}$ sub-group in silver has a greater stability than the $(3d)^{10}$ and $(5d)^{10}$ sub-groups in copper and gold, respectively.‡ We should therefore expect the perturbation of the $4d$ electrons in silver to be less marked than those of the $3d$ and $5d$ electrons in copper and gold, and hence for the additional cohesion to be least in silver. Reference

* N. F. Mott, *J. Inst. Metals*, 1944, **70**, 269 (discussion).
† J. Friedel, *Proc. Phys. Soc.*, 1952, [*B*], **65**, 769.
‡ Thus, silver gives rise only to univalent compounds, whereas copper and gold yield also divalent and trivalent compounds, respectively, indicating that in copper one $3d$ and in gold two $5d$ electrons can be perturbed sufficiently to take part in chemical combination.

to Figs. VI.31–35 will show that silver has a higher compressibility and coefficient of expansion, and a lower melting point and binding energy than either copper or gold, in agreement with the greater stability of the $(4d)^{10}$ sub-group.

Alloys of Copper, Silver, and Gold.

The lattice spacings and interatomic distances in copper, silver, and gold are shown in Table XII, and Fig. VI.10 shows the equilibrium diagrams of the systems copper–silver, copper–gold, and silver–gold. A continuous solid solution is formed in the system silver–gold, where the size-factor is very favourable. In the systems copper–silver and copper–gold the size-factors are almost identical, and are on the borderline of the favourable zone. Here it will be seen that the copper–silver alloys form a simple eutectiferous system, with comparatively restricted solid solutions. In the system copper–gold the size-factor is almost the same as in the system copper–silver, but as will be seen from Fig. VI.10, copper and gold are completely miscible in the solid state at high temperatures, although well-defined superlattices exist at low temperatures. The presence of these superlattices is undoubtedly due to the fact that the atomic diameters of gold and copper are on the borderline of the favourable zone, because this means that the formation of a solid solution is

TABLE XII.—*Lattice Spacings and Interatomic Distances of Copper, Silver, and Gold.*

Metal.	Lattice Spacings, kX.	Closest Interatomic Distance, kX.	Increase in Atomic Diameter, %.
Cu	3·6078 at 18° C.	2·5511	} 13
Ag	4·0779 at 25° C.	2·8835	} <1
Au	4·0704 ,,	2·8782	

accompanied by a considerable strain which may to some extent be relieved by a regular arrangement of the atoms. It is, however, clear from Fig. VI.10 that atoms of copper and gold tend to fit together more easily than atoms of copper and silver in spite of the size-factors being the same in the two systems. This is probably the result of the fact that the ion of gold is more easily polarizable than the ion of silver since, as explained above, the latter involves the relatively stable $(4d)^{10}$ sub-group. This means that the electron cloud of the gold ion can be more easily distorted, and so can fit more readily into the copper lattice than the ion of silver. This serves to emphasize the limitations of

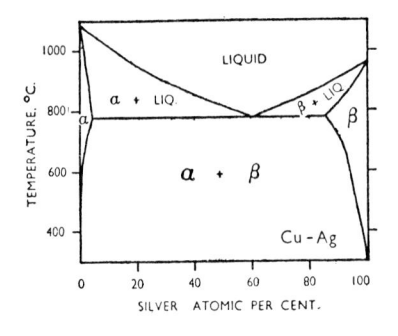

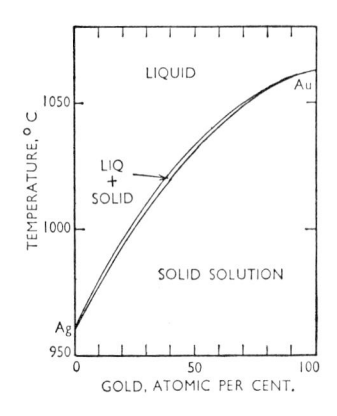

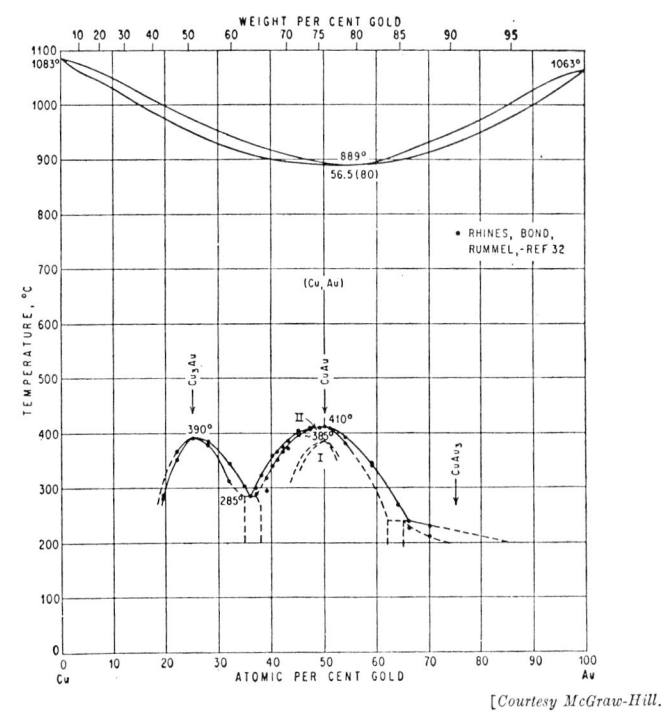

FIG. VI.10. The equilibrium diagrams of the systems Cu–Ag, Ag–Au, and Cu–Au. (*The last is after Hansen*)

attempts to represent the size of an atom by a single constant, and the polarization effects have been discussed by Hume-Rothery and Raynor.* Support for this view is given by the fact † that Young's modulus is lower for gold (8000 kg/mm^2) than for copper (13,000 kg/mm^2), although the bulk modulus is greater for gold (17,000 kg/mm^2 as compared with 13,500 kg/mm^2 for copper). This suggests that the resistance of gold to hydrostatic pressure is much greater than to a unidirectional force, in agreement with the view that the ion of gold is more easily deformable than that of copper.

The Theory of Copper and Silver Alloys.

In the alloys of copper, silver, and gold with one another, the atoms are all univalent, and the e.c. is always 1·0. At this electron concentration the face-centred and body-centred cubic structures would differ very little in energy if the valency electrons alone were concerned, and the face-centred cubic structure is preferred owing to the forces produced by the overlap of the ions.

When these metals are alloyed with elements of higher valency, the electron concentration is increased, and the study of these alloys has thrown much light on the principles underlying alloy formation.

In Section III.4 we have described the main characteristics of the equilibrium diagram of the system copper–zinc, where increasing zinc content results in the f.c. cubic α-phase being succeeded by the β, γ, ε, and η phases. Structurally analogous phases are found in other systems, the characteristic being that a phase of a given structure occurs at roughly the same electron concentration. Further, as shown below, the equilibrium diagrams are such that, if the size-factors are favourable and the electrochemical factors are small, some of the phase boundaries are almost superposed when the equilibrium diagrams are drawn in terms of electron concentration instead of in atomic percentages. These empirical relations suggest that, in the copper and silver alloys concerned, the equilibrium structures depend primarily on the electron concentration.

In the simplest theory, the ideas of which are due mainly to H. Jones, it is assumed that the solution of an element of higher valency in copper or silver can be regarded simply as increasing the number of electrons per atom, and that the alloy takes up the phase or mixture of phases which allows the electrons to be accommodated with the lowest Fermi energy.

We may consider two crystal structures which give rise to the two

* W. Hume-Rothery and G. V. Raynor, *Phil. Mag.*, 1938, [vii], **26**, 129.
† A. D. N. Smith, *J. Inst. Metals*, 1951–52, **80**, 477.

$N(E)$ curves shown in Fig. VI.11. If a given number of electrons is introduced into the two structures, the number of occupied states will be the same, and in Fig. VI.11 the shaded areas are equal. In Fig. VI.11(a) no electron has an energy greater than A, whereas in Fig. VI.11(b) the occupied states extend as far as B. It is clear, therefore, that a structure

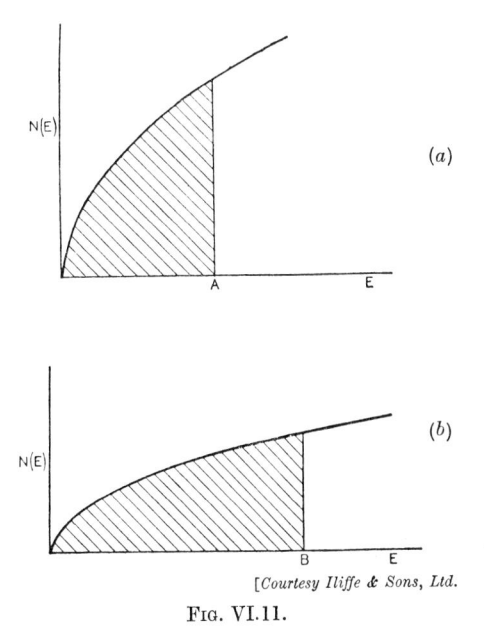

[Courtesy Iliffe & Sons, Ltd.

FIG. VI.11.

with a high $N(E)$ curve will accommodate a given number of electrons with a lower energy than one with a low $N(E)$ curve.

The general form of $N(E)$ curve for the face-centred cubic structure is as shown in Fig. VI.12, where the point A represents the stage at which the Fermi surface first touches the octahedral face of the zone (Fig. V.5),

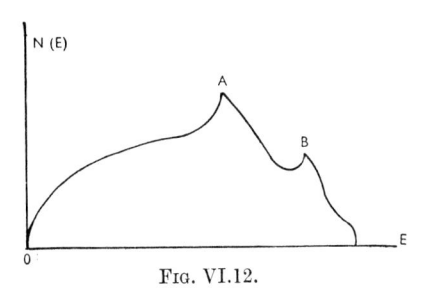

FIG. VI.12.

whilst the point B marks the stage at which the cube face is touched. The falls in the $N(E)$ curve after A and B mean that the number of electrons per unit-energy range is decreasing rapidly, and hence the increase in the energy for the addition of a given number of electrons increases rapidly beyond A, and again beyond B. If, therefore, we compare the crystal structure, say X, which gives rise to the $N(E)$ curve of Fig. VI.12 with another structure Y for which the $N(E)$ curve remains high beyond A, then, as far as the Fermi energy is concerned, the structure X will tend to become less stable relative to Y as the electron concentration increases beyond A, since the more stable structure will be that which accommodates the electrons with the lower energy.

The simple theory of Jones may be summarized by saying that, on increasing the electron concentration, there will be a general tendency for a phase to be stable at the point A, but to become increasingly less stable in the region of the fall from A to B because, on account of the large number of other possible crystal structures, there will probably be some other structure for which the $N(E)$ curve remains high, and which will therefore accommodate the electrons with a lower energy. The details of the process will depend on the relative forms of the $N(E)$ curves of the different possible structures, but, so far as the phase X is concerned, the steeper the fall in the $N(E)$ curve after the peak at A, the more rapidly will the phase tend to become unstable. Unless the energy contours are very distorted, the $N(E)$ curve will fall more rapidly the more nearly the Brillouin zone approximates to a sphere.

When expressed in this way, the ideas of the Jones theory are almost certainly correct for many alloys whose structures depend primarily upon electron concentration. The subject has been greatly confused by the fact that it was often applied with the assumption of spherical Fermi surfaces, although, as explained on p. 311, it has been known since 1937 that the Fermi surface for copper must be markedly distorted. It must also be realized that the change in Fermi energy relative to the bottom of the band is very different from the change in the total free energy of the system on going from a pure metal to a solid-solution alloy. The success of the Jones approach merely demonstrates that at a given alloy concentration the energy *differences* between various crystal structures correlate well with the differences in average band electron energies.

We may consider the alloys under the following headings:

(a) *The α/β-Brass Equilibrium.*

In 1934, Hume-Rothery, Mabbott, and Channel-Evans * showed that, where the size-factors were favourable, the maximum solubilities

* *Phil. Trans. Roy. Soc.*, 1934, [A], **233**, 1.

of B sub-Group elements in copper and silver corresponded to an electron concentration of approximately 1·40. If this rule were accurate, it would mean solubility limits of 40, 20, 13·3, and 10 at.-% of di-, tri-, tetra-, and pentavalent solutes, respectively. The actual values are about 40 at.-% (Cu–Zn, Ag–Cd), 20 at.-% (Cu–Ga, Cu–Al), 12 at.-% (Cu–Ge, Ag–Sn), and 7 at.-% (Ag–Sb), respectively, so that the rule becomes less accurate with increasing valency of the solute.

The exact form of an α-solubility curve must clearly depend upon the nature of the phase with which the α-phase is in equilibrium. In the general case, the solubility decreases with falling temperature, but Fig. III.19 (p. 119) shows that, in the system Cu–Zn, when the solubility limit of the α-phase is exceeded, the $\alpha/(\alpha + \beta)$ and $(\alpha + \beta)/\beta$ solubility curves have a characteristic form in which the solubility of zinc in the α-phase increases as the temperature falls. This is a characteristic of the equilibrium between the disordered f.c. cubic (α) and b.c. cubic (β) phases in this class of alloy, and may be termed the α/β *brass equilibrium*. Examination then shows that, provided the size-factors are small, the $\alpha/(\alpha + \beta)$ curves are roughly superposed if the equilibrium diagrams are drawn in terms of electron concentration. This is illustrated in Fig. VI.13 where several of these curves are drawn in terms of electron concentration. In the systems copper–indium and copper–tin, the size-factors are approaching the unfavourable zone, and the curves are displaced in the direction of lower electron concentration; the free-energy curves of Fig. VI.14 have shown how this may be understood if the relatively open b.c. cubic structure can accommodate the large atoms more readily than the close-packed f.c. cubic structure.

The general impression is clearly that the α/β brass equilibrium is controlled by the electron concentration, but that as the size-factor becomes unfavourable the effects are complicated by the strain energy resulting from the lattice distortion.

As explained on p. 328, the assumption of free electrons and spherical Fermi surfaces leads to the conclusion that the first peaks on the $N(E)$ curves for the α- and β-phases will occur at electron concentrations of 1·36 and 1·48, respectively. There is thus a region between these two points where, with increasing electron concentration, the $N(E)$ curve for the α-phase is falling, whilst that for the β-phase continues to rise, and hence by the argument of p. 310 the α-phase will become increasingly less stable than the β-phase. This gives an apparently very simple explanation of the fact that the change from the α to the β structure occurs at an electron concentration of about 1·4.

In 1937, H. Jones carried out more detailed calculations for the $N(E)$ curves of f.c. and b.c. cubic structures. The shapes of the curves

312   *Atomic Theory*

in the neighbourhood of the peaks depend upon the energy gaps, ΔE, at the surface of the zone, and the optical properties of copper indicate that $\Delta E = 4\cdot 1$ eV. By assuming this value for both f.c. and b.c. cubic structures, Jones obtained the $N(E)$ curves shown in Fig. VI.15. These

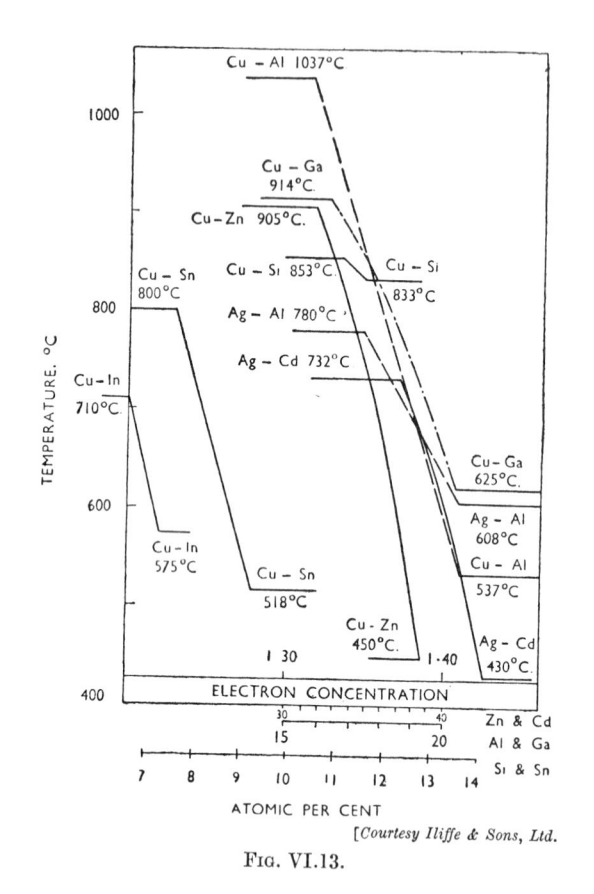

FIG. VI.13.

again show a region in which the curve for the face-centred cubic structure is falling whilst that for the b.c. cubic structure continues to rise, but the peak on the former curve is at an electron concentration of approximately 1·03, as compared with the value of 1·36 for the simple theory involving spherical Fermi surfaces. There is, however, a compensating change in the curve for the b.c. cubic phase, for which the peak on the $N(E)$ curve is at an electron concentration of about 1·20, as

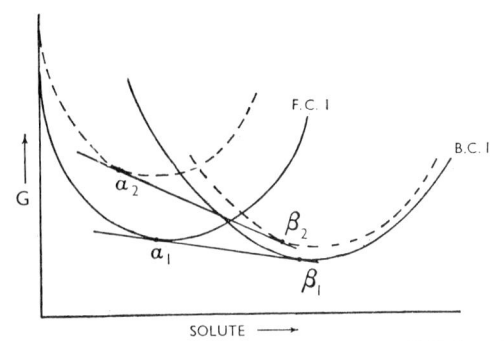

FIG. VI.14. Schematic free-energy/composition curves. F.C. 1 and B.C. 1 refer to face-centred cubic and body-centred cubic structures where the size-factors are very favourable. The corresponding dotted curves are for a case where the size-factor is increasing, and increases the free energy of the close-packed face-centred cubic solid solution more than that of the relatively open body-centred cubic structure. The points of contact of the common tangents show how the phase boundaries are displaced in the direction of lower solute concentration, and hence of lower electron concentration.

compared with the value of 1·48 for spherical Fermi surfaces. Fig. VI.16 shows Jones's calculations for the difference between the Fermi energies for f.c. and b.c. cubic structures as a function of the number of electrons per atom. The calculations refer to the absolute zero of temperature,

[*Courtesy Physical Society.*]

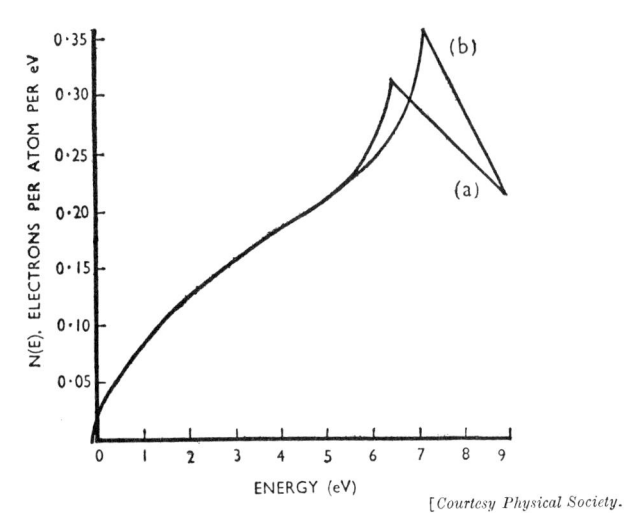

FIG. VI.15. Curves showing the number of states per unit energy range as a function of the energy (a) for the face-centred cubic, and (b) for the body-centred cubic structure. As explained in the text, these curves are of great historical interest, but may be incorrect for alloys.

and show that for e.c. 1·0–1·4 the lowest energy is given by the α-phase alone. From e.c. 1·4 to 1·44 the lowest energy results from a mixture of α- and β-phases, and at higher electron concentrations the β-phase by itself has the lowest energy. These calculations thus again account for an α/(α + β) phase boundary in the region of e.c. = 1·4.

The above calculations of Jones have produced a condition of great confusion. On the one hand they lead to the correct position of the α/β brass phase boundaries, and they suggest that the peak on the $N(E)$ curve for the f.c. cubic structure of copper is at an electron concentration of 1·08, in reasonable agreement with the work of Pippard

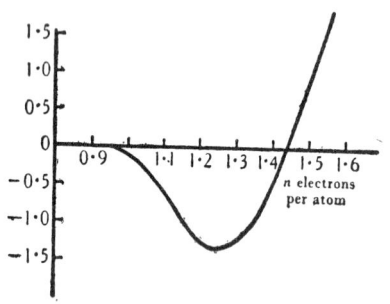

[Courtesy Physical Society.]

Fig. VI.16. The difference between the Fermi energies of the face-centred and body-centred cubic structures as a function of the number of electrons per atom.

referred to above (p. 304), according to which the Fermi surface has just touched the zone face at an electron concentration of 1·0. On the other hand, acceptance of the Jones calculations means the rejection of the simple correspondence between the electron concentrations of the peaks on $N(E)$ curves, and the compositions at which electron compounds occur, or phase boundaries are found.

This situation has been discussed in many places, and with the increased use of the idea of a pseudopotential (see p. 238) it is beginning to be possible to show that the correlation between the simplest (spherical-Fermi-surface) approach and a proper calculation of the relative energies of the α and β phases is not accidental.* In effect, the pseudopotential calculations show that large changes in the total electronic energy take place when the average wave-vector at the Fermi surface has a magnitude equal to the distance to the closest face of the Brillouin

---

* See "Phase Stability in Metals and Alloys" (edited by P. S. Rudman, J. Stringer, and R. I. Jaffee), especially the article by Heine. **1967**: New York and London (McGraw-Hill).

zone, in this case the (111) face. The change in energy when a particular part of the Fermi surface first makes contact with the zone is very small by comparison, as can be seen in the original calculations of Jones, represented in Fig. VI.16. (Notice that the region in question here is that at $n = 1\cdot03$, *not* $n \simeq 1\cdot0$, since the model implied a first contact at 1·03 electrons/atom.)

The theory based on the relative Fermi energies of two phases is not yet able to explain * the curious slope of the $\alpha/(\alpha + \beta)$ phase boundaries, although the conclusions of Zener referred to below (p. 322) do enable us to understand how the free energy of the b.c. cubic phase may become less than that of the f.c. cubic phase at high temperatures, and the increasing ranges of the β-phases at high temperatures may be understood in this way. The complete theory must clearly involve both electron concentration and lattice distortion and some of the principles have been discovered empirically by means of high-temperature X-ray techniques.

In equilibria of the α/β-brass type we are concerned with two phases, and the effect of lattice distortion is likely to be revealed as a relation between the composition and lattice spacings of the two phases, rather than as an effect on one phase boundary alone. We may note that the mean lattice distortion as revealed by X-rays is not a complete measure of the distortion process concerned. Thus, in the α/β equilibria in the systems copper–zinc and copper–tin, the lattice spacings of the α-phases in equilibrium with the β-phases at a given temperature are nearly the same, so that the mean lattice distortions of the α-phases are of the same order. But this distortion is produced by 35–40 at.-% of zinc in the system copper–zinc, and by less than 10 at.-% of tin in the system copper–tin. The distortion process is thus much more uniformly spread out in the system copper–zinc, and the data suggest clearly that a given mean lattice distortion, when distributed uniformly, affects the phase boundaries less than the same mean distortion proceeding from relatively few centres, where the local distortion must be correspondingly greater. To study this problem Andrews and Hume-Rothery measured the lattice spacings of the α- and β-phases in equilibrium with one another in the systems Cu–Zn, Cu–Al, Cu–Ga, Cu–In, and Cu–Sn at a standard temperature of 672° C. These results, together with the previously known equilibrium diagrams, enabled them to calculate (1) the electron concentration, *i.e.*, the number of electrons per atom, and (2) the volume per valency electron for the two phases

* The original paper of H. Jones (*Proc. Phys. Soc.*, 1937, **49**, 250) claimed to calculate the correct slope of the phase boundaries, but later work by C. Zener (*Phys. Rev.*, 1947, [ii], **71**, 846) showed that some mistake had been made.

which are in equilibrium at the temperature concerned. The results obtained are shown in Fig. VI.17, from which it will be seen that the electron concentrations of the two phases in equilibrium differ most

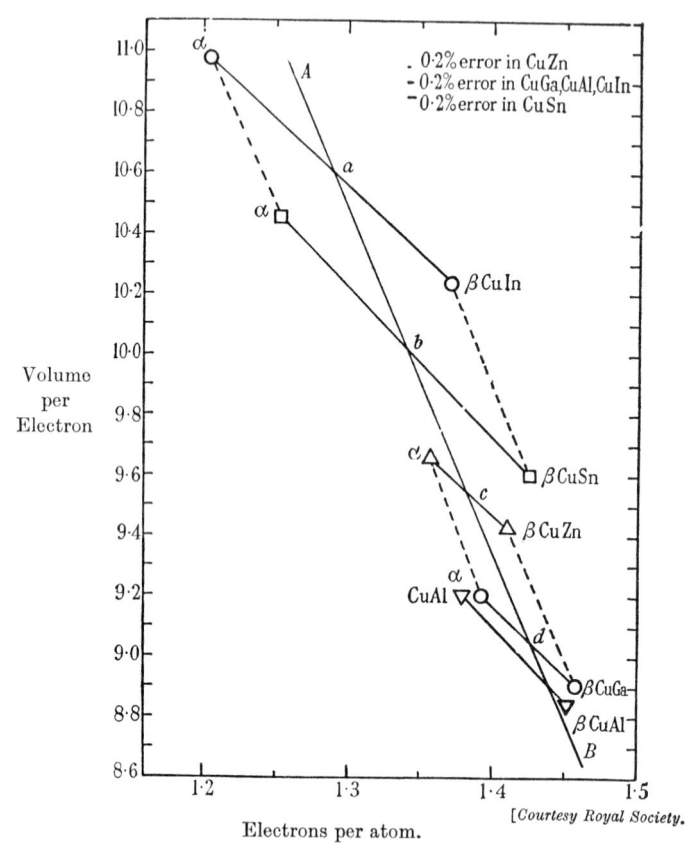

Fig. VI.17. In this diagram the abscissæ show the number of electrons per atom for the α (face-centred cubic) and β (body-centred cubic) phases in equilibrium with one another in the systems Cu–Zn, Cu–Ga, Cu–In, Cu–Sn, and Cu–Al, at a standard temperature of 672° C. The ordinates show the volume per valency electron as determined by high-temperature X-ray methods. The diagram shows the way in which the mean lattice distortion affects the electron concentration in the α/β brass kind of equilibrium.

greatly in the systems Cu–In and Cu–Sn, where the size-factors are most unfavourable. When plotted in this way, the lines joining the α and β points in each of the systems Cu–Al, Cu–Zn, Cu–Ga, Cu–In, and Cu–Sn form a series of five almost parallel straight lines, and the points a, b, c, d representing the mid-points of these lines in the systems Cu–Sn, Cu–In,

Cu–Ga, and Cu–Zn lie on the oblique straight line AB. For these last four systems, therefore, the effect of lattice distortion on the α/β-brass equilibrium is such that at 672° C there is a linear relation between the electron concentration and the mean volume per electron of the points lying midway between the $\alpha/(\alpha + \beta)$ and $(\alpha + \beta)/\beta$ boundaries in the equilibrium diagrams.

The data for the system copper–aluminium do not agree with this last relation, and it is to be noted that aluminium is by far the most electropositive of the elements concerned. Examination of a large number of equilibrium diagrams has shown that there is a general tendency for the simple principles of electron concentration to be obscured when the constituents of an alloy differ greatly in electrochemical properties, and this may be regarded as the result of the electrochemical factor.

(b) *The 3/2 Electron Compounds.*

At the electron concentration $3/2 = 1·5$, phases with the following structures are found (a prime indicates that the structure is ordered):

Body-centred cubic β- or β'-phases.
β-manganese phases, which may be denoted μ.
Close-packed hexagonal ζ- or ζ'-phases.

For the body-centred cubic structure, the first Brillouin zone is the rhombic dodecahedron of Fig. V.30, and the simple calculation given on p. 328 shows that a spherical Fermi surface touches the face of the zone at e.c. 1·48, in good agreement with the empirical discovery of the existence of β-phases at e.c. 1·5. This agreement may be fortuitous in view of the marked distortion of the surface from the spherical shown for many structures. Thus, the calculations of Jones referred to below suggest that the Fermi surface touches the zone of the body-centred cubic structure in copper alloys at e.c. 1·20, but this value is not completely established, because it depends on the assumption that the energy gap at the surface of the zone is the same as that indicated by the optical properties of copper with the face-centred cubic structure. The work of Cohen and Heine (for reference see p. 299) suggests that the solution of elements of higher valency in copper may make the Fermi surface more nearly spherical, and in this case the agreement of the empirical e.c. of 1·5 with the value of 1·48 required by a spherical Fermi surface might be understood, but there are difficulties in accepting this conclusion.

For the β-manganese structure, the first Jones zone * is shown in

* For the distinction between Brillouin zones and Jones zones, see p. 253.

Fig. VI.18, and is bounded by (221) and (310) planes. The zone is much more nearly spherical than that for the β-phases, and the completely filled zone accommodates 1·62 electrons per atom, whilst a spherical Fermi surface touches the zone at e.c. 1·41. These figures suggest that the fall in the $N(E)$ curve after the first peak will be much more rapid than for the $N(E)$ curve of the β-phase, for which the filled zone and the inscribed sphere correspond with e.c. of 2·0 and 1·48, respectively. This was regarded as a possible reason for the fact that the composition ranges of the μ-phases do not greatly exceed an electron concentration of 1·5, whereas those of the β-phases extend to considerably higher electron concentrations. This explanation may well be correct in view of the nearly spherical nature of the zone of Fig. VI.18, since there will

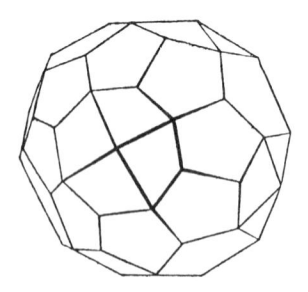

[Courtesy Clarendon Press.

FIG. VI.18. Jones zone for the β-manganese structure. The structure is cubic with 20 atoms in the unit cell. The Jones zone shown is bounded by (221) and (310) planes, and the complete zone corresponds to 1·62 electrons per atom. The inscribed sphere of the zone corresponds to 1·41 electrons per atom.

almost certainly be a marked fall in the $N(E)$ curve at electron concentrations slightly less than that of the completely filled zone (e.c. 1·62).

For the close-packed hexagonal or ζ-phases, the first and second Brillouin zones have the forms of Figs. V.31(a) and (b). The zones used by Jones (p. 252) have the forms of Fig. VI.19, and have curious characteristics. The volume of k-space bounded by planes of the type A, together with the horizontal planes at the top and bottom of these, constitute the first Brillouin zone of Fig. V.31. The structure factor is zero for the horizontal planes, and the energy gap vanishes across them. Within the volume of k-space bounded by planes of the types A, B, and C of Fig. VI.19(a), the energy varies continuously with the wave number for all directions of k-space. In the Jones zone, the energy discontinuities vanish along edges such as PQ and QR, and the smallest volume of k-space surrounded everywhere by an energy discontinuity is that of the Jones zone, with the addition on each of the A faces, of a small truncated prism, one of which is shown in Fig VI.19(a). The addition of these prisms results in the second Brillouin zone of Fig. V.31.

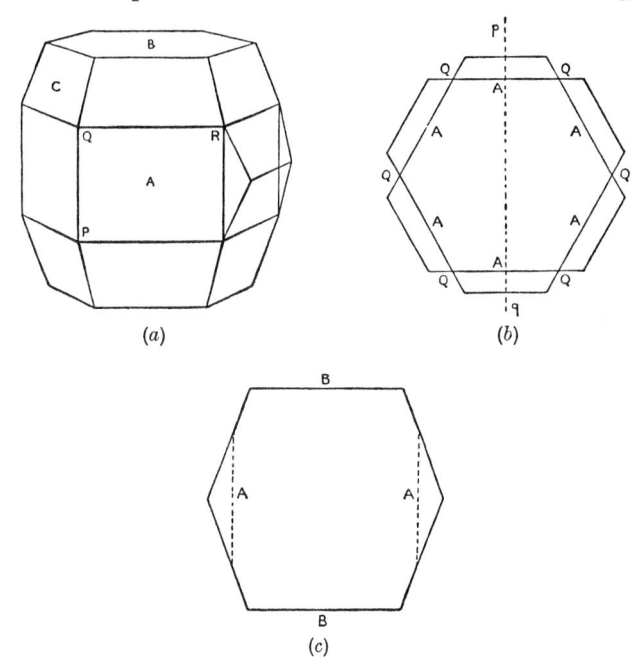

FIG. VI.19. (a) Jones zone with small prisms on one A face; (b) equatorial section with small prisms on all A faces; (c) vertical section through pq.

In Fig. VI.19 the ratio of the height of the zone to the length of the side of the base varies with the axial ratio of the crystal itself in real space. For zinc and cadmium, where the axial ratio $c/a = 1\cdot9$, the zone is short and squat compared with that for normal close-packed hexagonal structures with $c/a \sim 1\cdot6$. This can be understood in view of the relation between Brillouin zones and the reciprocal lattice.

The completed zone with the addition of the truncated prism on each A face contains states for 2 electrons per atom, regardless of the axial ratio (see p. 251). The incomplete zone of Fig. VI.19 contains a number of electrons equal to:

$$2 - \tfrac{3}{4}\left(\frac{a}{c}\right)^2\left(1 - \tfrac{1}{4}\left(\frac{a}{c}\right)^2\right)$$

and this indicates a number of electrons per atom varying from $1\cdot727$ for $c/a = 1\cdot57$ to $1\cdot749$ for $c/a = 1\cdot646$.

The dimensions of a crystal lattice may undergo interesting changes when the Fermi surface is in the region of a zone face. The typical $N(E)$ curve (Fig. V.8, p. 207) shows an upward inflection just before the Fermi surface makes contact with the zone face. When, therefore, an increase in the electron concentration results in the Fermi surface approaching a zone face, a lower energy may be produced if the crystal expands in the direction concerned, because the expansion will cause a contraction of the zone, and this will raise the $N(E)$ curve, and so accommodate the electrons with a lower energy, provided that the crystal contracts in some other direction so that the volume is maintained approximately constant. In the close-packed hexagonal structure, such changes are possible and have been discussed by Goodenough.* Apart from this, it has been shown by Jones † that if electron concentration produces an overlap across the face of a zone, the effect is to compress the zone, and so to expand the crystal in the direction concerned. Both these effects must be taken into consideration when dealing with the effect of electron concentration on lattice-spacing/composition curves.

There is no accurate knowledge of the form of the Fermi surface in the close-packed hexagonal ζ-phases. Considered empirically they exist at an electron concentration of $ca.$ 1·5. In binary alloys the a parameter always increases with increasing electron concentration, whilst the c parameter may increase or decrease according to the system concerned, but the combined effect is always to make the axial ratio c/a decrease with increasing electron concentration. Detailed work by Raynor and his collaborators has shown that in ternary systems such as copper–gallium–germanium the axial ratio is little affected if the composition is altered so as to maintain a constant electron concentration.

These results have been interpreted as indicating that in the ζ-phases the electron concentration is such that an overlap across the A faces of the zone has occurred, with the result that the a parameter increases with increasing electron concentration.

It is to be noted that in the system silver–aluminium, where the size-factor is very favourable,‡ the ζ close-packed hexagonal phase at e.c. 1·5 unites with the close-packed hexagonal phase of electron

* J. B. Goodenough, *Phys. Rev.*, 1953, **89**, 282.
† See T. B. Massalski, " Theory of Alloy Phases ", p. 63. **1956**: Cleveland. O. (American Society for Metals.)
‡ Using the interatomic distances in the crystals of the elements, the size-factor in the system silver–aluminium is − 5·5%, but the two atoms are such that no appreciable change in mean lattice spacing is observed when silver is dissolved in aluminium.

concentration $7/4 = 1.75$ to form the single wide phase field shown in Fig. VI.20. This suggests that there is no essential difference between the close-packed hexagonal electron compounds characterized by the

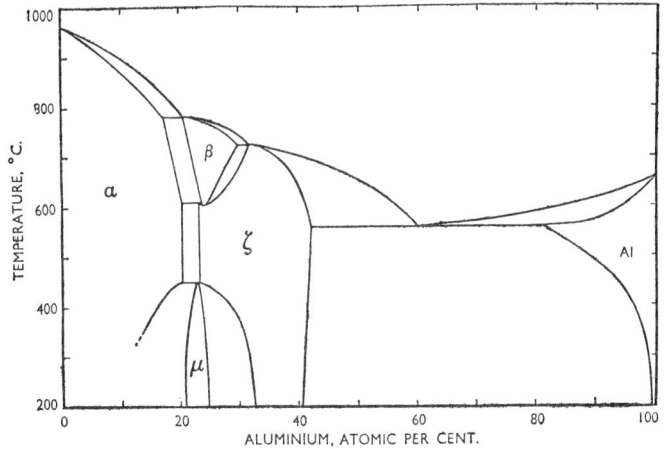

Fig. VI.20. The equilibrium diagram of the system silver–aluminium.

electron concentrations 1·5 and 1·75. The gap between the phases in most diagrams results from the fact that when the size-factors are not too favourable there is a general tendency for the γ-phase to be stabilized

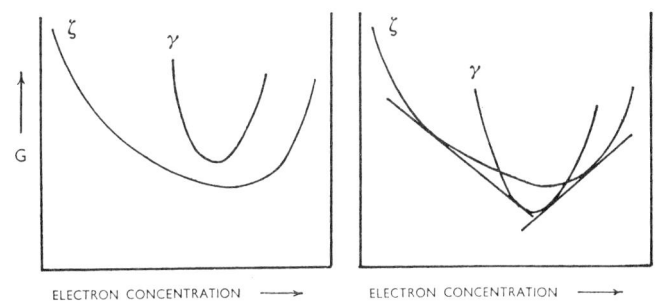

Fig. VI.21. To illustrate how increasing free energy of the γ-phase may result in the uniting of two close-packed hexagonal phases to form a single phase field.

i.e. for its free-energy/composition curve to fall to such an extent that it becomes lower than the free-energy curve for the close-packed hexagonal phase (see Fig. VI.21).

As explained above, the incomplete first Jones zone of the close-packed hexagonal structure can contain about 1·73–1·75 electrons per atom, the exact value depending on the axial ratio. In the close-packed hexagonal phases which occur at an electron concentration of 1·75, there must, therefore, be a large overlap into the second zone, and the variation of axial ratio with composition suggests that overlaps have occurred at both the A and C positions in Fig. VI.19.

(c) *The Relative Stabilities of the 3/2 Electron Compounds.*

There is at present no quantitative theory which explains why the body-centred cubic β-phases occur in some systems only, and are replaced by the μ- (β-Mn) or ζ- (c.p. hexagonal) phases in other systems. Systematic studies of the data have, however, shown that clear empirical principles exist in many cases.

Increasing valency of the solute favours the ζ or μ structures at the expense of the β-phases. Thus, in the systems copper–zinc, copper–gallium, and copper–germanium, the first has β-phases only at e.c. 1·5, the second has a β-phase stable at high temperatures, whilst the last has a ζ-phase only. This is a very simple case, but a systematic examination* has shown that the same general tendency is found in other systems, although the reason for this is not known.

Increasing temperature favours the β-phases at the expense of the ζ- or μ-phases. This principle was established empirically in 1940,* and it was later shown by Zener † that there is a tendency for body-centred cubic structures to be associated with an unusually large amplitude of vibration of the (110) [1̄10] shear-strain co-ordinate, and hence with an unusually large entropy (S). The criterion for the stability of a phase is for the free energy (G) to be low, where if U is the internal energy, $G = U - TS + PV$. A large value of the entropy implies a rapid decrease in free energy with rise of temperature, and the tendency for β-phases to be favoured by high temperatures can thus be understood.

Increasing size-factor favours the β-phases, and the general tendency is for the composition limits of these phases to be shifted in the direction of lower electron concentration as the size-factor increases. The phase boundaries in an equilibrium diagram depend on the relative forms of the free-energy/composition curves of different phases. The fact that the body-centred cubic phase is not close-packed suggests that it may accommodate large atoms more readily than the close-packed ζ-phase, and the tendency for the β-phases to be preferred with large size-factors can be understood in this way. Since the α-phase is also

* W. Hume-Rothery, P. W. Reynolds, and G. V. Raynor, *J. Inst. Metals,* 1940, **66,** 191.
† C. Zener, " Elasticity and Anelasticity of Metals ". **1948**: Chicago (University Press).

Electrons, Atoms, Metals, and Alloys

close-packed (f.c. cubic), we may expect its free-energy/composition curve to be affected more than that for the β-phase, and Fig. VI.14 shows how free-energy considerations will thus lead to a displacement of the copper-rich β-phase boundary in the direction of lower electron concentration. If the next phase in the equilibrium diagram is even less closely packed than the b.c. cubic structure, a similar argument would lead to a displacement of the high-electron-concentration phase boundary of the β-phase in the direction of lower electron concentration, and as explained below, the γ-phases which are adjacent to the β-phases but of higher electron concentrations may be regarded structurally as distorted body-centred cubic phases from which some atoms have dropped out with slight rearrangement of atomic positions. The general effects of size-factors on the compositions of β-phases can thus be understood.

Increasing electrochemical factor leads to the formation of ordered structures. Thus, in the system copper–zinc the β-phase has a disordered structure at high temperatures, and an ordered CsCl type of structure (Fig. III.23, p. 129) at low temperatures. In the systems silver–magnesium and gold–magnesium where the electrochemical factor is higher, the CsCl type of ordered b.c. cubic phase is stable up to the melting point, and no disordered β-phase exists. This tendency can be understood if increasing electrochemical factor leads to the two kinds of atom building up opposite charges, since in this case we shall expect a given atom to avoid those of its own kind, and this is one of the characteristics of most superlattices or ordered structures.

(d) *The γ-Brass Phases.*

The γ-brass type of structure occurs at an electron concentration of $21/13 = 1\cdot 615$, and may be regarded as based on a defect body-centred cubic structure. The unit cell of the b.c. cubic structure contains 2 atoms, and if we imagine a cell whose sides are three times as large as this, it will contain 27 small cells, and a total of 54 atoms. (See Fig. VI.22(a).) The γ-brass structure may be regarded as derived from this large cell by the dropping out of 2 atoms, and by slight displacement of the remaining 52 atoms as shown in Fig. VI.22(b).

The first Jones zone[*] of the γ-brass structure is shown in Fig. VI.23, and is bounded by two types of face corresponding to reflections from the (330) and (411) planes of the crystal. The completely filled zone contains 90 electrons per unit cell, and hence corresponds to an electron concentration of $90/52 = 1\cdot 73$, if all the lattice sites are occupied.

[*] Using the terminology of p. 251.

324   *Atomic Theory*

A spherical Fermi surface which touches the face of the zone contains 80 electrons per unit cell, and hence corresponds to an electron con-

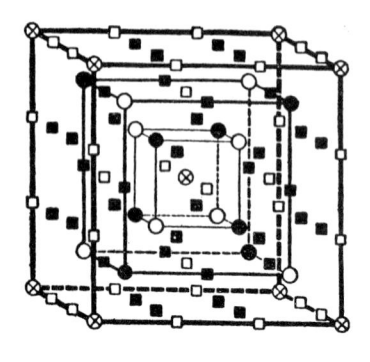

Fig. VI.22(a). The derivation of the structure of γ-brass from a simple cube-centred arrangement of atoms.

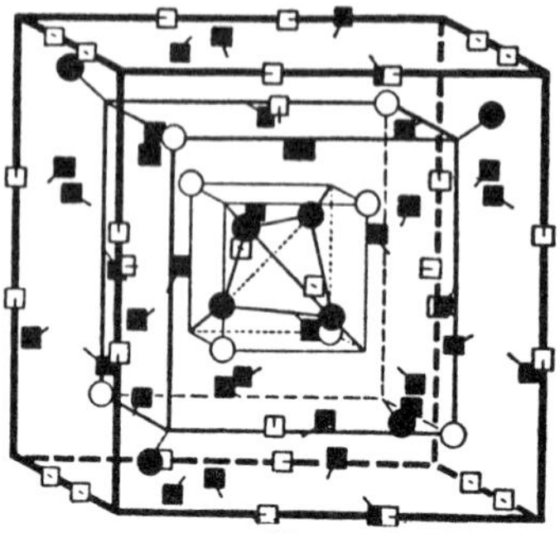

Fig. VI.22(b). The structure of γ-brass. (*After Bradley and Thewlis.*)

[*Courtesy Royal Society.*]

(a) The arrangement of atoms in a cubic cell formed by stacking together 27 unit cells of the body-centred cubic structure. This large cell corresponds to 54 atoms. If the atoms marked by crosses are now considered to be removed, the unit cell of the γ-brass structure is obtained by readjustment of the positions of the remaining atoms to give the structure illustrated in (b).

centration of $80/52 = 1 \cdot 54$. These figures were regarded as confirming the hypothesis that phases were stable in the region of a peak on the $N(E)$ curve, but this view must now be looked upon with caution until

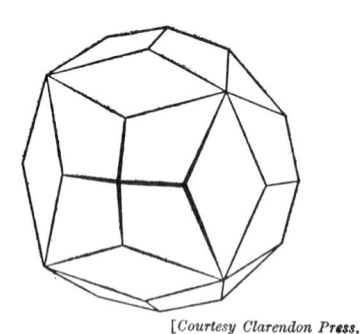

[*Courtesy Clarendon Press.*]

Fig. VI.23. Jones zone for γ-brass structure.

more is known of the distortion of the Fermi surface. The approximately spherical shape of the zone does, however, make it probable that, provided all the lattice sites are occupied, there will be a sharp drop * in the $N(E)$ curve just before the zone is completely filled at e.c. 1·73, and this agrees with the position of the copper–zinc γ-phase boundary on the zinc-rich side in Fig. III.19.

In the systems copper–gallium and copper–aluminium, the γ-phase extends to much higher electron concentrations, but experimental work † shows that, on increasing the percentage of the solute atoms, a critical stage is reached at e.c. 1·70 beyond which atoms drop out of the structure in such a way as to maintain a constant number of electrons per unit cell. This suggests that on reaching e.c. 1·70 the increase in energy accompanying the introduction of further electrons would be so great that, from the point of view of a low free energy, it is advantageous for atoms to drop out of the crystal structure rather than to allow the number of electrons per unit cell to increase. In this way the electron concentration (*i.e.*, the number of valency electrons *per atom*) can increase, whilst the number of electrons *per unit cell* remains constant.

Similar effects were found ‡ for the β-phase of the system nickel–aluminium, where the equiatomic alloy has the CsCl type of structure, and corresponds to an electron concentration of 1·5 (or 3 electrons per unit cell) if nickel is assumed to exert zero valency. On reducing the aluminium content of the alloy, normal substitution of nickel for aluminium takes place, and all the atomic sites remain occupied. But on increasing the aluminium content, the behaviour is quite different, and some of the nickel atoms drop out of the structure in such a way that the number of electrons per unit cell remains constant at approximately 3·0. Analogous effects have been found § for the b.c. cubic phase in ternary copper–aluminium–nickel alloys, and these results emphasize that the number of electrons per unit cell is the real characteristic of Brillouin-zone theory. The empirical discovery of the electron-concentration rules was possible only because most crystal structures are such that all lattice sites are occupied, in which case the number of electrons per unit cell is directly proportional to the number per atom, *i.e.*, to the electron concentration.

As explained above (Fig. VI.20) there is no γ-phase in the system silver–aluminium, where the size-factor is very favourable, and systematic examination ‖ shows that γ-phases are not formed where the size-

---

* This argument assumes that there is an energy gap between the first and second zone, or that the overlap is so slight that the valley in the $N(E)$ curve is near to the electron concentration for the completely filled zone.
† J. O. Betterton, Jr., and W. Hume-Rothery, *J. Inst. Metals*, 1951–52, **80**, 459.
‡ A. J. Bradley and A. Taylor, *Proc. Roy. Soc.*, 1937, [A], **159**, 56.
§ H. Lipson and A. Taylor, *ibid.*, 1939, [A], **173**, 232.
‖ W. Hume-Rothery, J. O. Betterton, Jr., and J. Reynolds, *J. Inst. Metals*, 1951–52, **80**, 609.

factor is small, or where it exceeds about 20%. Since the γ-structure is derived from a defect b.c. cubic structure, we may perhaps imagine that if all the atoms were of one size, there would be no reason for the dropping out of any of them from the b.c. cubic arrangement; whilst if they are of slightly different sizes, some of one kind may be dropped, with a slight rearrangement of the remainder to form the more stable γ-structure.

As with the β-phases, increasing size-factor displaces the composition limits of the γ phases in the direction of lower electron concentration. Increasing electrochemical factor displaces the composition limits of

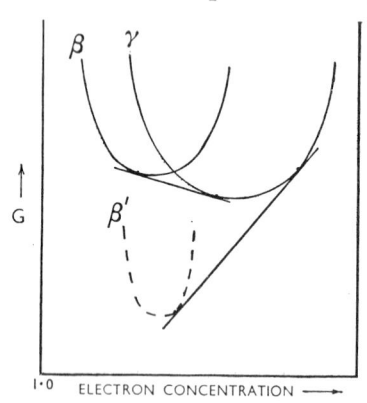

Fig. VI.24. To show how the composition of the γ-phase on the side of low electron concentration is displaced in the direction of higher electron concentration in cases where the β-phase is replaced by a very stable β'-phase (*i.e.*, a phase of low free energy).

the γ-phases in the direction of higher electron concentration. Since it is established empirically (p. 323) that the stability of the ordered β'-phases increases with increasing electrochemical factor, a free-energy curve diagram such as that of Fig. VI.24 enables us to understand how the γ-phase boundary on the side of low electron concentrations is shifted towards higher electron concentration by the presence of a very stable β'-phase. It is probable that increasing electrochemical factor lowers the free energies of the β'-phases more than those of the γ-phases, because the very simple CsCl structure enables an atom of one kind to surround itself with those of another.

(e) *The Liquid ⇌ Solid Equilibrium for α-Solid Solutions in Copper and Silver.*

When a solid solution is formed in copper, the liquidus and solidus curves may be raised, as in the system copper–nickel, or lowered, as in the system copper–zinc (Fig. III.19). The change in direction of the curves depends on the effect of the solute on the latent heat of fusion of the solvent. An increase in the latent heat of fusion corresponds to a raising of the liquidus and solidus curves, whilst a decrease produces the opposite effect. H. Jones * devised a method whereby the change in the latent heat of fusion could be calculated from the liquidus and solidus curves, and found that for solid solutions of cadmium, indium, tin, and antimony in silver, and of zinc, gallium, and germanium in copper, the change in the latent heat of fusion depended essentially on the electron concentration. A more detailed analysis by Hume-Rothery and Burns † showed how this simple electron-concentration effect was modified by lattice distortion, and by the electrochemical factor, the latter effect being pronounced in the system copper–arsenic owing to the electronegative nature of arsenic.

These results may be interpreted by assuming that the energy of the lowest electron state is sensitive to changes in atomic volume Ω, but relatively insensitive to changes in structure for a given Ω, whilst the Fermi energy is sensitive to changes in structure. In this case, if the change in volume on melting is almost independent of composition, the energy change on melting will depend mainly on the change in Fermi energy, and hence on the electron concentration. This simple relation between electron concentration and the liquid $\rightleftharpoons$ solid equilibrium holds only when solvent and solute are in the same row of the Periodic Table.

REFERENCES AND SUGGESTIONS FOR FURTHER READING.

Electron Theory of Solid Copper.
  K. Fuchs, *Proc. Roy. Soc.*, 1935, [A], **151**, 585; 1936, [A], **153**, 622.
  D. J. Howarth, *ibid.*, 1953, [A], **220**, 513.

Theory of α/β-Brass Equilibrium.
  H. Jones, *Proc. Phys. Soc.*, 1937, **49**, 250. This is the original paper and, as explained on p. 315 the work of Zener (*Phys. Rev.*, 1947, [ii], **71**, 846) indicates that some error has occurred as regards the effect of temperature.
  K. W. Andrews and W. Hume-Rothery, *Proc. Roy. Soc.*, 1941, [A], **178**, 464.
  M. H. Cohen and V. Heine, *Advances in Physics*, 1958, **7**, 395. This paper contains much of interest in connection with copper, silver, and gold, and their alloys.
  V. Heine, " Phase Stability of Metals and Alloys " (edited by P. S. Rudman, J. Stringer, and R. I. Jaffee), Section 2.1. **1967**: New York and London (McGraw-Hill).

Electron Theory of Close-Packed Hexagonal Copper–Zinc Phases.
  H. Jones, *Proc. Roy. Soc.*, 1934, [A], **147**, 396.

---

* H. Jones, *Proc. Phys. Soc.*, 1937, **49**, 242.
† W. Hume-Rothery and J. Burns, *Phil. Mag.*, 1957, [viii], **2**, 1177.

Electron Theory of γ-Brass.
H. Jones, *Proc. Roy. Soc.*, 1934, [A], **144**, 225.

Factors Affecting the Relative Stabilities of Body-Centred Cubic, Close-Packed Hexagonal, and β-Manganese Structures in Copper Alloys.
W. Hume-Rothery, P. W. Reynolds, and G. V. Raynor, *J. Inst. Metals*, 1940, **66**, 191.

APPENDIX.

The approximate positions of the first peaks on the $N(E)$ curves of the face-centred and body-centred cubic structures can be calculated as follows. As a first rough approximation we may assume that the electrons in the zone model have the E/k relations of free electrons, so that the Fermi surface of occupied states is a sphere whose volume is given by the equations of Section IV.1. The first peak on the $N(E)$ curve for the body-centred cubic structure occurs when the number of electrons is such that the Fermi surface just touches the Brillouin zone of Fig. V.30 (p. 251). If we use the free-electron approximation, the momentum associated with electron states on the spherical Fermi surface is given by equation IV (2) (p. 171), and for free electrons this momentum is equal to h/λ, where λ is the wavelength. We have, therefore, from equation IV (2):

$$\frac{4}{3}\pi p^3{}_{\text{max.}} = \frac{4}{3}\pi \frac{h^3}{\lambda^3} = \frac{Nh^3}{2V}$$

and hence:
$$\lambda = \left(\frac{8\pi V}{3N}\right)^{1/3}.$$

In this way we obtain for the electron states corresponding to the first peak on the $N(E)$ curve, an expression connecting the wave-length, λ, with the number of electrons, N, in a volume, V.

We have seen in Section V.1 (p. 200) that for the electron states on the surface of a Brillouin zone, the wave-lengths satisfy the Bragg relation:

$$n\lambda = 2d \sin \theta.$$

For the body-centred cubic structure, the first Brillouin zone is a rhombic dodecahedron (Fig. V.30) and the Bragg reflections are from the (110) planes of the body-centred cubic structure. The corresponding interplanar spacing, d, is equal to $\dfrac{a\sqrt{2}}{2}$, where a is the side of the unit cell. At the point where the Fermi sphere first touches the surface of the zone, the first-order reflection is clearly such that $n = 1$, and $\theta = 90°$ in the Bragg equation, and we may therefore write:

$$\lambda = a\sqrt{2}.$$

Now the unit cell of the body-centred cubic structure has a volume $V = a^3$, and contains two atoms, since the eight atoms at the corners are each shared with eight adjacent cells, and therefore each counts as $\frac{1}{8}$ for the particular cell considered, whilst the atom at the centre of the cell concerns the one cell alone. We may therefore write:

$$\frac{N_a}{V} = \frac{2}{a^3}$$

and
$$a = \left(\frac{2V}{N_a}\right)^{1/3},$$

where N is the number of atoms in volume V. It follows, therefore, that:

$$\lambda = \sqrt{2}\left(\frac{2V}{N_a}\right)^{1/3}.$$

Electrons, Atoms, Metals, and Alloys

In this way a second expression is obtained for λ, involving the number of atoms N_a in volume V. On equating the two expressions for λ, we have:

$$\frac{N}{N_a} = 2^{1/2} \times \frac{\pi}{3} = 1\cdot 48.$$

We conclude, therefore, that subject to these simplifying assumptions, the first peak on the $N(E)$ curve of the body-centred cubic structure occurs at an electron : atom ratio of $1\cdot 48$.

For the face-centred cubic structure, a similar calculation may be made. In this case the unit cell contains four atoms, and we may therefore write:

$$\frac{N_a}{V} = \frac{4}{a^3}.$$

The first peak on the $N(E)$ curve for the face-centred cubic structure corresponds to the (111) reflection, for which the spacing d is equal to $\frac{a\sqrt{3}}{3}$. In this case we may therefore write:

$$\lambda = \left(\frac{8\pi V}{3N}\right)^{1/3} = \frac{\sqrt{3}}{3}\left(\frac{4V}{N_a}\right)^{1/3}$$

and

$$\frac{N}{N_a} = \frac{\pi \times 3^{1/2}}{4} = 1\cdot 36.$$

It is thus very interesting to see how such a simple type of calculation leads to approximately correct values for the α/β phase changes in alloys of copper, although as explained above the true state of affairs is more complicated.

3. Some Metals of Higher Valency.

In earlier editions of this book the evidence that could be called upon or information about the electronic structures of polyvalent metals was wholly indirect, that is, it was drawn from experimental work of the type described in Section V.4 as "general." In recent years the various techniques of Fermi surface investigation have been successfully applied to a number of these metals, and the conclusions drawn from the older methods and from theoretical estimates have been in some cases confirmed and in others modified. The general indications of such developments are that for all the simple (*i.e.* non-transition) polyvalent metals the Fermi surfaces are rarely very different from those that would be suggested by the nearly-free-electron approach. It is important to emphasize, however, that this does *not* mean that their properties are nearly those predicted by a free-electron model, since not only do the finite energy gaps across zone faces cause the appearance or disappearance of pieces of Fermi surface, but also the character of electron states on the Fermi surface and their response to applied fields (as expressed in their velocities and effective masses) may nowhere resemble those of a free-electron Fermi sphere enclosing the same number of occupied states.

Magnesium, Zinc, and Cadmium.

The elements of Group II are metallic and are conductors of electricity, and those which are cubic must possess overlapping Brillouin zones, because as we have already explained (p. 221) the first Brillouin zones of simple translational lattices contain exactly two electrons per atom, and the crystals would be insulators if the zones were completely filled and no overlap occurred. Beryllium and magnesium crystallize in the close-packed hexagonal structure (Fig. III.12, p. 104); the axial ratios are 1·5682 and 1·6235 respectively, and are thus very nearly that required for close-packed spheres ($c/a = 1 \cdot 6330$). For zinc and cadmium the structures are considerably elongated in the direction of the c-axis (Fig. III.13), and the axial ratios are 1·8563 and 1·8859 respectively.

The Brillouin zones of the close-packed hexagonal structure have been described on p. 251. Great interest attaches to the Jones zone of Fig. VI.19, whose characteristics have been described on p. 252. The ratio of the height of the zone to the length of the side of the base varies with the axial ratio of the crystal, the variation being in the opposite direction to that of the structure of the crystal itself in real space. That is to say, in the case of zinc and cadmium, where $c/a = $ approx. 1·9, and the crystal structure is extended in the c-direction compared with the structure for close-packed spheres, the Brillouin zone is relatively short and squat compared with the zones for beryllium and magnesium, for which $c/a = $ approx. 1·6. The energy of an electron state at the point A is in general different from that of a state at the point B, and for the short squat zone of zinc and cadmium E_B is probably less than E_A, whilst with beryllium and magnesium it is probable that $E_A < E_B$. Since the complete zone contains exactly two electrons, and since all four metals are conductors of electricity, there must be an overlap into a second zone, and two kinds of overlap have to be considered.

(a) Overlap from the incomplete first zone across the faces A into the small truncated prisms (Fig. VI.19, p. 319). This will be termed the A overlap.

(b) Overlap from the complete first zone into a second outer zone. This may occur either at B in Fig. VI.19(a) or at the re-entrant corners (points Q in Fig. VI.19(b)). These may be termed the B and Q overlaps respectively, and the axial ratio is the main factor in determining which occurs first.

The calculations of Mott and Jones suggest that for beryllium and magnesium, and for other metals with an axial ratio less than 1·63, the A overlaps take place first, followed by the Q overlaps, but with two electrons per atom there is no overlap at B. For zinc and cadmium, with axial ratios of approximately 1·9, the B overlap takes place first, followed by the A overlap, but with two electrons per atom there is no overlap at Q. The axial ratio thus affects the order of the overlaps, but with electron concentrations of about 2·0, the total number of electrons outside the first zone does not depend greatly on the axial ratio.

There is an essential difference between the $N(E)$ curves for overlap into the outer zone at B and Q. The overlap at Q concerns electrons with wave numbers in six directions (perpendicular to the c-axis) whereas the overlap at B concerns electrons with wave numbers in only two directions (parallel to the hexagonal axis). For a given energy increase, therefore, there will be more electron states available in the case of the Q overlaps than for the B overlaps and, in consequence, the $N(E)$ curve for the Q overlaps will, after the initial vertically rising

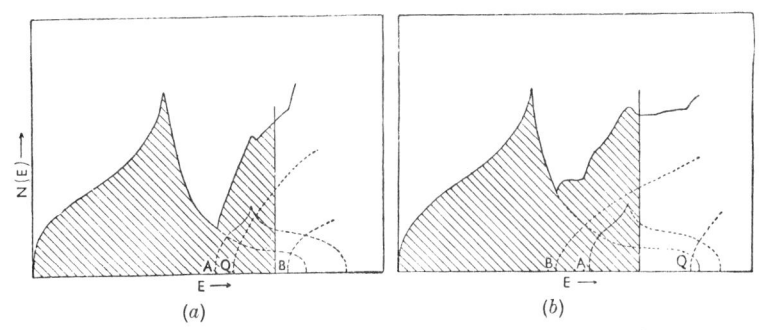

Fig. VI.25. $N(E)$ curves for (a) magnesium and (b) zinc and cadmium.

portion has been passed, rise more steeply away from the energy axis than that for the B overlaps. The exact form of the $N(E)$ curves depends, of course, on the details of the potential fields, and on the form of the Fermi surface.

The general qualitative forms of the $N(E)$ curves for magnesium, zinc, and cadmium were first given by Mott in 1938, and are shown in Fig. VI.25(a) and (b). For magnesium a somewhat similar curve (Fig. VI.26) has been obtained by Trlifaj * who used a more elaborate method of calculation, but in the absence of detailed curves for zinc and cadmium, it is convenient to discuss the effects in terms of Fig. VI.25, since although these are only approximations, their main features are almost certainly correct.

For magnesium, the condition of affairs is probably as shown in Fig. VI.25(a). The A overlaps occur first, and the small A zone in Fig.

* M. Trlifaj, *Czech. J. Physics*, 1952, **1**, 110.

VI.25(a) is that of the six small truncated prisms fitting on to the A faces of Fig. VI.19(a). The A overlap is followed by the overlaps at Q into the outer zone, but with two electrons per atom there is no overlap at B.

In contrast to this, the high axial ratio of zinc and cadmium, with the resulting short squat zone, is thought to produce the condition of affairs shown in Fig. VI.25(b). Here, the overlap has occurred first across the B faces of the zone, and then across the A faces but, with two electrons per atom, no overlap has occurred at Q.

The general electronic structure of Fig. VI.25(a) and Fig. VI.26 for magnesium is confirmed by Fermi surface studies.*

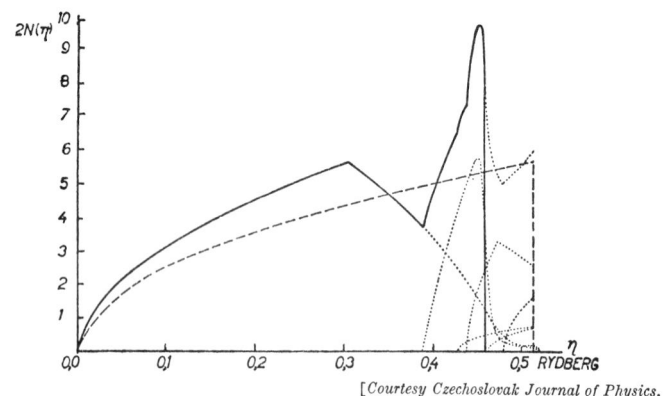

Fig. VI.26. $N(E)$ curve for magnesium according to Trlifaj. The full curve is obtained from the summation of the dotted curves, whilst the dashed curve is that of the free-electron theory.

The above conclusions have been thought to be confirmed by lattice-spacing/composition relations for solid solutions in magnesium. The work of Jones (p. 320) suggests that the effect of a Brillouin-zone overlap in a particular direction in k-space is to expand the lattice in the corresponding direction in real space. If, therefore, magnesium is alloyed with an element of higher valency with which it forms a solid solution, we shall expect the a parameter to change uniformly with composition, because the Brillouin-zone overlap has already occurred in pure magnesium. The c-parameter/composition curve should, however, show an abrupt change in direction at the point where the electron concentration becomes sufficiently great for the B overlap to occur. Changes of direction of this nature were found by Raynor and Hume-

* Experimental and theoretical work on the Fermi surface of zinc is described by D. F. Gibbons and L. M. Falicov, *Phil. Mag.*, 1963, **8**, 177.

Rothery, and by Raynor * whose curves for the a and c spacings of solid solutions of aluminium and indium in magnesium are shown in Fig. VI.27. Similar results were obtained for the solid solutions of all the trivalent solutes (Al, Ga, In, Tl), whilst for tetravalent solutes (Sn,Pb) the kinks in the c-spacing curves were at a lower atomic percentage of solute, as would be expected from the greater number of electrons per atom. It was also shown by Raynor * that when the axial ratios were plotted in terms of electron concentration, the points for the solid solutions of indium and tin lay on a single curve, whilst those for thallium and lead lay on a second curve, which differed very slightly from that for indium and tin.

[Courtesy Pergamon Press.

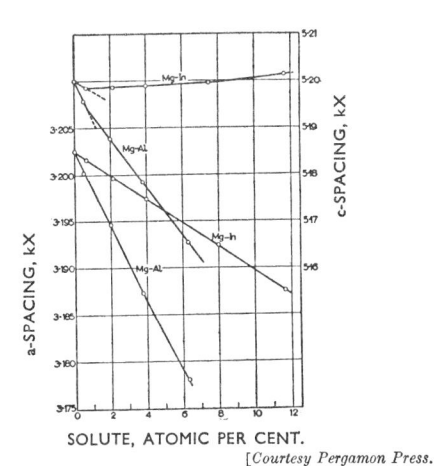

FIG. VI.27. The a and c lattice spacings for solid solutions in magnesium of aluminium and indium (20° C). (Raynor.)

This early work provided substantial confirmation for an electronic structure with zone overlaps of the form shown in the qualitative $N(E)$ curve of Fig. VI.25(a), and the agreement between the positions of the kinks in the curves for the Hall coefficients† and four c-lattice parameters was very striking. The situation has since been complicated by the failure of three later pairs of authors ‡ to confirm the kinks in the c-spacing/composition curves. The differences involved are very small,

* For a general summary of this work, see G. V. Raynor, "The Physical Metallurgy of Magnesium and Its Alloys." 1959: London (Pergamon Press).
† A. I. Schindler and E. I. Salkovitz, Phys. Rev., 1953, [ii], **91**, 1320.
‡ F. W. von Batchelder and R. F. Raeuchle, ibid., 1957, [ii], **105**, 59.
  D. Hardie and R. N. Parkins, Phil. Mag., 1959, [viii], **4**, 815.
  C. B. Walker and M. Marezio, Acta Met., 1959, **7**, 769.

and at present it is not possible to say which data are the more accurate.*

If the earlier work is accepted, the curves of Fig. VI.27 show clearly that lattice-spacing/composition relations may be profoundly altered if changing composition affects a Brillouin-zone overlap, and it follows that the apparent size of an atom in solid solution, as estimated by the extrapolation of the lattice-spacing/composition curve to 100% of the solute, may be quite different according as the particular lattice-spacing curve does or does not come immediately after a Brillouin-zone overlap. In this way it has been shown that the apparent atomic diameters of some elements when dissolved in copper or silver (no Brillouin-zone overlap), are much smaller than those estimated from extrapolations of the a-parameter/composition curves of solid solutions of the same elements in magnesium where a Brillouin-zone overlap is concerned.

Examination of Fig. VI.25 shows clearly that, whilst in magnesium the overlaps occur in the order AQB, the corresponding order for cadmium is BAQ. It is thus of great interest to examine what happens when cadmium is dissolved in magnesium. At high temperatures the two metals form a continuous series of solid solutions,† and since the atom of cadmium is somewhat smaller than that of magnesium, we shall expect that when cadmium is dissolved in magnesium both a and c parameters will decrease, provided that the zone overlaps are not affected. It is clear, however, that since there are only two electrons per atom in the whole series of alloys, and since the B overlap does not occur in magnesium and the Q overlap does not occur in cadmium, the substitution of cadmium for magnesium must eventually produce a stage at which the B overlap begins and the Q overlap dies away. From this point onwards we shall expect a marked increase in the axial ratio, since the c-parameter will tend to expand compared with the a-parameter. Fig. VI.28 shows the lattice spacings of magnesium–cadmium alloys at 310° C, and from this it will be seen how remarkably these expectations are fulfilled. The first substitution of cadmium for magnesium produces a decrease in both the a and c parameters, and the axial ratio is but little affected. In the region 60–70 at.-% of cadmium there is a clear change in the process, and the c-parameter increases rapidly, while the a-parameter decreases more rapidly than before. This is clearly in complete agreement with what is to be expected from the onset of the B overlap and the dying

* The work of Raynor involved the use of slowly cooled filings for the most dilute alloys, and the filings were prepared in argon. Hardie and Parkins and the American investigators used filings prepared in air, although the work of Raynor showed that such filings contained from 1 to 2 wt.-% of nitrogen and oxygen.

† Superlattices are formed at low temperatures.

away of the Q overlap, and there is a great fascination in seeing how these curious lattice-spacing relations find a logical explanation in terms of the zone theories.

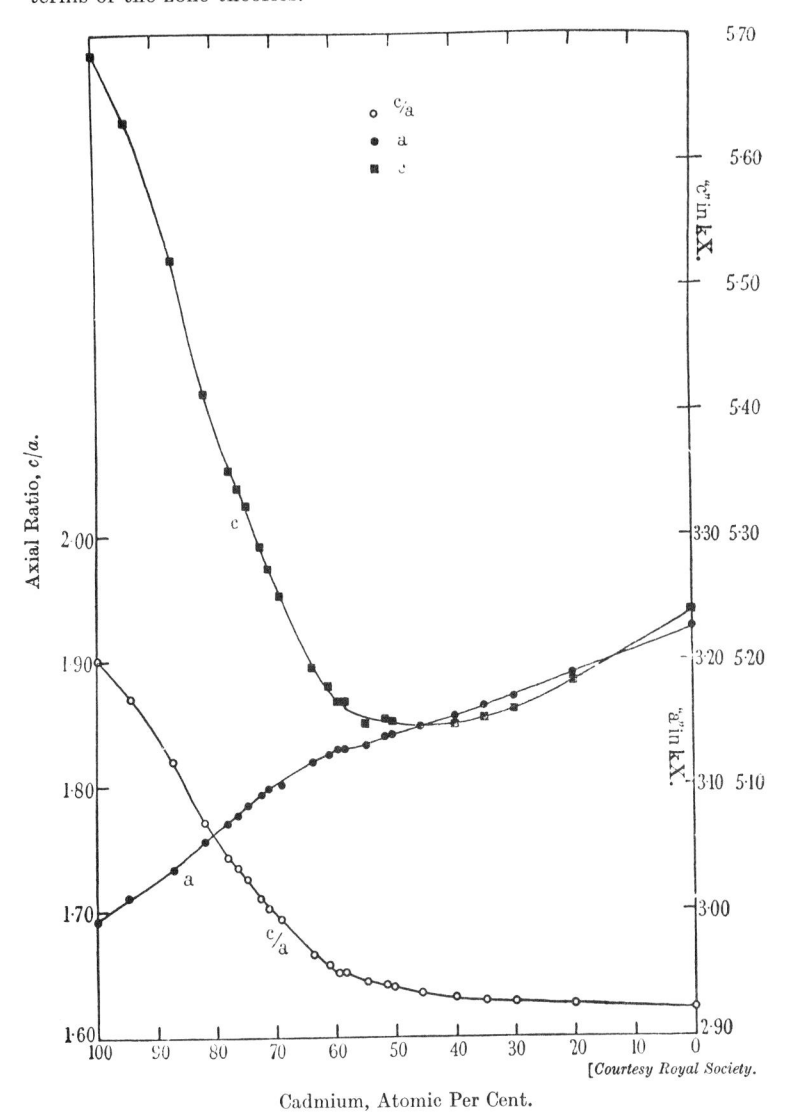

Fig. VI.28. The a and c lattice spacings of magnesium–cadmium alloys at 310° C, and also the axial ratio c/a.

The metal magnesium thus occupies a position of great interest from the point of view of electron theory, since not only are the predictions of the zone theory fulfilled, but the extreme closeness of approach of the B overlap, with two electrons per atom, means that the structure is near to a critical stage. This may even affect the mechanical properties, since if small amounts of elements of higher valency are present in solid solution, the crystal may be in a state in which a zone overlap may be produced by the application of tensile stress in the direction of the c-axis, and curious energy relations might be involved under alternating stresses.

Recent work[*] on zinc-based and cadmium-based alloys with copper or silver and silver, respectively (Figs. III.19 and III.23) have shown interesting changes in the variation of axial ratio (c/a) with electron concentration in the primary solid solution (η) which has c/a greater than the ideal value, and the ε phase (the 7/2 electron compound) which has c/a less than the ideal value ($\sim 1\cdot59$). In the former, c/a decreases rapidly with decreasing electron concentration, and the consequent expansion of the Brillouin zone in the c-direction produces a much more rapid decrease in the B overlap than would be expected simply from the decrease in volume enclosed by the Fermi surface. In the ε phase decreasing electron concentration produces a slow increase in c/a, which points to the absence of a B overlap; and in this phase the axial ratio correlates well with the electron concentration in a large number of alloy systems.

Atomic and Ionic Radii.

The preceding section refers to the energy characteristics of the electrons in the crystals of some of the elements of Group II, and it remains to consider briefly the nature of their electron-cloud patterns. When we compare an alkaline earth metal with the preceding alkali, we find that the alkaline earth metal has both a smaller *ion* and a smaller atomic diameter (as estimated from the closest distance of approach in the crystals of the elements). The contraction of the ion is, however, relatively smaller than the decrease in the atomic diameter, and the crystals of the alkaline earth metals are less " open " (see p. 285) than those of the alkalis. In contrast to this, when we pass from copper to zinc, silver to cadmium, or gold to mercury, the electron clouds of the underlying ions contract rapidly, whilst the atomic diameters, whether estimated from the closest distances of approach in the crystals of the elements or from the lattice distortions produced in solid solutions, increase. Zinc, cadmium, and mercury have thus more " open " structures

[*] T. B. Massalski and H. W. King, *Acta Met.*, 1962, **10**, 1171.

Electrons, Atoms, Metals, and Alloys

than copper, silver, and gold. It is found that 1 at.-% of cadmium expands the lattice of copper more than does 1 at.-% of silver, and this expansion is the result of the valency electrons, because the electron cloud of the *ion* of cadmium is smaller than that of silver. This principle is quite general, and in passing down a series of elements such as

$$\text{Cu} \longrightarrow \text{Zn} \longrightarrow \text{Ga} \longrightarrow \text{Ge} \longrightarrow \text{As}$$
$$\text{Ag} \longrightarrow \text{Cd} \longrightarrow \text{In} \longrightarrow \text{Sn} \longrightarrow \text{Sb}$$

the underlying ion with the outer group of 18 electrons shrinks rapidly, and its electron cloud plays a continually smaller part in determining the "size" of the atom. This point is illustrated clearly in Fig. VI.29, which shows both the *atomic* radii and the univalent *ionic* radii of some of the elements. The variation of the electron-cloud density of an ion in space cannot, of course, be adequately represented by a single number, but the univalent ionic radii of Pauling and Zachariasen * are the most satisfactory values to use when it is desired to represent the size of an ion by one constant. If the difference between the atomic radius (*i.e.*, one-half the closest distance of approach in the crystal of the element) and the ionic radius is looked on as a measure of the degree of "openness" of a metal, Fig. VI.29 shows clearly how the metals following an alkali metal become progressively less open with increasing valency, whilst in the elements following copper and silver the structures become progressively more open.

The rapid contraction of the ions in the elements following copper and silver leads to interesting solid-solubility relations in cases where the electron clouds of the ions are concerned in the process. When, for example, silver, cadmium, indium, tin, or antimony is dissolved in solid copper, the mean lattice distortions produced by 1 at.-% of the above elements are in the order:

$$\text{Ag} < \text{Cd} < \text{In} < \text{Sn} < \text{Sb}.$$

* So-called ionic radii were first deduced empirically in order to account for the interatomic distances in salts. These empirical ionic radii involve the charges on the ions, and are not measurements of the sizes of the ions if by "size" we mean the extent and density of the electron cloud. We may imagine, for example, that a univalent ion A^+ and a divalent ion B^{++} have exactly the same electron cloud pattern in free space. In an actual ionic crystal, the double charge on the B^{++} ion would result in its being pulled more strongly towards a given anion than the univalent A^+ ion would be at the same distance. The interatomic distances in salts would thus indicate a smaller ionic radius for the B^{++} ion than for the A^+ ion, even though in free space their electron cloud densities were the same. The so-called univalent ionic radii are attempts to express ionic radii free from this complication, and this work is due to Pauling (*J. Amer. Chem. Soc.*, 1927, **49**, 765) and Zachariasen (*Z. Krist.*, 1931, **80**, 137). The electron cloud density of an ion in space cannot, of course, really be represented by a single number, but where one constant has to be used, the univalent ionic radii of Pauling and Zachariasen are the most satisfactory.

338   *Atomic Theory*

Now when the size-factors are favourable and the electrochemical factor is small, we have seen (p. 311) that the maximum solid solubilities of the above elements in a univalent face-centred cubic solvent are of the order: 100 : 40 : 20 : 12 : 7–8. The univalent elements mix completely (*e.g.*, Ag–Au). The divalent and trivalent solutes dissolve up to a limiting electron concentration of approximately 1·4, whilst the

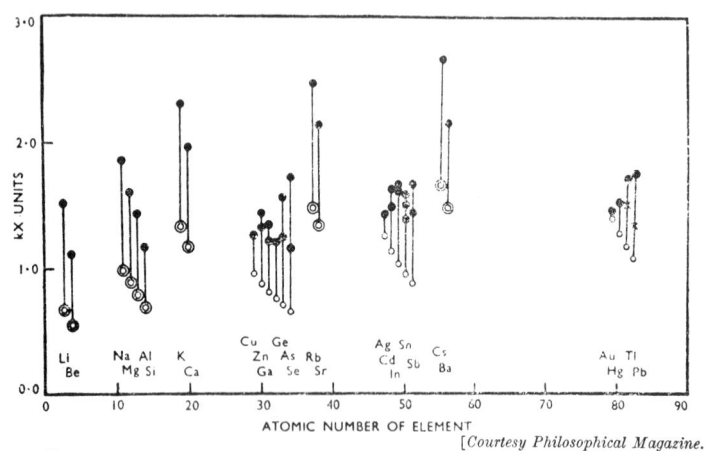

FIG. VI.29. Atomic radii and univalent ionic radii of the elements.

[*Courtesy Philosophical Magazine.*]

KEY.

● $\frac{1}{2}$ (interatomic distance in the crystal of the element).
◎ Zachariasen univalent ionic radii.
○ Pauling univalent ionic radii.
× Goldschmidt empirical ionic radii.

tetravalent (*e.g.*, Cu–Ge) and pentavalent (*e.g.*, Ag–Sb) solutes have slightly lower solubilities than those expected from a simple electron-concentration rule. The maximum solid solubilities of the above elements in copper are as follows:

Ag	Cd	In	Sn	Sb
5·4	1·7	10·9	9·3	5·9 at.-%

From these figures it will be seen that the solid solubilities of indium, tin, and antimony in solid copper approach more nearly to the normal values than do those of silver or cadmium, in spite of the fact that the mean lattice distortion per atom increases with the valency of the above solutes. We have thus the curious position that the mean lattice distortion of copper produced by one atom of indium is greater than that produced by one atom of cadmium, but that at the same time it is easier for the atom of indium to enter into solid solution in copper. The reason for this effect is almost certainly the rapid shrinking of the *underlying ion* which takes place with increasing atomic number of the above solutes. In the case of silver, as can be seen from Fig. VI.29, the atomic and ionic radii are of the same order, with the result that the introduction of the silver atom into the copper lattice involves the dense electron cloud of the silver ion. On passing to cadmium, indium, tin, and antimony, the underlying ions shrink rapidly, and their electron clouds play a diminishingly important part in the process, with the result that the solid solubilities approach more nearly to their normal values. This example serves to illustrate the dangers of attempting to represent the size of an atom by a single constant, and shows that in some cases it is necessary to consider not merely the characteristics of the valency electrons, but also those of the outermost shell of the ion.

Aluminium.

Aluminium is a trivalent metal with a face-centred cubic structure, and is a conductor because each complete zone contains two electrons. On passing from Na ⟶ Mg ⟶ Al ⟶ Si there is :

(1) A decrease in interatomic distance.
(2) A decrease in compressibility.
(3) A decrease in the coefficient of expansion.
(4) An increase in melting point.

We may, therefore, conclude that aluminium is a normal trivalent metal, with three electrons per atom taking part in the cohesion. The metal has, however, some curious characteristics which are not yet understood. The sequence of melting points for the above metals is Na (97° C), Mg (650° C), Al (660° C), Si (1420° C), so that the increase on passing from Mg to Al is very slight. This might be ascribed to some factor affecting the liquid rather than the solid phase, but the coefficient of expansion of aluminium is only slightly less than that of magnesium, and this is a property which depends only on the solid phase. A possible explanation of this is advanced later, but these properties of the metal are not really understood.

A major advance in the electron theory of aluminium was made by Heine,[*] whose work is a combination of theoretical calculation on the one hand, and conclusions based on measurements of specific heats (p. 259), the anomalous skin effect (p. 275), and the de Haas–van Alphen effect (p. 268), on the other. This approach has been extended by Harrison[†] and Ashcroft.[‡]

[*] V. Heine, *Proc. Roy. Soc.*, 1957, [A], **240**, 340, 354, 361.
[†] W. A. Harrison, *Phys. Rev.*, 1959, **116**, 555.
[‡] N. Ashcroft, *Phil. Mag.*, 1963, **8**, 2055.

The first Brillouin zone for aluminium is the truncated octahedron of Fig. VI.30, and all the lines of approach indicate that this zone is very nearly filled, and that there is a large overlap across the hexagonal faces, and a smaller overlap across the square faces.

The calculations show that, except in the immediate vicinity of the zone boundaries, the E/k relations are very nearly equal to those of the free-electron theory. To a first approximation we may regard the occupied states as almost filling the first zone, and spreading out to form large spherical caps on the hexagonal faces, and smaller spherical caps on the square faces. It is worth noting that the estimates of the band gaps are significantly smaller than those for magnesium and this can be associated with the relatively smaller ion cores (see Fig. VI.29) and the closer approach of energy states derived from the s and p atomic states. The bulk electrical transport properties of aluminium are in consequence not very different from what would be expected for a free-electron metal of valency 3.

The low-temperature specific-heat data (see p. 259) indicate that $N(E)$ at the Fermi level is about 1·6 times that for a free-electron gas. Much of the Fermi surface is in the second zone, and is not near to a zone face, and since this part of the surface will have a normal value of $N(E)$, it is concluded that there must be other parts with an abnormally high $N(E)$.

The data for the anomalous-skin effect, together with the theoretical work, indicate that the overlaps across the hexagonal and square faces of the zone have spread out from points such as L and X (Fig. VI.30) to such an extent that points such as K and U in the second zone are probably below the Fermi surface, although the latter has not yet reached points such as W in the second zone.

The de Haas–van Alphen effect in aluminium has been examined by Gunnerson,[*] who detected three periodic components of the oscillations, which correspond to three identical cushion-shaped pieces of the Fermi surface, with their principal axes perpendicular. The conclusion drawn is that, even in the first zone, the states are not quite all occupied, and that there are very small pockets of holes inside the corners W. These unoccupied states in the first zone account for a total of about $3·6 \times 10^{-3}$ holes/atom. There is also a set of low-frequency de Haas–van Alphen oscillations whose period is increased by the addition of a little magnesium in solid solution. From this it is concluded that a very small overlap has occurred into the third zone which may contain about 5×10^{-5} electrons/atom.

---

[*] E. M. Gunnerson, *Phil. Trans. Roy. Soc.*, 1957, [A], **249**, 299.

Electrons, Atoms, Metals, and Alloys

An alternative estimation of the Fermi surface in aluminium has been made by Harrison,* whose work is again a combination of calculation and deduction from experimental facts. In agreement with Heine, Harrison concludes that there is an overlap across both octahedral and square faces of the first zone, and that the fourth zone is empty. Harrison, however, concludes that the first zone is entirely filled, and that the overlap into the third zone is larger than that suggested by Heine.

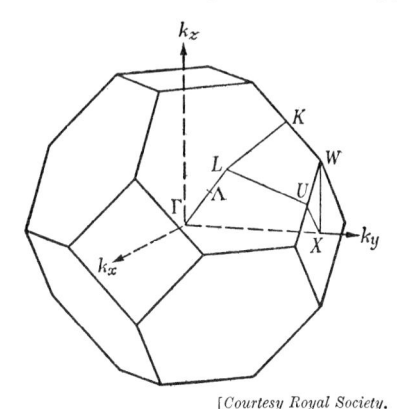

[Courtesy Royal Society.]
FIG. VI.30. First Brillouin zone for aluminium.

The electronic structure of solid aluminium is thus such that the first zone is almost filled, and that there is a large overlap across the hexagonal faces, and a considerable overlap across the square faces of the zone into the second zone, together with an exceedingly small overlap into the third zone. From the work of Jones (p. 320), we can thus understand why the atomic diameter of aluminium, estimated from the interatomic distances in the metal, is often greater than would be imagined from its behaviour when dissolved in some alloys where no zone overlap is concerned. In some cases, the apparent contraction of the aluminium atom in alloys is partly the result of a high electrochemical factor, but zone effects are often responsible for much of the contraction. Thus, aluminium dissolved in copper or silver expands the lattice as though the aluminium had an atomic diameter of 2·7 Å., as compared with an interatomic distance of 2·86 Å. in the pure metal. On the other hand, when dissolved in magnesium, where zone overlaps have occurred, the lattice is contracted as though the aluminium atom had an atomic diameter of 2·83–2·84 Å.

It is possible that the zone characteristics of aluminium referred to

* W. A. Harrison, *Phys. Rev.*, 1959, **116**, 555.

above are related to its large coefficient of expansion. On raising the temperature, electrons will be thermally excited from the lower to the higher zones. The expansion effect of Jones would then produce a greater expansion than if no zone overlaps were involved; this might also explain the abnormally low melting point.

The Apparent Atomic Diameters of Metals in Solid Solutions.

If one metal dissolves in another to form a primary solid solution, the lattice spacing/composition curve for dilute solutions is usually linear, and if it is extrapolated to 100 at.-% of solute the resulting lattice spacing may be called the apparent atomic diameter (A.A.D.) of the solute in the particular solvent concerned. As explained above, the A.A.D. of aluminium dissolved in copper is smaller than indicated by the interatomic distances in the crystal of pure aluminium, and this is partly due to the existence of a Brillouin-zone overlap in aluminium, but not in the alloy.

Much interest has been aroused in the problem of the factors affecting the A.A.D. values of the metallic elements when dissolved in one another. According to Vegard's law (1921) the lattice-spacing/composition curves for solid solutions of salts of the same structure are straight lines joining the values for the two pure salts, and similar relations, with corrections for the effects of co-ordination number, exist when the salts have different structures. In 1928 it was claimed that this law applied also to metallic solid solutions, but numerous exceptions were known, and a detailed examination by Hume-Rothery and Axon * has shown that in nearly all metallic solid solutions there are deviations from the so-called Vegard's law. The whole problem is very complicated, but in some cases the difference between the A.A.D. values for a given metal when dissolved in different solvents appears to be greatly influenced by the volumes per valency electron of the solvent and solute. We may illustrate this point by considering the solid solution of lithium in aluminium. Lithium, as an alkali metal, crystallizes in the body-centred cubic structure with atomic diameter (A.D.) † = 3·03 kX; the A.D. value for lithium in a hypothetical face-centred cubic structure would be about 3·12 kX. The Li^+ ion is very small compared with the closest distance of approach between two lithium atoms in the metallic crystal (see Fig. VI.1, p. 285) and the ions play a negligible part in determining the interatomic distances. The volume per valency electron, for which we may use the symbol V_e, is 21·5 kX³, so that the electron

---

* *Proc. Roy. Soc.*, 1948, [A], **193**, 1.
† The atomic diameter is taken as the closest distance of approach of the atoms in the crystal of the element.

gas of the valency electrons is very dilute. The metal has a high compressibility ($K = 8\cdot 9 \times 10^{12}$ dynes/cm.^{2}), which is due mainly to the valency electrons. We may therefore regard the crystal of metallic lithium as consisting of small Li$^+$ ions swimming in a dilute valency-electron gas, and the energy/volume relations of this gas may be looked on as giving rise to very weak attractive forces which hold the atoms together, and very weak repulsive forces which hold them apart.

In the face-centred cubic structure of aluminium the A.D. = 2·86 kX, so that the atoms are closer together than in lithium. With 3 valency electrons per atom, $V_e = 5\cdot 51$ kX^3 and the electron gas is about four times as dense as that in metallic lithium. The melting point of aluminium (660° C) is much higher than that of lithium (152° C), whilst its compressibility ($K = 1\cdot 37 \times 10^{12}$ dynes/cm^2) is much lower. If, therefore, we regard the interatomic distance as resulting from the balance between attractive and repulsive forces, then these are much stronger in aluminium than in lithium. If we considered a 1-dimensional chain of aluminium atoms, we could imagine one of these replaced by an atom of lithium with the latter retaining the attractive and repulsive forces characteristic of metallic lithium. In contrast to this, in a 2- or 3-dimensional lattice, a distortion round one atom affects the interatomic distances between other atoms. We can understand, therefore, how, when a lithium atom is substituted for one of aluminium, the much stronger forces of the aluminium lattice impress themselves on the lithium atom so that the latter no longer appears to possess the size characteristic of the very weak forces in metallic lithium. Actually, the A.A.D. value of lithium dissolved in aluminium is only 2·82 kX, and is thus nearly the same as that of aluminium itself, and since the ion of lithium is smaller than that of aluminium (2·86 kX), we can perhaps understand why the A.A.D. of lithium in aluminium is slightly smaller than 2·86 kX.

The above example of lithium dissolved in aluminium is one where the solvent has a smaller V_e value than the solute, whose A.A.D. value is thus made smaller. Axon and Hume-Rothery (*loc. cit.*) found that in many alloys this is a general principle, and that the converse also holds, *i.e.*, the A.A.D. value is increased if the V_e value of the solvent is larger than that of the solute. This simple principle is, however, often obscured by the Brillouin-zone effects referred to above, and also by a high electrochemical factor which tends to diminish the interatomic distances. Further complications are met with if the electron cloud of the *ion* of the solute is so large that ionic overlap occurs when a solute atom is introduced. From this it will be realized that it is only in exceptional cases that the so-called Vegard's law can be expected to hold for metals.

z

Other B-Group Metals.

Experimental information about the Fermi surfaces of most of the B-group metals is now available, and in all of them the Fermi surface, when expressed in terms of the reduced-zone scheme (see Appendix I(5), p. 252) has many different sheets of rather complicated geometries. There is also a growing amount of theoretical work (mainly by Heine and his co-workers) on the application of pseudopotential methods (p. 238) to the evaluation of such data in terms of a small number of parameters, which specify the pseudopotential and which can be discussed in terms of the electronic structures of the different atoms. Detailed discussions of these developments are beyond the present scope of this book, but it is worth noting that they make reasonable (while not rigorously predicting) the departures from the three basic metallic crystal structures observed in mercury, indium, gallium, and tin. Heine emphasizes that the number of nearest neighbours and their relative directions in these complex structures has nothing to do with directed covalent bonds in the conventional sense. In covalent bonds the atomic arrangements make large constraints on the wave-functions of the bonding electrons; the complex structures arise from distortions that minimize the constraints set on the band electrons by the pseudopotentials associated with the lattice sites.

References to some of this work are given below.

REFERENCES AND SUGGESTIONS FOR FURTHER READING.

Electron Theory of Magnesium, Cadmium, and Close-Packed Hexagonal Lattices.

N. F. Mott and H. Jones, " The Theory of the Properties of Metals and Alloys," p. 161. **1936**: Oxford (Clarendon Press).

H. Jones, *Proc. Roy. Soc.*, 1934, [A], **144**, 225.

W. Hume-Rothery and G. V. Raynor, *J. Inst. Metals*, 1938, **63**, 227.

W. M. Lomer and W. E. Gardner, *Progress Mat. Sci.*, 1969, **14**, 143.

Electron Theory of B-Group Metals.

V. Heine in " The Physics of Metals " (edited by J. M. Ziman. **1969**: Cambridge (University Press).

General Lattice-Spacing Relations.

H. J. Axon and W. Hume-Rothery, *Proc. Roy. Soc.*, 1948, [A], **193**, 1.

G. V. Raynor, " Progress in Metal Physics." Vol. I, p. 1. **1949**: London (Butterworths Scientific Publications).

Lattice-Spacing Relations in Solid Solutions in Magnesium.

W. Hume-Rothery and G. V. Raynor, *Proc. Roy. Soc.*, 1940, [A], **177**, 27.

This paper describes the work showing how the apparent sizes of atoms in solid solution are affected by Brillouin-zone overlaps.

G. V. Raynor, *ibid.*, 1942, [A], **180**, 107.

W. Hume-Rothery and G. V. Raynor, *ibid.*, 1940, [A], **174**, 479 (Mg–Cd alloys).

Effect of Ionic Radius on Solid-Solubility Relations.

W. Hume-Rothery and G. V. Raynor, *Phil. Mag.*, 1938, [vii], **26**, 143.

4. The Transition Elements.

General.

The transition metals of the Long Periods are those in which an incomplete group of 8 electrons expands into one of 18 by the building up of a sub-group of 10 d-electrons. The methods of soft X-ray spectroscopy have shown that in the crystals of these elements the ns, np, and $(n-1)d$ electron states of the free atoms have broadened into overlapping bands, so that the electrons concerned are in hybrid (spd) states. The electron theory of these metals is thus very complicated and we shall deal here only with some of the more simple theories. The transition metals are of interest because they include the metals of highest melting points, and also the ferromagnetic metals, iron, cobalt, and nickel, in the First Long Period.

TABLE XIII.—*Crystal Structures of the Transition Elements.*

In all cases the allotropic forms are shown in the order of the temperature ranges in which they are stable, the form stable at the highest temperature being at the top.

K	Ca	Sc *	Ti	V	Cr	Mn	Fe	Co	Ni	Cu
b.c.c.	b.c.c. c.p.h. f.c.c.	f.c.c. c.p.h.	b.c.c. c.p.h.	b.c.c.	b.c.c.	δ b.c.c. γ f.c.c. β complex α complex	δ b.c.c. γ f.c.c. α b.c.c.	β f.c.c. α c.p.h.	f.c.c.	f.c.c.

Rb	Sr	Y	Zr	Nb	Mo	Tc	Ru	Rh	Pd	Ag
b.c.c.	b.c.c. c.p.h. f.c.c.	c.p.h.	b.c.c. c.p.h.	b.c.c.	b.c.c.	c.p.h.	c.p.h.	f.c.c.	f.c.c.	f.c.c.

Cs	Ba	La	Hf	Ta	W	Re	Os	Ir	Pt	Au
b.c.c.	b.c.c.	β f.c.c. α c.p.h.	b.c.c. c.p.h.	b.c.c.	b.c.c.	c.p.h.	c.p.h.	f.c.c.	f.c.c.	f.c.c.

			Th	Pa	U					
			f.c.c.	b.c. tetragonal	γ b.c.c. β complex α complex					

* It is not known which allotrope is stable at the higher temperature.

Crystal Structures.

Table XIII shows the crystal structures of the transition metals, from which it will be seen that allotropy is common. Iron is unique in having the body-centred cubic structure as the form stable at both high and low temperatures, with the face-centred cubic modification stable in the intermediate range 910°–1389° C. In all other cases, if the body-centred cubic structure exists, it is the form stable at the highest temperature. The reason for this is probably the same entropy effect as that described on p. 322 in connection with the stability of the β phases in copper alloys.

In the Second and Third Long Periods, close-packed structures appear in Group IIIA and then, on passing along these Periods, the different structures follow a regular sequence c.p. hex. → b.c. cubic → c.p. hex. → f.c. cubic, with the b.c. cubic structure reaching its maximum stability in Groups VA and VIA. In the First Long Period, the behaviour of the earlier Groups is the same, but manganese behaves abnormally, with complicated structures in the α and β modifications which are stable at low temperatures. In this Period, the b.c.c. structure extends as far forward as Group VIIIA (Fe), whilst the f.c.c. structure extends as far back as Group VIIA (Mn). Cobalt, like rhodium, crystallizes in the f.c.c. structure, but there is also a low-temperature modification with the c.p. hexagonal structure. This, however, differs from the c.p. hexagonal structures of ruthenium and osmium in the later Periods, in that the axial ratio (1·623) is nearly that for the close-packing of spheres, whereas for the elements of Group VIIIA the axial ratios are very much lower (Ru, 1·582; Os, 1·579).

In the Fourth Long Period, the crystal structures are quite different, and it is probable that in the actinides some of the 5f electrons are involved in the metallic bonding.

Atomic Diameters.

Fig. VI.31 shows the atomic diameters of the elements of the Long Periods as defined by the closest distances of approach of the atoms in

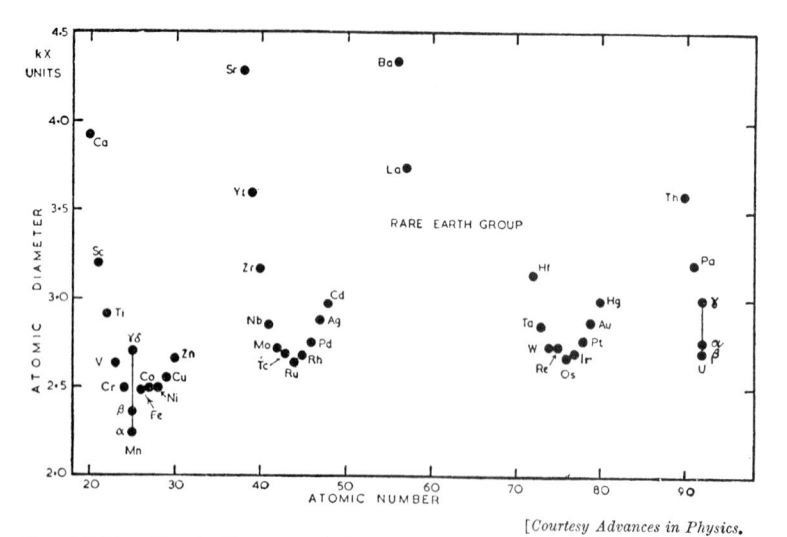

[*Courtesy Advances in Physics.*]

FIG. VI.31. Atomic diameters of elements of Long Periods as defined by closest approach of atoms in the crystals of the pure metals.

the crystals of the pure metals. In each Period there is a progressive decrease in atomic diameter on passing from Group IA to VIA, after which the relation changes and there is an increase on passing to copper, silver, and gold. The two later Long Periods are characterized by the fact that ruthenium and osmium in Group VIII have smaller atomic diameters than the preceding elements in Group VIA (molybdenum and tungsten), whereas in the First Long Period the atomic diameters of chromium and iron are almost identical.

Compressibilities.

The compressibility is a measure of the ease with which the atoms of a solid can be squeezed together, and since at zero pressure the process is reversible, the reciprocal of the compressibility, which may be called the incompressibility, is a measure of the difficulty in pulling the atoms apart. Consequently, strong bonding is indicated by a high incompressibility and a low compressibility. The compressibilities of the transition elements are shown in Fig. VI.32, and suggest that in all three Long Periods the strength of bonding increases with the group number as far as Group VA, after which a difference exists between the First and later Periods. In the Second and Third Long Periods, the compressibilities continue to diminish and reach minimum values at ruthenium and iridium, respectively. In the First Long Period, the compressibility of α-manganese is anomalous, as would be expected from its abnormal crystal structure, but otherwise there is little change on passing from V → Cr → Fe → Co → Ni. The value for vanadium may be affected by impurity, but the compressibility of chromium is clearly greater in relation to that of titanium than would be expected from the values for the corresponding elements in the later Periods.

In general, for the elements of any one sub-group, the compressibility increases with the atomic volume. Fig. VI.32 shows, however, that in spite of their smaller atomic volumes, the compressibilities of titanium, vanadium, and chromium are slightly greater than those of zirconium, niobium, and molybdenum respectively, and there is thus a strong suggestion that the cohesion for the middle and later elements of the First Transition Series is weaker than would be expected from the behaviour of the analogous elements in the Second and Third Transition Series.

Coefficients of Thermal Expansion.

Fig. VI.33 shows the coefficients of expansion of the transition elements, and these exhibit the same general trends as the compressibilities, although there are differences in detail. The coefficient of

348   *Atomic Theory*

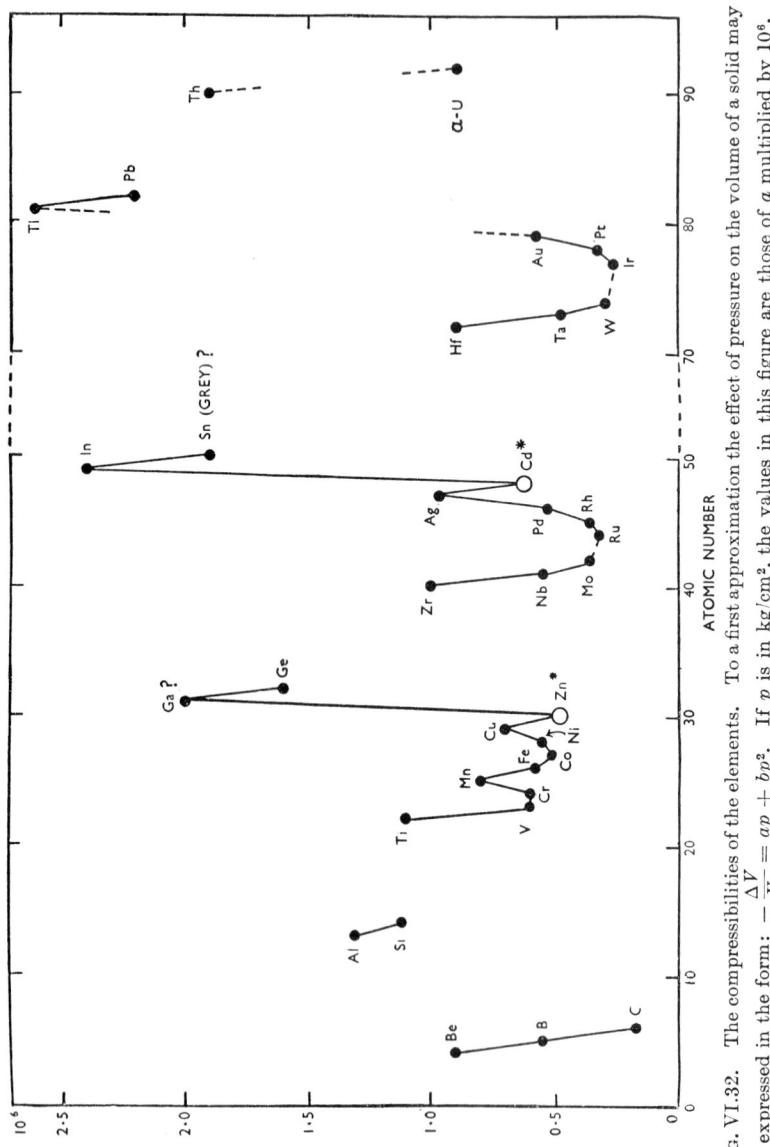

FIG. VI.32. The compressibilities of the elements. To a first approximation the effect of pressure on the volume of a solid may be expressed in the form: $-\dfrac{\Delta V}{V_0} = ap \pm bp^2$. If p is in kg/cm², the values in this figure are those of a multiplied by 10^6.

For footnote see opposite page.

Footnote to Fig. VI.32.

* For zinc and cadmium, the values plotted (large open circles) are 3 times the linear compressibility in the direction of the close-packed hexagonal planes, in order to give an indication of the strength of bonding in these planes. Owing to the high value of the axial ratio, the volume compressibility of the crystal as a whole is much greater.
Values of $a \times 10^6$ for elements of higher compressibility are as follows: Na 15, K 29, Rb 31, Cs 40, Mg 3, Ca 6, Sr, 8 Ba 10.

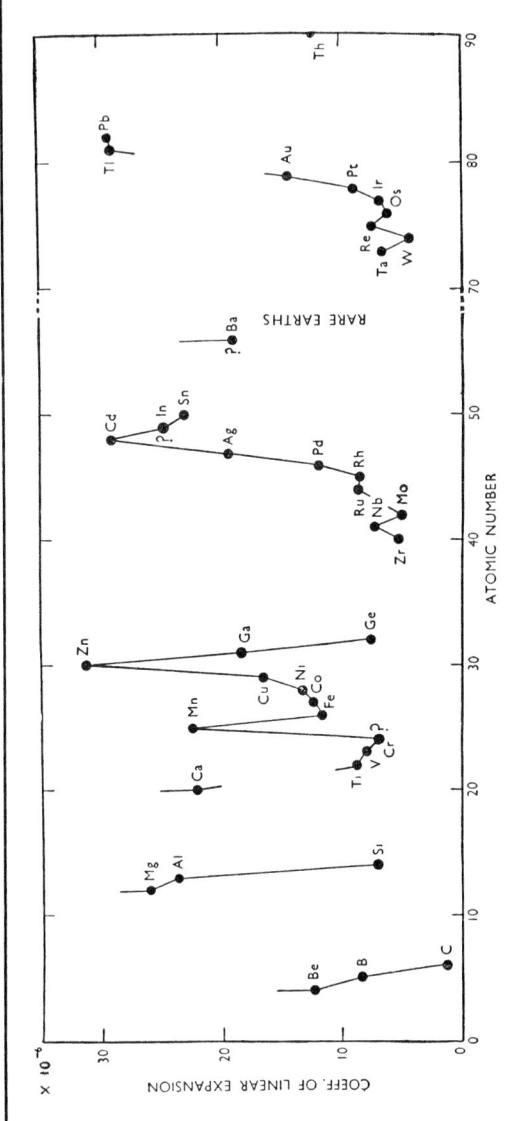

Fig. VI.33. The coefficients of linear expansion of the elements.

Values for elements with higher coefficients of expansion are as follows: Li 56, Na 71, K 83, Rb 90, Cs 97 $\times 10^{-6}$.

expansion is a measure of the resistance to increasing amplitude of thermal vibration and consequently strong bonding is indicated by a low coefficient of expansion. There is again a difference between the First and later Periods. In particular, the coefficients of expansion of iron and cobalt are relatively much higher compared with those of titanium, vanadium, and chromium, than would be expected from the values for the corresponding elements in the later Periods.

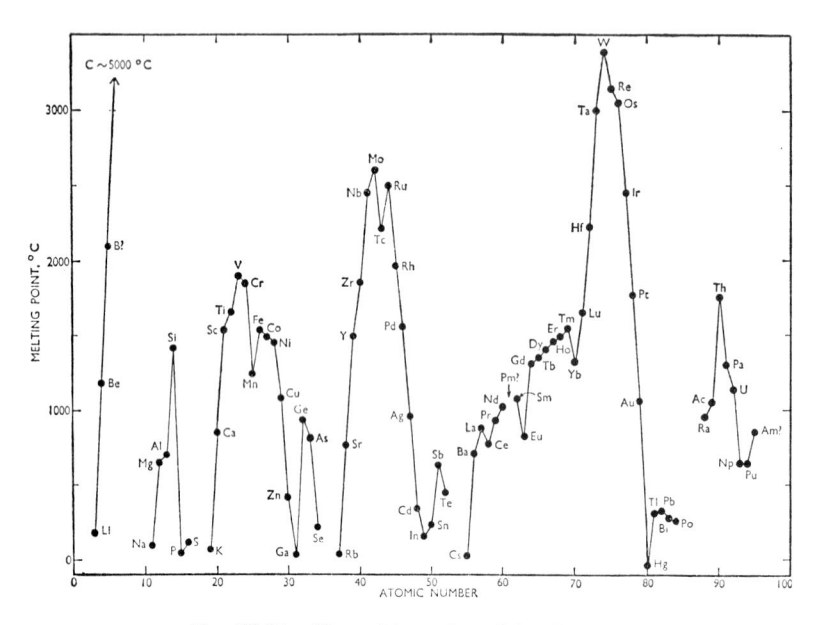

FIG. VI.34. The melting points of the elements.

Melting Points.

The atomic diameters, compressibilities, and coefficients of expansion are properties of the solid phase alone. In general, a high melting point indicates a strong resistance to thermal agitation, and hence a strong atomic bonding, but as the melting point by definition involves equilibrium between solid and liquid phases, it is not such a satisfactory property to study. Fig. VI.34 shows the melting points of the transition elements, and emphasizes the great difference between the Second and Third Periods, where there are spectacular maxima at Group VIA (molybdenum and tungsten), and the First Period where, in spite of the normal crystal structure of δ-manganese, its melting point is anomalous, and there are much smaller differences in the values for the elements from titanium to nickel.

Binding Energies.

Fig. VI.35 shows the heats of sublimation or binding energies of the transition elements. By definition, the heat of sublimation is the energy

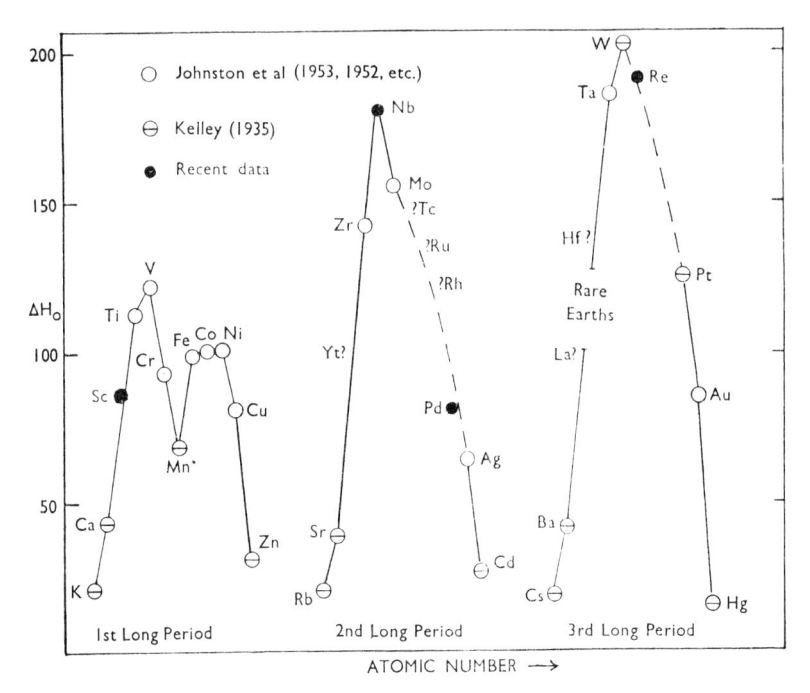

FIG. VI.35. The heats of sublimation of the elements at $0°$ K. (ΔH_0, kcal/g mole)

* The differences between the values for the different forms of manganese are very small.

required to remove an atom from the solid to the vapour state, and although it is a definite physical constant of the element, it involves the ground state of the free atom, and is not a constant of the solid state alone. High binding energies are generally accompanied by low compressibilities (cf. Figs. VI.32 and VI.35), but there is no simple relation between the two quantities, because the compressibility is a measure of the energy required to produce a small displacement of the atoms from their equilibrium position in the solid, whereas the binding energy is the

352   *Atomic Theory*

work required to remove an atom from the solid to the vapour phase. Fig. VI.35 again emphasizes the differences between the elements of the First and of the later Periods.

Electronic Specific Heats.

As explained below (p. 354) theories of the electronic structures of the transition metals imply that the d-electrons give a narrow band in which the density of states is as much as ten times larger than that in the bands of simple metals. From the considerations of p. 259 we therefore expect large values of the electronic specific heat. These are, in fact, found experimentally and are clear evidence that the d-electrons must be described by Bloch wave-functions rather than atomic wave-functions in these metals. The variation in the electronic specific heat coefficient (γ) along any row of transition metals shows that large variations in $N(E)_F$ as a function of E take place; but this variation cannot be used to give directly the $N(E)$ vs. E curve for any given metal, since changes in crystal structure and total band width are taking place. These results do support, however, the type of d-band structure shown in Fig. VI.40.

In ferromagnetic or antiferromagnetic metals and alloys new effects in the specific heat arise from magnetic ordering; and when, in an alloy, the ordering temperature falls to the range where electronic specific heats are normally measured the interpretation of the results is difficult.

Magnetic Properties.

Some of the magnetic properties of metals are dealt with in Section VII.1, but we may refer here to the *saturation moments* which have played an important part in the development of the theory. When a ferromagnetic substance is magnetized, the saturation moment at the absolute zero represents the greatest degree of magnetization that can be obtained. The data are usually expressed in Bohr magnetons per atom, and the accepted values for iron, cobalt, and nickel are 2·22, 1·71, and 0·61 μ_B respectively. These represent the number of unpaired electron spins per atom in the metallic crystal, and are not whole numbers, and this means that we cannot regard these ferromagnetic metals as consisting of an array of identical atoms or ions, each having the same number of electrons missing from the $(3d)^{10}$ sub-group. We have either to regard the ferromagnetic metal as consisting of atoms in more than one electronic state, or to adopt a model in which we lose sight of the individual atoms, and consider the whole assembly as giving rise to a band containing the number of unpaired spins required to give the above values for the saturation moments per atom. The latter viewpoint is adopted in the collective-electron theories of Mott and Jones, of Stoner, and of Wohlfarth, whilst the former approach underlies the hypothesis of Pauling which has been extended and modified by several investigators (see pp. 360–370).

When an element such as copper, zinc, or aluminium dissolves in nickel, the magnetic saturation moment is reduced by v Bohr magnetons for each atomic substitution, where v is the valency of the solute. Since the saturation moment of nickel is 0·6 μ_B, this means that the graph connecting saturation moment with composition is a straight line giving a zero moment at 60 at.-% of copper; whilst in the nickel–aluminium alloys the ferromagnetism is destroyed three times as quickly. In contrast, when the same solutes are dissolved in iron, for which the saturation moment is 2·22 μ_B, the effect is to reduce the saturation moment by 2·22μ_B for each atomic substitution, regardless of the valency of the solute. The data for cobalt are less certain, but the results resemble those for nickel rather than those for iron. An interesting review of these effects has been given by Stoner [*] and their interpretation is discussed later.

The Electron Theory of Transition Metals: Qualitative Principles.

For the alkali metals at the beginning of each Transition Series, the electronic energy levels of the free atoms are in the order $ns < (n-1)d < np$, where n is the quantum number of the valency electrons (*e.g.*, for potassium, $n = 4$). In copper, silver, and gold at the end of the Transition Series the order of the levels is $(n-1)d < ns < np$. For the free atom, therefore, the general effect of passing along the Transition Series is for the $(n-1)d$ level to move downward through the ns level.

When the atoms assemble to form metallic crystals, each sharp energy level broadens out into a band. For the alkali metals (potassium, rubidium, caesium), it is a reasonable approximation to regard the outer electrons [†] as existing in nearly pure s-like states. In the alkaline earth metals, the d-bands have dropped to such an extent that some of the outer electrons are properly described by hybrid (spd) functions. Further passage from Group IIA → IIIA → IVA . . . results in a continuation of this process, the proportion of the d function in the lower-energy range of the hybrid band increasing with the number of the Group.

There is general agreement that, up to the stage at which the breaks in the sequence of physical properties (pp. 348–351) occur, all the

[*] E. C. Stoner, *Rep. Progress Physics*, 1946–47, **11**, 43.
[†] For brevity we use the term "outer electrons" to describe all the electrons outside the core with the configuration of the preceding inert gas. By (spd) hybridization, we mean the hybrids formed by ns, np, and $(n-1)d$ functions.

outer electrons are involved in the cohesion, and all are in hybrid (spd) states. Beyond this stage, the position is less clear, and several different and conflicting methods of approach have been used. In the *collective-electron theories*, all the outer electrons are regarded as being in energy bands of the whole crystal, and as contributing to the cohesion. From this point of view the breaks in the sequence of physical properties are ascribed to an increase in the repulsive part of the wave function. In the *Pauling hypothesis*, the breaks in the sequences of physical properties are assumed to result from the entry of some of the outer electrons into atomic or non-bonding d-orbitals, whose wave functions are associated with individual atoms, and do not belong to the crystal as a whole. These atomic d-orbitals are distinct from the remaining d-orbitals, which are regarded as forming hybrid (spd) bonding orbitals, whose resonance gives rise to the metallic bond.

At the end of the Transition Series, on passing from Groups IB → IIB → IIIB . . ., there is fairly general agreement * that the completely filled $(nd)^{10}$ sub-groups shrink rapidly, and that the d-electrons become purely atomic in nature, and cease to play a part in metallic cohesion.

Band Theories.

The first and most simple theory of the transition metals is that of Mott and Jones,† who considered metals such as iron, cobalt, and nickel at the end of the Transition Series. For these, the $4p$ fraction of the hybrids was ignored, and the outer electrons were regarded as existing in two distinct, but overlapping bands derived from the $(3d)$ and $(4s)$ states of the free atoms. A similar picture was advanced for the later elements of the Second and Third Long Periods, with appropriate increases in the quantum numbers.

The d-band contains 10 electrons per atom, and the s-band only 2, and a number of facts indicate that the d-band is much higher and narrower than the s-band, so that in a simplified form the $N(E)$ curves for the two bands are as in Fig. VI.36.

From this viewpoint, the diamagnetism of copper indicates that the d-band is full, and the position is as shown in Fig. VI.36(a), where the complete sub-group of ten $(3d)^{10}$ electrons forms part of the Cu^+ ion. The weak diamagnetism of copper is the result of the weak paramagnetism (p. 191) of the $(4s)$ electrons being outweighed by the

---

* In one development of the Pauling theory (p. 363), some of the $3d$ electrons of zinc and gallium were regarded as contributing to the metallic cohesion, but this is not generally accepted.

† N. F. Mott and H. Jones, " The Theory of the Properties of Metals and Alloys."

diamagnetism of the Cu⁺ ion, which contains only completed groups of 2, 8, or 18 electrons. In solid nickel the d-band is not completely filled and the position in the paramagnetic state is as illustrated in Fig. VI.36(b), where there are holes or unoccupied states in the ($3d$) band. For ferromagnetism to be shown there must be a permanent predominance of electrons with one spin, and the model presented is one in

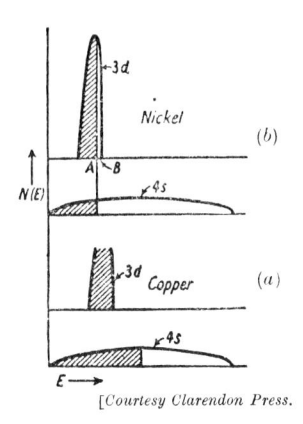

FIG. VI.36. Density of states in nickel and copper.

which the ($3d$) band of Fig. VI.36(b) is regarded as consisting of two bands, one containing electrons of positive spin and the other of negative spin. In the ferromagnetic state one of these two half-bands is regarded as completely filled, whilst the other contains a number of unoccupied states or " holes ", as in Fig. VI.37. This means a predominance of positive over negative spins equal to the number of holes in the ($3d$) band, and we shall consider later (Section VII.1) the process which makes this condition of affairs more stable (i.e., of lower energy) than the normal distribution with equal numbers of electrons of each spin. If we accept the simple model of Fig. VI.37, the saturation moment at the absolute zero will be equal to the number of holes in the ($3d$) band and, since we know the total number of electrons in the ($3d$) and ($4s$) bands, the number in the latter can be estimated. In this way the results shown in Table XIV were obtained for the metals concerned.

In the original theory of Mott and Jones the cohesive forces were regarded as due to the ($4s$) electrons alone. The relatively small number of these, particularly for iron, made this view difficult to accept and, as explained below (p. 356), additional cohesive processes were later suggested. In one approach,* the outermost (3d) electrons were regarded as giving rise to attractive forces of a van der Waals type.

In Mott and Jones's model the electrical conductivity of the transition elements is due mainly to electrons in s-states, and the abnormally large resistance is due to the high $N(E)$ curve for the electrons in d-states, since, as we explain later (p. 408), this gives a high probability of electrons being scattered so as to undergo $s \to d$ transitions which do not occur in the univalent metals.

TABLE XIV.

Metal	Total No. of Electrons Outside the Argon Shell.	Saturation Moment, μ_B	No. of Holes in (3d) Band	No. of Electrons in:	
				(3d) Band	(4s) Band
Fe .	8	2·2	2·2	7·8	0·2
Co .	9	1·7	1·7	8·3	0·7
Ni .	10	0·6	0·6	9·4	0·6

In the model of Fig. VI.37, the saturation moment of nickel is directly related to the number of holes in the d-band. This kind of theory was naturally attractive to those concerned with the magnetic properties of metals, and much work was carried out in which these properties were interpreted in terms of two-band assumptions, in which idealized band forms were assumed. These were usually such that the s-band was parabolic, whilst the head of the d-band was parabolic in the reverse direction, and was thus of the form $N(E) \propto (E_\omega - E)^{1/2}$, where E_ω is the energy at the head of the band. Fig. VI.39 shows the kind of model used, and explains some of the terms that may be encountered. As a result of this work it was gradually realized that the condition of affairs in Fig. VI.37, although reasonably correct for nickel, did not apply to iron, because here the forces stabilizing the ferromagnetic state were not great enough to fill one-half of the d-band completely, so that the distribution is of the kind shown in Fig. VI.38, where there are unoccupied states in both halves of the (3d) band. If y is the number of holes in the lower half-band, and x that in the upper half-band, the saturation moment is $(y - x)$, and the number of electrons in the (4s) band is greater than for the model of Fig. VI.37, and the total number of holes in the (3d) band is greater than the saturation moment. From a study of the magnetic properties of alloys, Coles † has concluded that for iron the approximate distribution on the basis

* N. F. Mott, *Phil. Mag.*, 1953, [vii], **44**, 187.
† See W. Hume-Rothery and B. R. Coles, *Advances in Physics*, 1954, **3**, 149.

Electrons, Atoms, Metals, and Alloys

of a two-band model is 0·9 s-electrons per atom, and 7·1 d-electrons, of which 4·65 are of + and 2·45 of − spin. This removes part of the difficulty referred to above in connection with the strength of cohesion, although it is still not clear why the cohesion in copper is so much weaker than that in iron.

Figs. VI.37 and VI.38 are, of course, purely diagrammatic and drawn

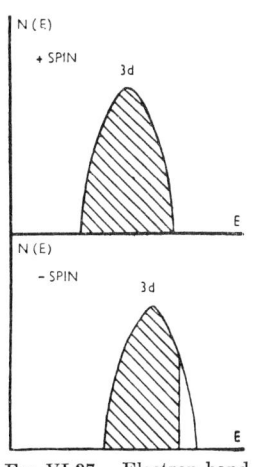

FIG. VI.37. Electron-band condition for ferro-magnetism in nickel.

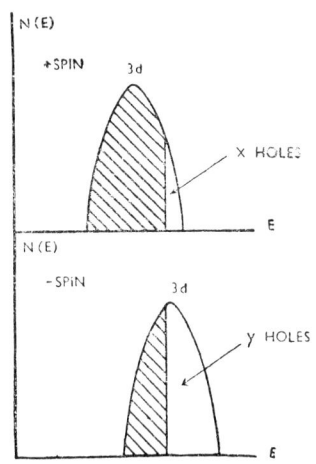

FIG. VI.38. Electron-band condition for ferromagnetism in iron.

so as to show the distribution of spins; they are not to be taken as suggesting that two separate bands exist. The fact that the saturation moments are not whole numbers is to be interpreted as indicating that, for example, in nickel, there are in the crystal as a whole on the average 0·6 electrons per atom in (4s) states. If instead of thinking of the crystal as a whole, we consider the electron clouds round the individual atoms, then in the theory of Mott and Jones it is imagined that atoms may exist with different numbers of electrons in the (3d) shell, e.g., $(3d)^8$, $(3d)^9$, $(3d)^{10}$. The electronic configuration of an atom is regarded as continually changing from one electronic state to another, but Mott and Jones imagine that when an atom exists in a high valency state, e.g. $(3d)^8$, it does so for a sufficient time for it to act as a positive ion attracting the neighbouring (4s) electrons, and in this way giving rise to an additional cohesive force which is one of the causes of the high binding energies of the transition metals. This view was later withdrawn by Mott on the grounds that, in copper–nickel alloys, it would require the existence of superlattices which have not been observed. It is perhaps significant that in iron–cobalt and in iron–nickel alloys superlattices are formed in spite of the closely similar chemical properties and atomic diameters of the two metals, and it is possible that for these alloys the original suggestion of Mott and Jones may be correct in some ways, even though the division into independent s- and d-bands is wrong.

The difference between the distributions represented by Figs. VI.37 and VI.38 may offer an explanation of the different effects of elements such

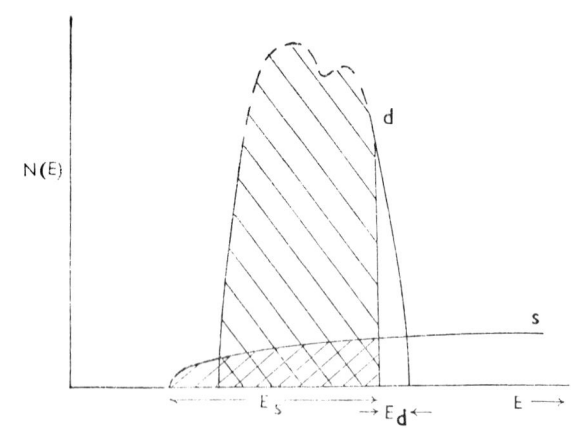

[Courtesy Advances in Physics.

FIG. VI.39. Overlapping d-band and s-band model. (Shading indicates occupied states.) Number of occupied s-band states $= n_s$; number of unoccupied d-band states $= n_d$; relative magnetization (ζ) and degeneracy temperature (T_0) are defined by $\zeta = q_f/n_{0d}$ and $kT_0 = \epsilon_{0d}$, where q_f is the observed intensity of magnetization in Bohr magnetons per atom, k is Boltzmann's constant, and zero suffixes refer to 0° K.

as aluminium on the saturation moments (p. 352) of nickel and cobalt on the one hand, and of iron on the other. For nickel and cobalt (Fig. VI.37), where the one half-band is filled, entry of electrons into the d shell must reduce the saturation moment. In iron (Fig. VI.38), with holes in both halves of the ($3d$) band, there is the possibility of filling the holes in pairs and so producing an effect which is equivalent to the removal of one iron atom for each atomic substitution. This explanation would be valid only if the $N(E)$ curve were nearly horizontal for a considerable range.*

The d-states of a given quantum number are divided into 5-sub-sub-

* This point and many other aspects of the electronic structures of transition metals are discussed in detail in an excellent recent review by Mott (*Advances in Physics*, 1964, 13, 325).

Electrons, Atoms, Metals, and Alloys

groups, each holding not more than 2 electrons (one of each spin). In a perfectly free atom, these electron states are degenerate, but in the field of a cubic crystal they split into two groups, containing 2 and 3 electron states, respectively. The energy ranges of these overlap, and the result is to produce an $N(E)$ curve with a double peak, provided that the d-band is regarded as a separate band. Many calculations have been made of bands of this kind, and the results of Fletcher and Wohlfarth[*] on nickel are shown in Fig. VI.40.

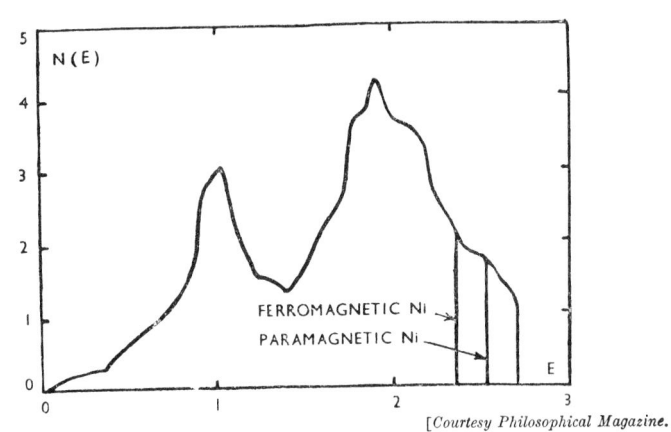

[Courtesy *Philosophical Magazine*.

FIG. VI.40. Density of states curve for the $(3d)$ band in nickel. The abscissae show the electronic energy in eV. measured from zero at the bottom of the band. The ordinates show the density of states $N(E)$ such that the total area under the curve corresponds to five states per atom of either spin. The curve is derived from five sub-bands, of which two are primarily responsible for the left-hand peak, and three for the right-hand peak, although extensive overlapping of the sub-bands occurs. The curve is due to Fletcher and Wohlfarth.

The concept of separated s- and d-bands is of great historical interest, and led to much valuable interpretation and generalization of results, and stimulated experimental research work in many directions. In recent years it has, however, become increasingly clear that even though some of the d-electrons may exist in separated atomic d-orbitals, most of the electrons concerned in the cohesion of the transition metals are in hybrid (spd) states, so that, even for the transition metals of the later Groups, the simple separation into s- and d-bands is unjustified.

More recent theoretical calculations of band structures for transition metals are more realistic, in so far as they take into account the hybridization between states derived from s-levels and those derived from d-levels. The results imply Fermi surfaces with many different sheets

[*] G. C. Fletcher and E. P. Wohlfarth, *Phil. Mag.*, 1951, [vii], **42**, 106.

for most transition metals, and some of them correspond to s-like character in one part of the zone and d-like character in others. The experimental data now becoming available concerning Fermi surfaces of various transition metals seem to be in reasonable accord with the theoretical calculations. It is also of interest to observe that in a given column of the Periodic Table some features of the Fermi surface seem to appear for all three transition rows.

The successful application of the indications of such theoretical and experimental work to the discussion of such properties as the electronic specific heat, the magnetic susceptibility, and even the superconducting behaviour of transition metals, seems to imply a strong justification for the general collective-band viewpoint of the electronic structure of transition metals. It is clear, however, that with the possible exception of nickel and palladium at the end of the first groups a well-defined band of s electrons cannot be regarded as providing the dominant carriers of current in the electrical properties.

The Hypothesis of Pauling and its Modifications.

The theories of the preceding section are collective-electron theories in the sense that the assembly of electrons is considered as a whole without reference to individual atoms. An alternative approach to the electron theory of solids is based on the use of H.L.H. functions (p. 241), in which each electron is described by a wave function localized around one particular pair of atoms. It is this kind of theory that underlies the ideas of Pauling referred to on p. 162, in which metallic cohesion is ascribed to resonance between all the structures which might be obtained by allotting the electrons to definite one-electron or two-electron bonds between the various atoms. We shall consider later some of the conditions which determine whether this approach is justified.

The Pauling hypothesis has developed over many years, and we may deal first with the original paper * in which Pauling assumes that bond formation in the transition elements is due to electrons in s-, p-, and d-states, and not only to electrons in s-states. The d-orbitals alone are not well suited to bond formation, but hybridization between s-, p-, and d-orbitals may lead to very stable hybrid (spd) orbitals, such as those formed in ferricyanide ions $Fe(CN)_6'''$. The ($3s$) and ($3p$) orbitals are too deep down in the atoms to be available for bonding, and in the elements of the First Long Period there are nine orbitals available for hybrid-bond formation, five ($3d$), one ($4s$), and three ($4p$) orbitals.

* L. Pauling, *Phys. Rev.*, 1938, [ii], **54**, 899.

The facts described in section (1) show clearly that there is no continuous increase in the strength of bonding on passing from potassium to cobalt, as would be the case if all the orbitals filled up in succession as the number of outer electrons increased from one to nine. This led Pauling to conclude that only some of the d-orbitals are available for bonding, and that the remainder, which may be called *atomic d-orbitals*, are associated with individual atoms, and do not form bonds. In accordance with the general principles (p. 51), the electrons in the atomic d-orbitals tend as far as possible to remain unpaired, and it is these unpaired electrons which give rise to the saturation magnetic moments. The observed saturation moments are 2·2, 1·7, and 0·6μ_B for iron, cobalt, and nickel, respectively, and increase to a maximum of 2·44μ_B for an alloy of cobalt and iron. This led Pauling to conclude that there are 2·44 atomic d-orbitals per atom, and that the remaining 2·56 d-orbitals are bonding orbitals which, together with the one s- and three p-orbitals, give rise to 6·56 hybrid (spd) bonding orbitals per atom.

For the elements of the First Long Period, the picture given by Pauling is that, on passing from K → Ca → Sc → Ti → V, there is a continuous increase from one to five in the number of electrons per atom used for bond formation, in accordance with the sequences of physical properties discussed above.

In chromium Pauling considers that 5·78 electrons per atom are available for bond formation, and that the remaining 0·22 electrons per atom enter the atomic ($3d$) orbitals. From this stage onwards he assumes that the number of bonding electrons remains constant at 5·78 per atom, and that each step along the Periodic Table results in one more electron entering the atomic d-orbitals, where the electrons remain unpaired as long as this is possible.

In this way the scheme shown in Table XV is obtained. The atomic saturation moment increases on passing from chromium to a point between iron and cobalt, and then decreases, because only 2·44 atomic d-orbitals are assumed to exist. The same general scheme is assumed to hold for the corresponding elements of the later Periods.

The above description refers to the first paper by Pauling,* and since there are in all 9 hybrid (spd) orbitals, the assumption of 2·44 atomic d-orbitals leaves 6·56 hybrid (spd) orbitals for bonding. Examination of Table XV shows that only 5·78 orbitals are regarded as being used, and in a later paper † Pauling emphasizes this point, and regards the 0·78

---

* *Loc. cit.*
† L. Pauling, *Proc. Roy. Soc.*, 1949, [A], **196**, 343.

unused orbitals as "metallic orbitals" whose existence is a necessary characteristic of the metallic state.

It should be emphasized that the work of Pauling is hardly entitled to be called a theory of the transition metals. The numerical values given in Table XV are deliberately chosen to agree with the

TABLE XV.—*Distribution of Electrons in Bonding Orbitals According to Pauling's Hypothesis.*

Element	Total No. of Electrons Outside the Argon Shell	No. of Electrons in Atomic d-Orbitals		No. of Electrons in Bonding Orbitals
		+ Spin	− Spin	
K	1	...	...	1
Ca	2	...	...	2
Sc	3	...	...	3
Ti	4	...	...	4
V	5	...	...	5
Cr	6	0·22	...	5·78
Mn	7	1·22	...	5·78
Fe	8	2·22	...	5·78
Co	9	2·44	0·78	5·78
Ni	10	2·44	1·78	5·78
Cu	11	2·44	2·44	...

experimental data, and the work should be regarded not as a theory or explanation, but rather as an interpretation of the facts, and an attempt to deduce from them the numbers of electrons per atom concerned in the bonding process. The original Pauling hypothesis or interpretation is undoubtedly correct in stressing that on passing along the Long Rows of the Periodic Table, a break in the electronic process occurs in the region of Groups V–VII. It is also correct in emphasizing that the bonding electrons are in hybrid (spd) bands, and not in independent s- and d-bands as was assumed in the preliminary theories of Mott and Jones. It is unsatisfactory in that it gives no reason for the difference between the properties of the elements in the First and later Long Periods, and because the Pauling view of a constant number of bonding electrons per atom from Group VI to Group VIIIC does not agree with the physical properties referred to on pp. 348–351.

In the Pauling concept, the metallic bond results from the resonance of electrons between different configurations, under conditions in which the number of electrons per atom is less than the number of neighbours to which the atom is bound. There are thus too few electrons to form normal co-valent bonds, and fractional bonds are formed. This concept was developed further by Pauling,* according to whom the effect of bond order and resonance can be combined in a single equation of the type:

$$R(1) - R(n) = 0{\cdot}3 \log n \quad \ldots \quad \text{VI (10)}$$

where $R(1)$ is the radius of a normal single co-valent bond, and $R(n)$ is the radius of the atom in a structure where v single bonds resonate among N positions, and $n = v/N$. This equation is empirical, and is based on data for multiple bonds of carbon; its application to metals is a matter on which opinions differ greatly.† In cases where v and n are known, the equation can be used to calculate $R(1)$, and in this way Pauling produced tables of single-bond radii for all the metals including those of the Transition Series; for the latter the numbers of bonding electrons were assumed to be those in Table XV.

The value of this work is difficult to assess but, in general, caution is needed before single-bond radii, or related quantities, calculated by equation VI(10), are used to a degree of refinement greater than 0·1 Å. The equation agrees excellently with the change of atomic radius during the allotropic change of iron from $\alpha \rightleftharpoons \gamma$, if it is assumed that the same number of electrons per atom are concerned in the bonding in the two structures.

In later developments of the Pauling views, it was considered that $(3d)$ electrons were concerned in the metallic cohesion in copper, zinc, and gallium, and that the valencies ‡ of these metals were 5·44, 4·44, and 3·44, respectively, instead of the usually accepted valencies of 1, 2, and 3. This work is not generally accepted,§ and the later developments of the Pauling hypothesis are in some ways less convincing than the original paper of 1938, which undoubtedly contains much truth.

Modifications of the Pauling Scheme.

In 1951, Hume-Rothery, Irving, and Williams ‖ suggested an alternative scheme of valencies of the transition metals, which is in better agreement with the physical data, and is also more closely related to the valencies of inorganic chemistry. The data of Figs. VI.31 to

---

* L. Pauling, *J. Amer. Chem. Soc.*, 1947, **69**, 542 ; *Proc. Roy. Soc.*, 1949, [A], **196**, 343, 350.
† The history of the equation is extremely complicated, and is discussed fully by W. Hume-Rothery and B. R. Coles, *Advances in Physics*, 1954, **3**, 149.
‡ The term " valency " here means the number of electrons per atom concerned in metallic bonding.
§ See W. Hume-Rothery and B. R. Coles, *loc. cit.*, and W. Hume-Rothery, *Ann. Rep. Chem. Soc.*, 1949, **46**, 52.
‖ W. Hume-Rothery, H. M. Irving, and R. J. P. Williams, *Proc. Roy. Soc.*, 1951, [A], **208**, 431.

VI.35 show that iron, cobalt, and nickel resemble each other much more closely than do the corresponding triads ruthenium, rhodium, palladium, and osmium, iridium, platinum in the later Periods, whilst there is a smaller change in the physical properties on passing from Fe → Cu than from Ru → Ag or Os → Au. All the physical properties emphasize that the strength of bonding is greater for molybdenum and tungsten than for chromium, and that manganese, even in its γ (f.c.c) and δ (b.c.c) modifications, has an abnormally low strength of bonding.

In the inorganic chemistry of these elements, it is a general principle that, in any one Group, the predominant valencies increase on passing down the Periodic Table. In Group VI, for example, the trivalent chromic compounds are more stable than the hexavalent compounds, whereas for molybdenum and tungsten the valency of 6 is the most important. Now, the valencies of an element are an indication of the numbers of electrons per atom that can be perturbed sufficiently to take part in the various types of molecular cohesion. It is, therefore, reasonable to expect the predominant valencies to be related to the cohesion in a metallic crystal which is itself a gigantic molecule. Hume-Rothery, Irving, and Williams, therefore, argued that the Pauling scheme was correct in giving molybdenum and tungsten a valency of almost 6, but that for chromium the number of bonding electrons was probably smaller. For brevity we shall call this the "modified scheme".

On passing from Group VI to Group VII, the atomic diameters show a marked expansion on passing from Cr → Mn (γ and δ), and relatively small changes on passing from Mo → Tc or W → Re. In the modified scheme this is regarded as the result of the great stability of the divalent manganous salts, in which the Mn^{++} ion contains the $(3d)^5$ configuration where the $(3d)$ shell is exactly half-filled; an extra stability is known to be acquired by sub-groups which are completely filled or exactly half-filled. The corresponding rhenium salts are, however, comparatively unstable, and as this element shows an array of high valencies, its small atomic diameter and high melting point can be understood. The modified scheme, therefore, regards γ- and δ-manganese as structures in which the divalent state plays a prominent part in determining the electron configuration. This view has received some support from work on the paramagnetism of solutions of copper in γ-manganese, but neutron-diffraction data (p. 392) show that other states must also be concerned in the resonance process.

For the elements of Group VIII, the text-book by Sidgwick [*] gives

[*] N. V. Sidgwick, "The Chemical Elements and Their Compounds." **1950**: London (Oxford University Press).

the valencies in the form of Table XVI where the degree of underlining increases with the importance of the valency in the chemistry of the elements concerned. In the First Long Period, the predominant valencies of iron, cobalt, and nickel are 3 and 2, and there is a relatively small drift towards lower valencies on passing from Fe → Co → Ni.

TABLE XIV.—*Valencies of Elements of Group VIII (Sidgwick).*

Element	Valency	Element	Valency	Element	Valency
Fe	3 2 ion	Ru	8 7 6 5 4 3 2	Os	8 6 4 3 2
	6 3 2 covalent				
Co	4? 3 2 ion	Rh	5 4 3 2?	Ir	6 5 4 3 2 (1)
	2 3 covalent				
Ni	3 2 1 ion	Pd	4 3 2	Pt	(6) 4 3 2 (1)
	3 2 covalent				

In the later Periods, both ruthenium and osmium show an array of high valencies, and on passing from Ru → Rh → Pd and from Os → Ir → Pt, the drift towards lower valencies is more pronounced. The correspondence between the valencies of inorganic chemistry, and the physical constants of Figs. VI.31–VI.35 is striking.

The picture of the transition metals given by the modified scheme is as follows:

Groups IA → IIA → IIIA → IVA → VA. A steady increase in the number of bonding electrons per atom, as in the original Pauling scheme.*

Groups VIA → VIIA → VIIIA. In the Second and Third Long Periods, the number of bonding electrons per atom increases to 6 in molybdenum and tungsten. At this stage electrons begin to enter the atomic orbitals, but the number of bonding electrons per atom remains high on passing from Mo → Tc → Ru, and from W → Re → Os, in general agreement with the Pauling scheme.

In the First Long Period the data suggest strongly that for chromium the number of bonding electrons is less than 6, and decreases on passing to manganese, and then increases for iron. Neutron-diffraction results have, however, disproved the suggestion that α-iron contains a high proportion of the $(3d)^5$ configuration.

* Hume-Rothery, Irving, and Williams concluded that the number of bonding electrons per atom in vanadium was less than 5, but more recent data have shown that the melting point is higher than they imagined.

Groups VIIIA → VIIIB → VIIIC. In the Second and Third Long Periods there is a rapid decrease in the number of bonding electrons per atom on passing through Group VIII, whilst in the First Long Period the decrease is very much less pronounced.

A detailed study of binding energies has been made by Baughan * who proposes a valency scheme in general agreement with that of Hume-Rothery, Irving, and Williams, although there are minor differences for the elements of Group VIII.

Directed Bonding in Transition Metals.

In the alkali metals at the beginning of each Transition Series there is one outer electron per atom, the atomic volume is large, and the volume per bonding electron is also large. On passing along the series, the continuous decrease in atomic diameter, and the increase in the number of bonding electrons, mean that the volume per bonding electron becomes very small. The atomic diameters of the transition metals of the middle and later Groups are of the same order as the normal co-valent radii given by Pauling for various kinds of (spd) bonds. These facts have led several investigators to conclude that, for the metals in the middle of the Transition Series, the bonding forces very closely resemble those in normal co-valency crystals such as the diamond.

In the crystal of the diamond, each atom has four close neighbours, and as there are four orbitals available (one s, and three p) every atom can form normal co-valent bonds with its four neighbours. In the crystals of the transition metals, this is not possible. If all the s-, p-, and d-orbitals were used, there would be nine orbitals available, and normal co-valent bonding would require a structure in which each atom had nine close neighbours. No metal crystallizes in such a structure, and as explained above, the body-centred cubic structure, in which each atom has 8 close neighbours, reaches its maximum stability in Group VI, and not in Group VIIIA. If the s- and d- orbitals alone were used, six orbitals would be available, but again transition metals of Group VI crystallize in the b. c. cubic structure with 8 close neighbours.

Considerations such as these have led to the conclusion that if directed co-valent bonding exists in transition metals, the bonds will either be fractional bonds, or will not include some of the electrons. For the face-centred cubic structure, the picture suggested is that of Fig. VI.41(*a*), in which an atom is bonded to its closest neighbours by d^3 orbitals of the type shown, whilst the d^2 non-bonding orbitals are in the directions of the cube axes, and do not overlap sufficiently to cause

* E. C. Baughan, *Trans. Faraday Soc.*, 1954, **50**, 322.

bonding. For the body-centred cubic structure the picture is that of Fig. VI.41 (b); in this the d^2 non-bonding orbitals are again in the directions of the cube edges, but do not overlap sufficiently to produce bonding.

This interpretation is open to the criticism that bonding by d-orbitals

d^3 BONDING ORBITALS d^2 NON-BONDING ORBITALS

(a)

d^3 BONDING ORBITALS

d^2 NON-BONDING ORBITALS

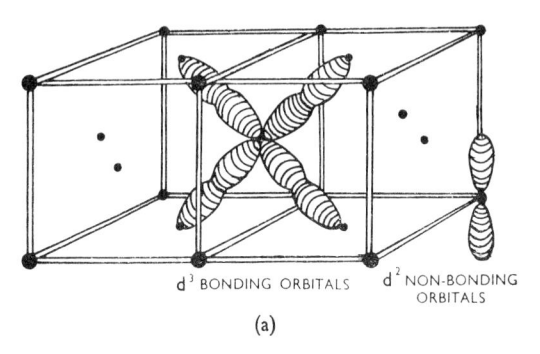

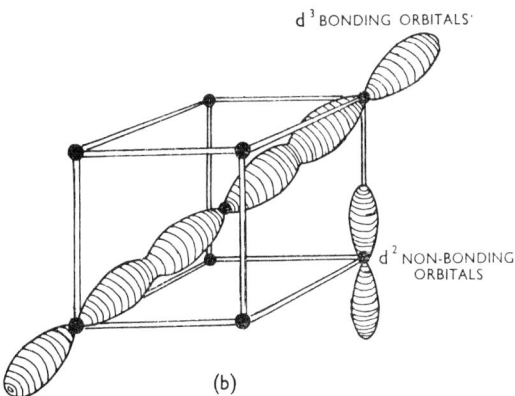

(b)

FIG. VI.41. Models to illustrate bonding in (a) face-centred cubic structure and (b) body-centred cubic structure. Only one set of d^2 non-bonding orbitals is shown.

alone is unlikely, since it is known that hybridized (sd) or (spd) bonding is preferred in chemical compounds. Concepts such as those of Fig. VI.41 may well be correct for non-bonding orbitals, but for bonding orbitals hybridized states are far more probable.

An alternative explanation has been given by Altmann, Coulson, and Hume-Rothery,* according to whom the structure of a metallic

* S. L. Altmann, C. A. Coulson, and W. Hume-Rothery, *Proc. Roy. Soc.*, 1957, [A], **240,** 145. For brevity these authors are referred to as A., C., and H.-R.

368 Atomic Theory

crystal results from resonance between many electronic configurations. Some of these, such as the ionic ones, may have no marked directional properties, although they may be of great importance in determining certain properties of the crystal. Other hybrids may have such strong directional properties that they determine the structure, even though they form only a small proportion of the total wave function, and do not play an important part in determining other properties of the metal. A., C., and H.-R. then investigate which of the hybrid bond orbitals of quantum chemistry have the directional properties required to account for the three typical metallic structures.

In the transition metals, orbitals of the s, p, and d types are available for bonding. The electron clouds of all these (i.e. the values of ψ^2)

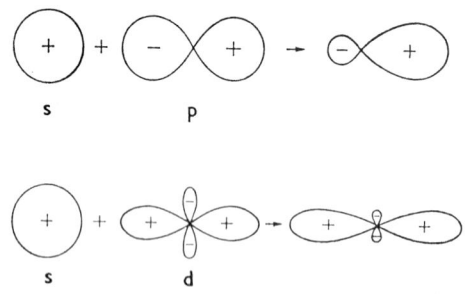

Fig. VI.42. All the orbitals should be imagined as centred around the same nucleus, and the diagram shows how combination of the functions occurs.

[Courtesy Royal Society.]

possess central symmetry, and are *gerade*, in the sense of p. 136. The wave functions themselves (i.e. the values of ψ) are *gerade* (g) for s- and d-orbitals, but are *ungerade* (u) for p-orbitals (see Fig. VI.42).

For the pure orbitals, ψ^2 is always *gerade*, irrespective of the g or u character of ψ. For hybrid orbitals, ψ^2 is *ungerade* for combinations involving u-orbitals. In such combinations, the electron cloud is biased on one side of the atom (see Fig. VI.42), and this favours strong bonding, and strong directional properties.

In hybrids made from combinations of *gerade* orbitals, the electron clouds have *gerade* distribution (see Fig. VI.42), so that the electron has an equal chance of finding itself on either side of the central atom. The bonding is weaker than in the first case, but directional properties still continue.

For the body-centred cubic structure, it is not possible to form 8 hybrids with cubic symmetry from s-, p-, and d-orbitals, but 4 hybrids of tetrahedral symmetry can be formed by (sd^3) hybridization. Since

Electrons, Atoms, Metals, and Alloys

the (sd^3) hybrids are *gerade*, they give rise to 8 equivalent directions of bonding (Fig. VI.43(*a*)). The (d^4) hybrids which are directed towards the corners of the base of a tetragonal pyramid are also *gerade*, and so give 8 bonds as shown in Fig. VI.43(*b*). Further bonding to the 6 second-closest neighbours of the b.c. cubic structure can be obtained by

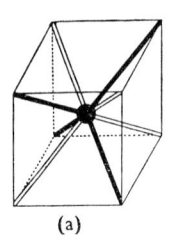

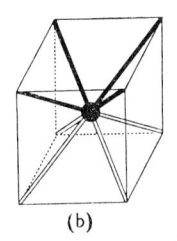

[Courtesy Royal Society.]

Fig. VI.43. Diagrammatic representation of : (*a*) the (sd^3) and (*b*) the (d^4) hybrids. One-half of each hybrid is represented with a solid black line so as to emphasize the original symmetry of the hybrids.

the use of the *gerade* (d^3) hybrids, which form bonds of π, rather than of σ type.

According to A., C., and H.-R., the total bonding in the b.c. cubic structure may be written:

$$(sd^3)^m(d^4)^n(d^3)^q.$$

The proportion of the (sd^3) hybrid is $m/(m+n+q)$, and it does not follow that $(m+n+q) = 1$, because the structure may contain other resonating configurations which may be important in determining some properties of the crystal, although unimportant as regards the directional characteristics.

By considering the directions of the bonds in the different structures, A., C., and H.-R. conclude that the three typical structures of the transition metals are determined by the following hybrids:

body-centred cubic	$\left.\begin{array}{c}sd^3\\d^4\end{array}\right\}$	closest neighbours (pure g)
	d^3	second-closest neighbours (pure g)
close-packed hexagonal	sd^2	neighbours in basal plane (pure g)
	$\left.\begin{array}{c}pd^5\\spd^4\end{array}\right\}$	neighbours in planes above and below (mixed g and u)
face-centred cubic	p^3d^3	4 equivalent sets in resonance (mixed g and u)
	sd^5	(pure g)

When this scheme is examined, it will be seen that, on the average, the d weight will be lower for the c.p. hexagonal structure (where it ranges from 2/3 to 5/6) than for the b.c. cubic structure (where it varies from 3/4 to 1). We thus expect the different structures to appear in the sequence: c.p. hex.—b.c. cubic—c.p. hex., with the b.c. cubic structure appearing at the point where the proportion of d function is greatest. Further, the (p^3d^3) hybrid * has the lowest d weight of all, and we can thus understand why the f.c. cubic structure appears at the beginning and end of each Transition Series.

As explained above (p. 353), the $(n - 1)d$ level lies between the ns and np levels at the beginning of the Transition Series, and so we can understand why the c.p. hexagonal structure persists as one allotropic form at the beginning of the Transition Series (Group IIA and IIIA) but drops out at the end. For the above scheme requires hybridization of p-orbitals with only three of the d-orbitals for the f.c.c. structure in the (p^3d^3) hybrid, but with four or five of the d-orbitals for the c.p. hexagonal structure, and the latter process will be less likely at the end of the Transition Series where the d-levels are becoming lower and lower in energy.

The above scheme also suggests that if one or more of the d-orbitals become atomic and non-bonding, and are completely filled, the c.p. hexagonal structure will be prejudiced at the expense of the other two. For the f.c.c. and b.c.c. structures can be formed by hybrids involving only three of the d-orbitals, whereas the hybrids for the c.p. hexagonal structure require four or five d-orbitals. In this way, if the views of Hume-Rothery, Irving, and Williams are accepted (p. 363), we can understand why the c.p. hexagonal structure appears in technetium, ruthenium, rhodium, and osmium, but not in manganese and iron.

Alloys of Transition Metals.

In the absence of any quantitative theory of transition metals, it is only natural that there is as yet no satisfactory theory of their alloys. Systematic experimental work has, however, revealed some general principles.

The *size-factor principle* referred to on p. 120 applies to alloys of transition metals, and a difference of about 15% between the atomic diameters separates the zones of favourable and unfavourable size-factor. For alloys of iron, this means that all the elements from vanadium to copper in the First Transition Series are of favourable size-factor, whilst titanium is on the borderline of the unfavourable zone.

In the alloys of copper with the succeeding elements Zn, Ga, Ge . . .,

* In the original paper Altmann, Coulson, and Hume-Rothery did not consider the (sd^5) hybrid, and part of their argument was incorrect for this reason, but has been corrected above.

and of silver with Cd, In, Sn . . ., it is a general principle that, for a given atomic percentage of solute, the depression of the liquidus and solidus curves increases with the valency of the solute, i.e. increases with the distance between solvent and solute in the same row of the Periodic Table. Fig. VI.44 (a), which is due to Hellawell and Hume-Rothery,* shows the slopes of the δ liquidus and solidus curves in iron-rich alloys of metals of the same Period. Manganese behaves

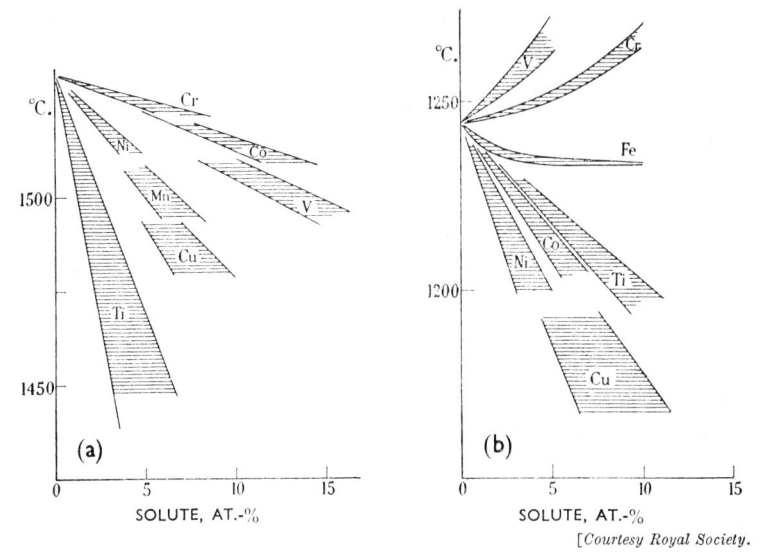

FIG. VI.44. Slopes of δ liquidus and solidus in : (a) iron-rich and (b) manganese-rich alloys.

anomalously, but for the other elements the depressions produced by equal atomic percentages are in the order $Ti > V > Cr$ and $Co < Ni < Cu$. There is, thus, a clear relation between the depression of the liquidus and solidus, and the distance between solvent and solute in the Periodic Table. In this case the borderline size-factor of titanium reinforces what appears to be an electronic effect. Work by Sutton and Hume-Rothery † shows that the lattice distortions produced by equal atomic percentages of the above solutes follow the same sequences.

Fig. VI.44 (b) shows the corresponding liquidus and solidus curves for

* A. Hellawell and W. Hume-Rothery, *Phil. Trans. Roy. Soc.*, 1957, [A], **249**, 417.
† A. L. Sutton and W. Hume-Rothery, *Phil. Mag.*, 1955, [vii], **46**, 1295.

solutions of the same elements in δ-manganese. In this case, all the
elements are of very favourable size-factor, except titanium, for which
the size-factor is considerable (+10%). It is well known that an
increasing size-factor tends to depress the liquidus and solidus curves,
and this effect is apparent for titanium. For the remaining transition
metals, the depressions of the liquidus and solidus curves follow a clear
sequence on passing from V → Cr → Fe → Co → Ni, and it is only on
reaching copper that this tendency is interrupted.

This tendency for the alloys of transition metals to show relations
which depend on the distance between solvent and solute in the Periodic
Table, has received remarkable confirmation in the case of some
intermediate phases in the binary alloys. We have seen above (p. 345)
that on passing from Group VI to Group VIII in the Second and Third
Long Periods, the crystal structures of the elements follow the sequence
b.c. cubic → c.p. hexagonal → f.c. cubic. Work by several investigators * has shown that intermediate phases with close-packed hexagonal
structures are formed in systems such as molybdenum–rhodium, where
one metal is of lower and the other of higher Group Number, than in
those (Groups VIIA and VIIIA) for which the c.p. hexagonal structure
appears in the pure metals. In view of the controversial state of the
theory of transition metals, it is convenient to describe these relations
in terms of the Group Numbers of the elements, which may be regarded
as increasing from IV (Ti, Zr, Hf) to VIIIC (Ni, Pd, Pt). All diagrams
are drawn with the Group Numbers increasing from left to right, so
that, on passing from left to right, the number of electrons outside the
inert-gas shell increases.

Fig. VI.45 is drawn in this way, and shows that, with very few
exceptions, the compositions of the c.p. hexagonal or ε phases for any one
metal of Group V or VI move from right to left as the Group Number of
the second metal increases from Group VII → VIIIA → VIIIB →
VIIIC. This change in composition is in the direction that would be
expected if the c.p. hexagonal structure were favoured by a characteristic average Group Number. This is equivalent to a characteristic
electron : atom ratio, where the number of electrons is that outside the
inert-gas shell. A correlation with the number of vacancies of the
Pauling atomic orbitals is also possible.

Apart from the intermediate c.p. hexagonal phases, alloys of these
transition metals give rise to phases with the characteristic σ structure,†

---

* A summary is given by C. W. Haworth and W. Hume-Rothery, *Phil. Mag.*,
1958, [viii], **3**, 1013.

† For a discussion of the σ structure see : W. Hume-Rothery, R. E. Smallman,
and C. W. Haworth, " The Structure of Metals and Alloys," p. 243. **1969**: London (Institute of Metals).

and also to structures of the α-manganese and β-tungsten types. These are also included in Fig. VI.45, from which it will be seen that, although the compositions of the β-tungsten phases are usually near to the atomic

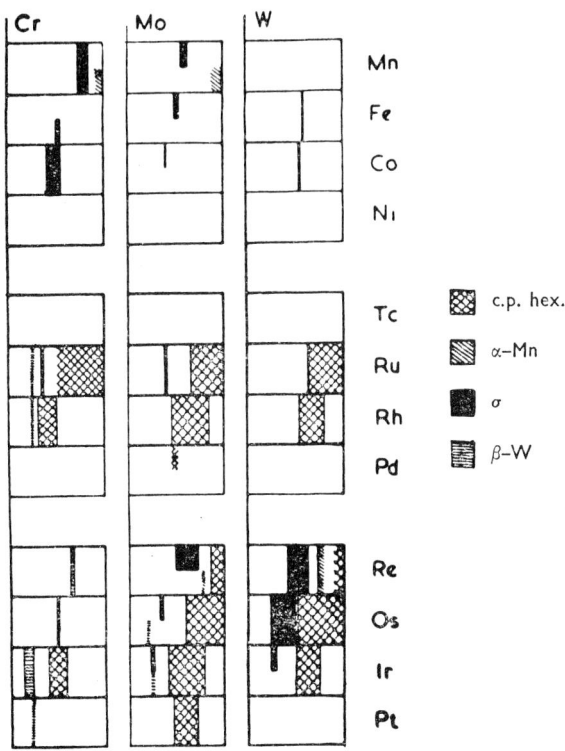

FIG. VI.45. Approximate composition limits of the different phases in the alloys of Cr, Mo, and W with the sequences of elements (Mn, Fe, Co, Ni), (Tc, Ru, Rh, Pd), and (Re, Os, Ir, Pt). In the system Cr–Ni a σ phase at ~30% Ni has been observed (H. J. Schuller and P. Schaub, Z. *Metallkunde*, 1960, **51**, 81) in thin films prepared by evaporation. This agrees with the principle described in the text, but the σ phase has not yet been observed in bulk metal.

ratio A_3B, where A is the element from Group V or VI, the compositions of the other phases show the same general shift as is found for the c.p. hexagonal phase. For the σ phases, this effect has been discussed by many writers but, although there is general agreement that the shifts are in the direction to be expected from a constant electron : atom ratio—an interpretation in terms of Pauling vacancies is also possible—

no satisfactory scheme has yet been advanced which accounts for all the composition ranges in terms of either a single electron : atom ratio, or a single ratio of Pauling vacancies to atoms.

Examination of Fig. VI.45 shows, however, that in every system of the metals concerned, on passing from left to right, the phases occur in the characteristic order :

b.c. cubic → β-W → σ → α-Mn → c.p. hex. → f.c. cubic

In all systems, one or more of the above phases are absent, but the characteristic order is retained. Although, at present, there is no quantitative understanding of these effects, there is a very strong suggestion that electronic factors control the structures of these alloys, in cases where the size-factors are favourable.

The Problem of the Transition Metals.

The transition metals continue to provide one of the central problems in the theory of metals. Much progress has been made in recent years with band-structure calculations, and we are even beginning to acquire some direct experimental evidence about Fermi surfaces, but a major difficulty persists. It is clear that significant modifications of the $(3d)$ atomic wave-functions take place in going from the free atom to the metal crystal, and that significant contributions to cohesion are associated with these modifications. It is also clear that some electrons in the metal will possess characteristics not too dissimilar from those of the s-electrons of copper. On the other hand, over much of the volume of an atomic cell the d-functions will be little changed from those of the atom, while in certain directions in k-space symmetry factors will make possible strong hybridization between d- and s-wave-functions.

In consequence, it is natural for people interested in structure and cohesion to base arguments on the behaviour of transition metals in chemical compounds where the atomic character of the d-functions and the existence of hybridization are widely recognized. It is equally natural for people who recognize the Bloch-function character of all the outer $(d + s)$ electrons, and the existence of a Fermi level in an energy range where d- and s-functions exist, to concentrate on a collective-electron viewpoint (see p. 354), in which a band structure and a density of states are used as a basis for discussions of the low-lying excitations involved in the typically metallic properties of the transition metals. (In fairness to this band-structure approach, it must be pointed out that it has not in recent years been restricted to the simpler, separate s- and d-band models described above, but care has been taken to include properly the hybridization between the two bands.) It is clear that this second viewpoint is the valid one for discussions of Fermi surfaces, transport properties, and most magnetic properties.

It is equally clear that a large part is played in the total energy of the system by excitations of a type more easily associated with those in atoms, and containing important contributions from correlations between electrons on a given atom. This makes it likely that attempts to rationalize the variations in cohesion and structure in the transition metals (on which band theory is almost silent) in terms of quasi-chemical concepts have some validity.

The metallurgist has to concern himself with both aspects and he must draw upon both physical and chemical concepts, whilst treating with reservations the comments of the physicist on the chemist's viewpoint and *vice versa*.

5. Rare-Earth Metals and Actinides.

The table of the electronic structures of atoms in their ground-state configurations in Section II.1 shows that the elements between La and Hf correspond to the gradual filling of the $(4f)$ electron shell. For all these elements the $(5s)$, $(5p)$, and $(6s)$ shells are full, so that the $(4f)$ states are well shielded from external influences and, in accordance with the principal quantum number of 4, their wave-functions have negligible amplitude in regions where the $(6s)$ electron density is still quite significant. As a result the $(4f)$ electrons take no part in chemical bonding in molecules or in metallic cohesion in the metals and their alloys. (It should be noted, however, that while few of the isolated atoms possess $(5d)$ electrons, the usual valence state in ionic compounds is that of the 3^+ ion.) In the pure metals there are normally two $(6s)$ electrons and one $(5d)$ electron involved in the band structures, so that they can be regarded as early transition elements with the addition of a partly filled $(4f)$ electron shell; and this is borne out by the ease with which they form solid solutions, not only with La ($4f^0$, $6s^2$, $5d$) and Lu ($4f^{14}$ $6s^2$ $5d$) but also with scandium and yttrium, which are ($4s^2 3d$) and ($5s^2 4d$), respectively. This is slightly modified in certain of the rare-earth metals by the well-known tendency of elements to make small readjustments to their configurations, if they can thereby achieve the stabilizing influence of a $(4f)$ shell that is empty, exactly half-full, or full. There is thus strong evidence to suggest that Eu and Yb in the metallic state have two conduction electrons and $(4f)$ shell occupations of 7 and 14, respectively. In cerium, the same tendency shows itself in the well-known low-temperature (or high-pressure) collapse of the crystal structure to a state in which the single $(4f)$ electron tends to be promoted to act as a fourth conduction electron. This point is returned to below.

The crystal structures and interatomic distances of the rare earths accord with this description.* Because, however, of the increase in nuclear charge along the series the atomic diameter of the 3-valent configuration does not remain constant but decreases steadily from ~ 3.7 Å at lanthanum to ~ 3.4 Å at lutetium; this is known as the laut havide contraction. The lower-valent states of Eu and Yb are indicated by their atomic diameters of 3·96 and 3·86 Å, while the low-temperature form of Ce has a diameter of 3·42 Å compared with the high-temperature value of 3·64 Å.

Until the work of F. H. Spedding and his collaborators at Ames, Iowa, had made available relatively large amounts of the different rare earths in a fairly pure state, little was known of the properties of these metals, but in recent years a great deal of data has accumulated concerning the rare-earth metals in both polycrystalline and single-crystal form and their alloys both with one another and with other metals.†

Not unexpectedly, wide ranges of solid solution are found in alloys of the rare earths with one another and with Sc and Y. The same is true for alloys with thorium; but solid solubility in or for other metals is normally very small, although the later elements dissolve to some extent in palladium and gold, and slightly in silver.

Intermetallic compounds of rare earths with a large number of other metals have been found to exist, and many of them have interesting magnetic properties, especially those of the formula RM_2 and RM_5, where R is a rare-earth metal and M is Fe, Co, or Ni. There have recently been indications that the RCo_5 alloys may prove to be useful permanent-magnet materials.

The complex magnetic structures of the rare-earth metals are discussed briefly in Section VII.1 (see also the article by Elliot, *loc. cit.*†). Neighbouring (4f) shells do not interact directly, but there exists a strong coupling via the conduction electrons. The paramagnetic moments found at high temperatures, unlike the (3d) elements, contain contributions from both orbital angular momentum (L) and spin angular momentum (S), since the influence of the crystalline field on the deeply buried (4f) shells is too weak to quench the former. In the first half of the series L and S couple in opposition and in the second half in parallel.

---

* For the details see W. Hume-Rothery, R. E. Smallman, and C. W. Haworth, " The Structure of Metals and Alloys," 5th edn., p. 109. **1969**: London (Inst. Metals).

† A useful survey is provided by K. A. Gschneidner, " Rare-Earth Alloys." **1961**: Princeton (Van Nostrand). The magnetic properties are discussed by R. J. Elliot in Vol. IIA of " Magnetism " (edited by G. T. Rado and H. Suhl). **1965**: New York and London (Academic Press).

The Anomalous Behaviour of Cerium.

Reference has already been made to the change found in cerium to a modification of much smaller atomic volume at low temperatures (below $-178°$ C) or high pressures (> 7000 kg/cm^2). This phase, which is face-centred cubic, $a = 4\cdot84$ Å, seems to possess only a very small amount of $(4f)$ character, unlike the high-temperature (expanded) f.c.c. phase (γ-cerium, $a = 5\cdot161$ Å). The situation is further complicated by the fact that a double-hexagonal c.p. β-cerium phase (La structure), with an atomic volume close to that of γ-cerium, seems to be the thermodynamically stable one between approximately room temperature and $-178°$ C; by the fact that considerable hysteresis is found in all the phase transformations, and that they are greatly modified by thermal cycling and by cold work. In dilute alloys of cerium in metals like Y and La it does not seem meaningful to ascribe to it a definite number (0 or 1) of $(4f)$ electrons, and one should probably think of its configuration in terms of a virtual bound state (see Section VI.6).

Actinide Metals.

Developments in atomic energy have made available for metallurgical study a number of the elements, following thorium in the Periodic Table, where the $(5f)$ electron shell is partly occupied, and these show some analogies with the rare-earth series. The actinide metals are rich in complex allotropic modifications, and these seem to be associated with less well-defined f-electron shell configurations. (As in the first and second transition series, where the latter seem to have d-electrons with less well-defined atomic character, so also the $(5f)$ electrons in the metallic state seem to be associated with finite energy-band widths, unlike the strictly atomic character of $(4f)$ states in rare earths.) In spite of a considerable amount of experimental work the electronic nature of plutonium metal is still a subject of discussion.*

The behaviour of actinides in alloys is correspondingly variable; thus, when dissolved in palladium, uranium behaves rather like molybdenum with six outer electrons, but in many compounds it shows the magnetic character associated with localized f-electrons.

6. Models for Dilute Alloys.

In our discussions of the changes, on alloying, of the degree of occupation of the Brillouin zones in alloys based on such metals as copper

* See, *e.g.*, " Plutonium, 1965 " (Proceedings of the Third International Conference on Plutonium) (edited by A. E. Kay and M. B. Waldron). **1967**: London (Inst. Metals).

or magnesium, we have not discussed the details of the local modification in space of the electron density required to screen the modified potential around the impurity atoms (see Fig. VI.46(a)). This question has been the source of much theoretical work in recent years, especially by Friedel and his co-workers.*

In a simple situation such as a dilute solution of zinc in copper the extra positive charge on the ion core of zinc is efficiently screened by a pile-up of conduction electrons in the atomic cell containing the zinc atom, although oscillations in the charge density extend to a greater distance and are important for some properties. The electron states contributing to this screening are drawn from the whole of the occupied range of the conduction band and the density-of-states curve is little changed from that of pure copper (see Fig. VI.46(b)).

On the other hand, we know that when one arsenic atom is introduced into the semiconductor germanium the extra electron occupies a state localized in space around the impurity atom and corresponding to a sharp local level of the energy distribution in the region between the valence band and the conduction band that is "forbidden" for pure germanium. Calculation and experiment show that this localized state is rather extensive in space and the corresponding wave-function has a finite amplitude in a volume containing a large number of the germanium atoms.

Such a sharp level cannot arise for an impurity in a metal (at least, not above the bottom of the conduction band) because it would mix or hybridize with conduction-electron states of similar energy. This mixing is, as we have seen, very strong when the would-be local state has a character similar to that of the conduction-electron states, as is the case for a ($4s$) state on a zinc atom in copper. For local states of rather different character, however, such as a ($3d$) state on manganese dissolved in copper, the symmetry requirements of the state make the conduction-band mixing rather weak, and the ($3d$) shell on the manganese atom preserves to some extent its localized atomic character, although some broadening in energy, perhaps as much as 1 eV, takes place (see Fig. VI.46(c)). The resultant situation has been described by Friedel as a *virtual bound state* and the properties of such states are of great importance for the properties of dilute alloys. When the local-state/conduction-electron mixing is weak, features of the atomic configuration of the impurity (such as the possession of a magnetic moment) tend to persist, although the magnetic coupling of the conduction electrons to the local state tends to be dominated by such mixing as does exist. When the mixing is stronger, the magnetic moment on the impurity can

* A valuable review is given by J. Friedel, *Nuovo Cimento, Suppl.*, **1958, 7,** 287.

disappear, and we can thus understand how iron can apparently carry a magnetic moment when dissolved in gold but not when dissolved in aluminium, where the larger energy range of the conduction electrons makes greater mixing feasible.

When a transition metal is dissolved in a transition metal a wide range of situations is possible. If host and impurity are close neighbours in the Periodic Table mixing of their electron states is easy, the impurity

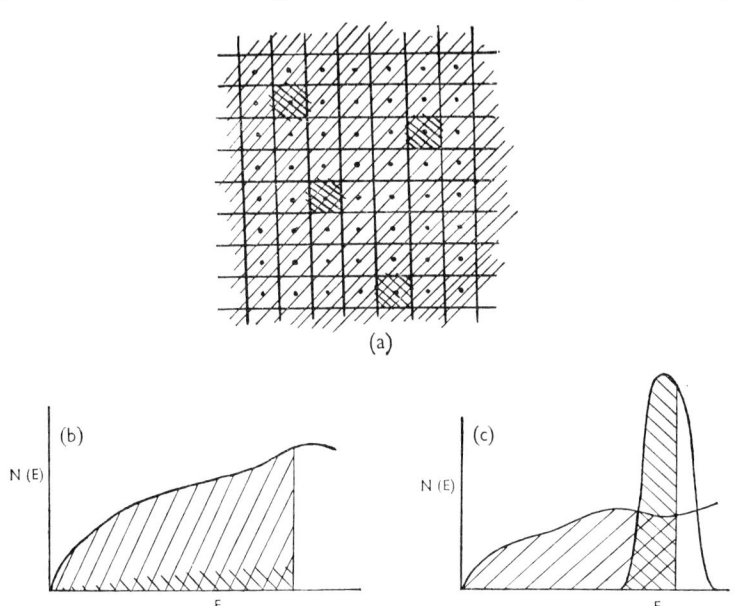

Fig. VI.46. Distribution of screening charge cloud around solute atoms in a dilute alloy:
  (a) in the space lattice;
  (b) in *energy*, for a simple solute in a simple solvent, e.g., Li–Mg, or Cu–Zn;
  (c) in *energy*, for a transition-metal solute in a simple solvent, e.g., Al-Fe.
Reverse hatching represents the extra electrons associated with the solute.

is efficiently screened within its atomic cell, and the distribution of allowed states will remain rather like that of the pure host metal, as in Cu–Zn, the *occupation* of these states taking into account the change in average conduction-electron concentration. In such a situation a description of the band structure of the alloy in the terminology of Section VI.4, p. 354, is appropriate, but it should be remembered that if the alloy is ferromagnetic the collective-band description does *not* imply that host and solute sites carry the same magnetic moment. In fact, in nickel–cobalt alloys, which seem to be well described in this way, neutron-scattering studies show that the increase in magnetic moment (of $\sim 1\mu_B$ per atom) associated with the addition of cobalt to nickel is well localized on the lattice sites occupied by the cobalt atoms.

In conclusion, it can normally be assumed that in an alloy of two simple metals (*i.e.* not containing partly filled d or f shells) the outer electrons will occupy a common band structure and a Fermi surface will exist* in the Brillouin zone with a shape somewhat modified from that of the pure host metal. (The same situation may hold for two transition metals if they are fairly close neighbours in the Periodic Table.) The electron density in space will not, however, be uniform, and local charge accumulations (or depletions) will exist to provide screening (or anti-screening) of the solute atoms. When, however, a transition metal or rare-earth metal is dissolved in a simple metal the electrons in the partly filled d or ($4f$) shell will remain fairly well localized not only in *space* but also in the *energy* distribution. Normally, this localization in energy will be very sharp for ($4f$) electrons, which mix only weakly with conduction-electron states of the host, but much less sharp ($\frac{1}{2}-1$ eV) for d-electrons, which mix fairly strongly. An alternative view of this last situation is to regard the conduction electrons as entering a local d-state as they enter the impurity potential but remaining there for only a fairly short period of time before they depart again as normal conduction electrons.

REFERENCES AND SUGGESTIONS FOR FURTHER READING.

General Theory of Transition Elements.

N. F. Mott and H. Jones, " The Theory of the Properties of Metals and Alloys." **1936** : Oxford (Clarendon Press).
F. Seitz, " The Modern Theory of Solids." **1940** : New York (McGraw-Hill Book Co., Inc.).
L. Pauling, *Phys. Rev.*, 1938, [ii], **54**, 899.
W. Hume-Rothery, *J. Inst. Metals*, 1944, **70**, 229.
W. Hume-Rothery and B. R. Coles, *Advances in Physics*, 1954, **3**, 149.
N. F. Mott, *Advances in Physics*, 1964, **13**, 325.

---

* There is, for an alloy, a slight theoretical difficulty about the definition of a Fermi surface in k-space, since an electron in the presence of a scattering potential will not persist indefinitely in a state of given k. Since there is now a spatial extent associated with an electron which is of the order of the mean free path rather than of the order of the dimensions of the specimen, the uncertainty in k is much larger than that discussed on p. 169 but still much smaller for most dilute alloys than k_{Fermi}. We can therefore think of the Fermi surface in k-space as slightly blurred by scattering, although there is still, at $0°$ K, a well-defined cut-off in *energy* at the Fermi level.

PART VII. SOME PHYSICAL PROPERTIES.

1. Some Magnetic Properties.

General.

In Sections VI.4 and VI.5 we have referred to the ferromagnetism of iron, cobalt, nickel, and some of the rare-earth metals and we may now describe briefly the theories of this subject. The phenomenon of ferromagnetism has been known for a very long time, but it was not until the middle of the 19th century that it was shown conclusively by Faraday that other substances exhibited magnetic properties, and that in general three classes could be distinguished. If a specimen, in the form of a rod, is placed in a non-uniform magnetic field (e.g. between the poles of a magnet), the substance concerned is said to be *diamagnetic* if the rod tends to set itself at right angles to the field, and *paramagnetic* if it tends to set itself parallel to the field. Diamagnetic substances are repelled by a magnet, whilst paramagnetics are attracted. In diamagnetic and paramagnetic substances, the magnetic effects are only shown in the presence of an external field, but the third class of *ferromagnetic* substances may become permanently magnetized, although the magnetism may be destroyed by heating to a sufficiently high temperature, and above this temperature, which is known as the Curie Point (see p. 386), they become paramagnetic.

The magnetic susceptibility κ is a means of representing the magnetizability of a substance, and is defined so that, if the intensity of magnetization is I in a magnetic field of strength H, then

$$I = \kappa H.$$

The intensity of magnetization refers to unit volume, and the mass susceptibility χ is equal to $\frac{\kappa}{\rho}$ where ρ is the density, whilst the gram-atomic susceptibility is equal to the mass susceptibility multiplied by the atomic weight.

The susceptibility of diamagnetic substances is negative, whilst that of paramagnetic substances is positive and decreases with rise of temperature (see p. 386 below). The susceptibility of ferromagnetic substances is, of course, positive, and is very large numerically, until the field becomes so strong that the specimen approaches saturation.

Recent work, particularly by Néel, has led to the recognition of a fourth class of substance known as *antiferromagnetics*. At high temperatures, these are paramagnetic, with susceptibilities which increase as the temperature is lowered, until a critical temperature known as the *Néel point* is reached,* below which the susceptibility decreases with falling temperature. Antiferromagnetics are thus characterized by a maximum in the susceptibility/temperature curve. A further class of substance known as *ferrimagnetics* was recognized by Néel and is referred to on p. 395.†

Magnetic Domains.

The above description refers to the ferromagnetic state, which persists in the absence of an applied field. Ferromagnetic substances are, however, often found in an apparently unmagnetized state, and the cause of this has only recently been discovered. The first suggested explanation was that the substance consisted of a number of crystals, each magnetized in one direction, but with their orientation distributed at random, so that there was no resultant magnetic moment. This explanation was disproved by the discovery that large single crystals of iron or its alloys could also be obtained in the unmagnetized state. It has now been shown that a single crystal consists of a number of domains, each magnetized in one direction, and that in the non-magnetic state the effects in the different directions neutralize one another, so that there is no resultant magnetic moment. The application of an external field causes the different domains to be magnetized in the same direction, with the production of a resultant moment which may be retained as a permanent magnetization when the external field is withdrawn. The existence of these domains is revealed by the formation of definite patterns when the surface of a ferromagnetic crystal is covered with a suspension of a magnetic oxide of iron. Some of the patterns studied in the early history of the subject were undoubtedly influenced by the straining of the surface of the crystal during the polishing operation, but this defect has been overcome, and the existence of magnetic domains in ferromagnetics is conclusively established.

The reason for the stability of a domain structure is that its formation may reduce or avoid the existence of free magnetic poles in a crystal and so lower the energy. If a single crystal of a ferromagnetic substance has

---

* In some of the early literature this is called the *Antiferromagnetic Curie Point*, cf. Kittel, *loc. cit.*, p. 435.

† Useful surveys of the different types of magnetic order are given by:

C. Kittel, "Introduction to Solid State Physics," 2nd edn. **1956**: New York (John Wiley and Sons); London (Chapman and Hall).
H. Morrish, "Physical Principles of Magnetism." **1965**: New York and London (John Wiley).

Some Physical Properties

all its elementary magnets pointing in the same direction, free poles will exist at the surface, but this may be prevented by the formation of a domain structure * such as those of Fig. VII.1, where the magnetic flux is closed through the specimen.

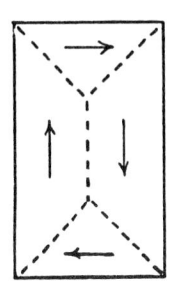

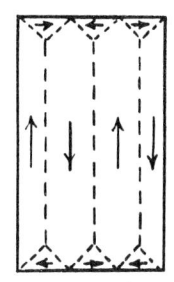

[Courtesy Pergamon Press.]
FIG. VII.1. Idealized domain structures.

Magnetic Susceptibilities.

The gram-atomic susceptibilities of the elements are shown in Fig. VII.2. The exact experimental values are very sensitive to the presence of ferromagnetic impurities (e.g. iron), but the following general trends are now well established:

(1) The inert gases and the preceding elements which crystallize according to the $(8 - N)$ rule are diamagnetic. The diamagnetism is generally weak, the gram-atomic susceptibilities being of the order of 0 to 50×10^{-6}. Antimony and bismuth are more strongly diamagnetic, for reasons which are discussed later (p. 399).

(2) The metals preceding the $(8 - N)$-rule elements, and also the alkali and alkaline earth metals,† are either weakly paramagnetic or weakly diamagnetic, the gram-atomic susceptibilities lying within the range $+ 50 \times 10^{-6}$ to $- 50 \times 10^{-6}$.

(3) The transition elements from Group IIIA onwards are more strongly paramagnetic, and there is a curious alternation of relatively high values in the Groups of odd number, and low values in those of even number. This may be seen by comparing the points in Fig. VII.2 for the sequences Sc → Ti → V → Cr → Mn, Y → Zr → Nb → Mo, and Hf → Ta → W → Re. In the transition elements of Group VIII the para-

---

* A useful description is given by U. M. Martius, " Progress in Metal Physics ", Vol. III., p. 140. **1952**: London (Pergamon Press). See also C. Kittel, *loc. cit.*, p. 414; L. F. Bates, *J. Inst. Metals*, 1953–54, **82**, 417.

† The paramagnetism of barium is abnormally high.

384 Atomic Theory

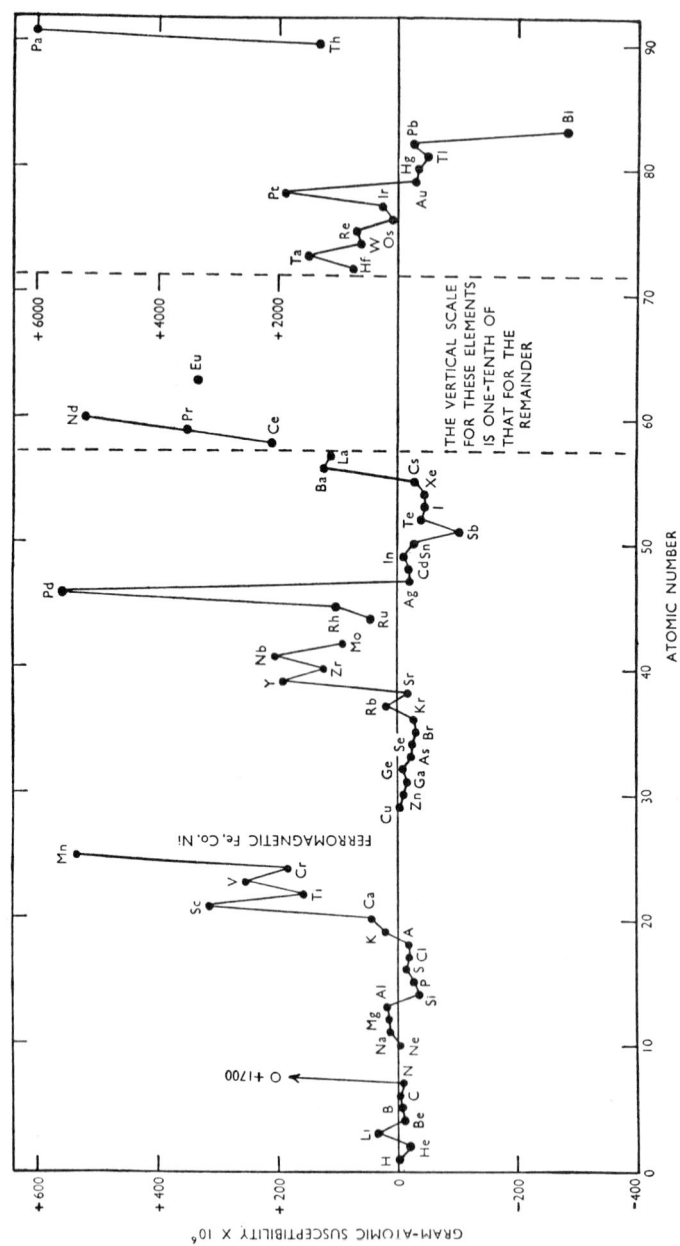

FIG. VII.2. Magnetic susceptibilities of the elements.

magnetism becomes weaker on passing to the later Periods. In the First Long Period all the elements of Group VIII are ferromagnetic, although the ferromagnetism of nickel is destroyed at a relatively low temperature (353° C). In the Second Long Period, palladium is strongly paramagnetic and there is a marked increase on passing from Ru → Rh → Pd, for which the values are systematically higher than for the corresponding elements Os → Ir → Pt in the next Period.

(4) Many of the Rare Earth Metals are very strongly paramagnetic, the gram-atomic susceptibilities being of the order of ten times those of the transition elements. In Fig VII.2 the vertical scale has been changed by a factor of 10 for the Rare Earths.

(5) Ferromagnetism is confined to iron, cobalt, and nickel in the First Long Period, and to a few of the Rare Earth Metals (Gd, Dy, Er, and Ho) which are ferromagnetic at low temperatures. Since ferromagnetic substances become paramagnetic at a sufficiently high temperature, the facts suggest strongly that ferromagnetism is either an extreme form of, or closely connected with, paramagnetism, and that diamagnetism is distinct.

General Theoretical Ideas.

Most theories of ferromagnetism involve the assumption that ferromagnetic substances contain a number of elementary magnets, and that the magnetic state corresponds with these elementary magnets all taking up the same orientation. The general characteristics of the phenomena can be explained if the ordered or orientated state is the result of a " co-operative process " in which the energy required to change the orientation of one elementary magnet increases with the perfection of the orientation of all the other elementary magnets. If, for example, all the elementary magnets are pointing in the same direction, then a change in the orientation of any one elementary magnet will be opposed by the order of the whole assembly,* whereas the less perfect the order of the assembly, the smaller will be the average restraining force exerted on any one elementary magnet. In this way it is possible to show that on raising the temperature, the magnetism will be destroyed, at first slowly, and then with a sudden rush, over a comparatively narrow range of temperature, at the *Curie Point*.† Above the Curie point, the substance becomes paramagnetic,

* The effect of any one elementary magnet is, of course, mainly due to its immediate neighbours, and to a lesser extent to those farther away.
† The general nature of the process is very like that of the destruction of the long-range order of a superlattice in a solid solution.

and its susceptibility varies with the temperature, according to a relation of the type: *

$$\chi = \frac{C}{\theta - \bar{\theta}_0},$$

where θ is the temperature concerned, and $\bar{\theta}_0$ the temperature of the Curie point. Many ordinary paramagnetic substances (i.e. those which are paramagnetic and do not exhibit ferromagnetism at a low temperature) have susceptibilities which vary with the temperature according to a relation of the type:

$$\chi = \frac{C}{\theta + A}$$

and if we write $-A = \bar{\theta}_0$, then $\bar{\theta}_0$ may be called the negative Curie temperature of the material. This temperature-dependence of paramagnetic susceptibility means that little is to be gained by the detailed comparison of susceptibility values referring to some arbitrary (e.g. room) temperature.

Many considerations led to the conclusion that the elementary magnets must either be electrons revolving in orbits, or the spinning electrons which had been found to explain the finer details of atomic structure. In either case, the ferromagnetic body would contain something which was revolving, and the known properties of gyroscopes led to the prediction of the various gyromagnetic phenomena, in which, according to the nature of the experiment,† rotation of a ferromagnetic substance produces magnetization, or alternatively the sudden magnetization of the substance tends to produce rotation. If we use the symbol ρ to denote the *gyromagnetic ratio*, i.e. the ratio of the angular momentum and the magnetic moment of the elementary magnetic gyrostat, it can be shown that for a spinning electron $\rho = m/e$, whilst for an electron in a circular orbit $\rho = 2m/e$. In the general case, $\rho = 1/g \cdot 2m/e$, where g is the so-called *Landé splitting factor*, and equals 2 for a spinning electron and 1 for orbital motion. Experiment then showed that for ferromagnetic metals for which reliable results could be obtained, g was nearly equal to 2, from which it was concluded that the elementary magnets were spinning electrons, and it was assumed that the same applied to the strongly paramagnetic elements.

* It should be noted that the $\bar{\theta}_0$ in this equation, the paramagnetic Curie point, is not really exactly the same as the ferromagnetic Curie point, at which the ferromagnetism disappears, but for the present purpose this difference may be ignored.

† An interesting account is given in "Modern Magnetism," by L. F. Bates. **1951**: Cambridge (University Press).

Some Physical Properties 387

On the other hand, in non-metallic ferromagnetic substances the effect of the orbital moment may be much greater.

As explained on p. 356 the saturation moments of ferromagnetics are usually expressed in Bohr magnetons (μ_B) per atom. Since the maximum component in a given direction of the total angular momentum of an atom is $J\hbar$ the saturation magnetic moment per atom is $gJ\mu_B$, or $2S\mu_B$, if S is the spin angular momentum quantum number. Since $S = \frac{1}{2} \times$ the number of parallel spins, the saturation moment in Bohr magnetons per atom is the number of unpaired parallel spins per atom for a spin-only ferromagnet.

In a semi-classical treatment of the susceptibility of a ferromagnet above its Curie temperature, the constant C in the expression given above is equal to $N\mu^2/3k$, where k is Boltzmann's constant, N the number of atoms, and μ the magnetic moment per atom. A quantum-mechanical treatment shows that this is not in fact equal to $gJ\mu_B$, the saturation moment, but to $g\sqrt{J(J+1)}\mu_B$, or $2\sqrt{S(S+1)}\mu_B$, for spin-only atoms. The value of μ derived from the constant C is therefore referred to as the *effective moment* μ_{eff}, sometimes denoted $p_{eff}\mu_B$, where p_{eff} is the *effective magneton number*. It should be noted that five unpaired spins with no orbital contribution will give $S = 5/2$ and $\mu_{eff} = p_{eff}\mu_B = 2\sqrt{S(S+1)}\mu_B = 5\cdot9\mu_B$. With 1, 2, 3, and 4 parallel spins p_{eff} is 1·73, 2·8, 3·9, and 4·9 respectively. These *effective magneton numbers* are at first rather confusing, because they mean that except for $n = 1$, n parallel spins give an effective magneton number of about $(n + 1)$. In cases where orbital angular momentum is involved, as well as spin momentum, the effective magneton number may be very much greater than the number of unpaired electron spins.

In the general theory of Parts IV and V we have seen that the electrons at the absolute zero occupy the $N/2$ lowest energy states, each state containing two electrons of opposite spins. In metals, where the elementary magnets are spinning electrons, it is clear that any preponderance of electrons of a given spin must increase the Fermi energy. Ferromagnetism can, therefore, exist only if some other factor more than counteracts this effect, and the theories of ferromagnetism have taken two main lines. On the one hand, various simplified models, such as the two-band model of Mott and Jones (p. 354), have been used to systematize the magnetic data. This work has attained considerable success, at the expense of not paying much attention to the justification of the simplifying assumptions. On the other hand, many attempts have been made to explain the reasons why the elementary magnets set themselves parallel in some substances, and not in others. This work has been comparatively unsuccessful, and there is at present no real understanding of the ultimate cause of ferromagnetism. We may now describe briefly some of the main approaches to this problem.

The Heisenberg Theory of the Exchange Integral.

In 1928 it was suggested by Heisenberg that the process responsible for the stabilization of the ferromagnetic state was the exchange or interchange energy, to which reference was made in connection with the structure of the hydrogen molecule (Part III, p. 144). In the H_2 molecule the exchange integral is negative, and the most stable state is that in which the electrons have opposite spins, and produce a normal co-valent linkage. Heisenberg suggested that, in some cases, the exchange integral might be positive, and that the ferromagnetic state might then be the stable condition. This is an arbitrary assumption for which no proof has yet been given, but if it is accepted the following factors are important in determining whether ferromagnetism is shown :

(1) The atoms must have an incomplete shell of electrons.

(2) The incomplete shell must have a relatively low electron cloud density near the nucleus. This means that the second quantum number l must be large, and in practice only d and f sub-groups have to be considered.

(3) The atoms must not be too close together or the electron clouds overlap so greatly that the exchange integral is negative, and normal cohesion, with electron pairs of opposite spins, is encountered.

(4) The atoms must not be too far apart, or the exchange integral becomes very small, and the resultant energy term is too small to stabilize the ferromagnetic state.

If we denote by D the ratio of the distance between two atoms in a crystal, and the " diameter of the orbits " of the overlapping electron shells,* the relation between the exchange integral, A, and the value of D is regarded as of the general form shown in Fig. VII.3. Iron, nickel, and cobalt are thought to possess values of D corresponding to the part of the curve where A is relatively large and positive. The dependence of ferromagnetism on a critical value of D is thought to be shown by the fact that, although manganese is not ferromagnetic (because the atoms are too close together), some compounds of manganese are ferromagnetic. Thus one of the manganese nitrides is ferromagnetic, and it has been

---

* The expression " diameter of the orbit " is used to denote twice the distance from the nucleus to the point at which the outer part of the electron shell concerned has its maximum electron cloud density.

suggested that this is the result of the manganese atoms being pushed apart by the atoms of nitrogen which exist in interstitial solid solution. The same explanation may account for the ferromagnetism of the Heusler alloys, which consist of copper, manganese, and aluminium in proportions approximating to the composition Cu_2AlMn, whilst ferromagnetism is also shown by some alloys of chromium with tellurium or platinum. The general conclusion to be drawn from these considerations is that new ferromagnetic alloys would probably be found if transition elements of Groups VI–VIII were alloyed in such a way as to obtain crystal structures with suitable interatomic distances.

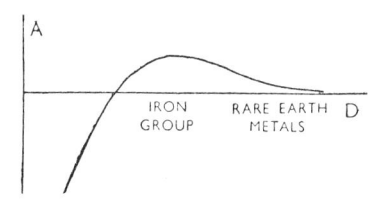

FIG. VII.3. In this diagram D is the ratio of the distance between two atoms to the " diameter " of the incompletely filled shell of electrons. If D is small, the exchange integral A is negative, and normal cohesion with electron pairs of opposite spin is obtained. If D is very large, the exchange integral becomes too small to have any appreciable effect. With a critical value of D, the exchange integral is thought to be positive and large enough to produce ferromagnetism.

In rare-earth metals no overlap between the $(4f)$ wave-functions of neighbouring atoms occurs and A is consequently zero. Ferromagnetism in these metals is ascribed to an indirect coupling of the $(4f)$ moments via their exchange interaction with the conduction electrons. In these metals the orbital contribution to the magnetic moment is not quenched as it is by the crystalline field in the $(3d)$ ferromagnets and the saturation moment is $gJ\mu_B$ per atom.

It should be noted that if the interatomic distances in the crystal of an element correspond with a point to the left of the maximum on the curve, then compression of the metal decreases the value of A, and so decreases the magnetization. Conversely, destruction of the magnetization tends to produce a contraction, and it is thought that the low coefficient of expansion of alloys such as Invar is due to the alloy corresponding to a point to the left of the maximum in Fig. VII.3, so that the normal thermal expansion is counteracted by a contraction due to the destruction of the magnetization.

Since the theories we have described regard the elementary magnets as consisting of spinning electrons in incompletely filled shells, it was

natural at first to associate the saturation moment at low temperatures *
with the number of holes in the d-shells in the atoms. For this purpose
the saturation moment must be expressed in Bohr magnetons per
atom, and the fact that the observed numbers are not whole numbers
implies that on the average some of the atoms are in one state, and
some in another. This point of view implies, of course, that we
associate the electrons with individual atoms, but this does not imply
that a given atom is permanently connected with a given number
of electrons. We should say rather that there is a certain probability
of an atom having an incomplete d-shell with a magnetic moment, and
this " hole " in the d-shell moves through the lattice, and in the " spin-
wave " theory of Bloch the motion of the hole in the d-shell is associated
with a ϕ wave ($=$ spin-wave) in the same way that the motion of a free
electron is associated with a Ψ wave. Just as $\Psi^2 d\tau$ is the probability of
finding an electron in the volume element $d\tau$ so $\phi^2 d\tau$ is the probability
of finding a hole in the d-shell in the volume element $d\tau$; in the latter case $d\tau$ should be larger than an atom. The spin-wave can
then be associated with a momentum, a kinetic energy, and a velo
city, but the details of this theory are too mathematical for the present
book.

For nickel the assumption that the number of holes in the d-shell
is equal to the saturation moment at low temperatures is almost
correct, and detailed calculations have been made by Stoner † and
by Wohlfarth.‡ From Fig. IV.11 (p. 191) it is clear that if a stable state
is to be obtained, the increase in Fermi energy resulting from an excess of electrons of one spin must be more than counterbalanced by
the energy of the positive exchange integral, the details of the process
depending on the form of the $N(E)$ curve. Stoner has examined this
problem in detail for a $N(E)$ curve whose head has a standard parabolic
form. This work shows that for ferromagnetism to occur, the exchange
integral must not merely be positive but must exceed a certain critical
value. Further, the assumption that the number of holes in the d-shell
is equal to the saturation moment at the absolute zero is valid only if
the exchange integral exceeds a second and larger critical value. With

* At low temperatures in a strong field, all the elementary magnets point in
the same direction. For the relation between this saturation intensity and the
results of observations at higher temperatures, reference may be made to " Magnetism and Matter," by E. C. Stoner. **1934** : London (Methuen & Co., Ltd.), and
" Modern Magnetism," by L. F. Bates. **1951** : Cambridge (University Press).

† E. C. Stoner, *Proc. Roy. Soc.*, 1938, [A], **165**, 372 ; this paper deals with the
general problem of magnetization and saturation moments.
  Proc. Roy. Soc., 1939, [A], **169**, 339 ; deals with the thermal effects associated
with ferromagnetism.

‡ E. P. Wohlfarth, *Proc. Roy. Soc.*, 1949, [A], **195**, 434 : deals with nickel and
nickel–copper alloys.

Some Physical Properties

intermediate values of the exchange integral it is possible for ferromagnetism to occur with a saturation moment less than the number of holes in the d-shell. For nickel, it is concluded that the exchange interaction is almost, but not quite, sufficient to produce complete saturation at the absolute zero, and so the number of holes in the d-band is slightly greater than 0·6 per atom.

For iron, the exchange interaction is not sufficient to produce complete saturation at the absolute zero and the position resembles that of Fig. VI.38 (p. 357). From a study of the magnetic properties of iron alloys, Coles * concludes that in pure iron there are about 0·35 paired holes per atom in the d-shell. The number of electrons with positive spin is thus $5 - 0·35 = 4·65$, and as the observed saturation moment is 2·2, the number of electrons with negative spin is $4·65 - 2·2 = 2·45$ per atom. The total number of d-electrons is thus 7·1 per atom, and hence the number of ($4s$) electrons per atom is about 0·9. The case of cobalt has not yet been considered in detail, but the data indicate that it resembles nickel rather than iron.

Most of the above work has been in terms of the single d-band theories of Mott and Jones. In the early Pauling hypothesis (p. 360), where the saturation moment is regarded as associated with the atomic d-orbitals, the exchange process concerns these alone, and the remaining d-electrons are in hybrid (spd) bonding orbitals.

In the above description we have assumed throughout that ferromagnetism is the result of a positive exchange integral. This assumption has some theoretical basis, but is without any real proof, either experimental or theoretical. Wohlfarth † has made calculations using a simplified model in which only the radial factor of the ($3d$) functions is used, and has shown that this leads to a negative value of the exchange integral regardless of the interatomic distances. This may not, however, necessarily be the case ‡ when the angular factors of the d functions are taken into account. There is still uncertainty as to the sign of the exchange integral, and useful references to some of the discussions will be found in Vol. 25 of the *Reviews of Modern Physics*.

The uncertainty regarding the exchange integral has led some authors to doubt the usefulness of Fig. VII.3. It has also been suggested that, in the absence of any quantitative calculation, the concept of a positive exchange integral is little more than a figure of speech, and that it might be replaced by the term " energy arising from Coulomb interaction ".

---

* B. R. Coles, Thesis (Oxford University), **1951**.
† E. P. Wohlfarth, *Nature*, 1949, **163**, 57.
‡ E. P. Wohlfarth, private communication.

This difficulty in connection with the exchange integral has led to alternative suggestions regarding the cause of ferromagnetism. In particular, Zener * has advanced a theory of the transition metals assuming that the exchange integral is always negative and will tend to an antiferromagnetic alignment of the electron spins. This assumption is made dogmatically and is no less arbitrary than the accepted view that the exchange integral may become positive at a critical value of R/r. If Zener's assumption is accepted, the increase in Fermi energy resulting from the excess of positive spins in the ferromagnetic state has to be counterbalanced by some process that lowers the energy, and this process is assumed to be a spin coupling between the conduction electrons and the d-electrons, the conduction electrons serving to carry the spin from one atom to the next. Unfortunately, this work has included many developments which are generally regarded as unacceptable; in particular, Zener assumes that the electronic configuration in the solid metal is essentially the same as in the free atom, and this is not only most improbable, but is almost certainly in contradiction to the neutron-diffraction data referred to below. The underlying ideas that the exchange integral is always negative and that coupling between conduction and core electrons is responsible for ferromagnetism are, however, favoured by many physicists. Thus, Pauling † has advanced a theory in which the saturation moment of iron is regarded as arising from a structure in which each atom has 2 electrons in atomic ($3d$) orbitals, giving rise to a saturation moment of 2 μ_B. The remaining 6 electrons ‡ are conductivity electrons in hybrid (spd) orbitals, and the interaction between these and the atomic ($3d$) orbitals is regarded as responsible for an additional moment of 0·2 μ_B, thus making a total saturation moment of 2·2 μ_B. The quantitative development of this idea depended on the assumption that as many as 6 electrons/atom could be treated as particles of a free-electron gas, and was open to serious criticism, but the underlying idea that ferromagnetism results from interaction between conductivity and core electrons has been adopted by several later authors.

Neutron Diffraction and Antiferromagnetism.

Antiferromagnetic substances (p. 382) are characterized by their Néel temperatures, at which the susceptibilities attain maximum values. In 1932 Néel advanced the idea that, in such substances, the exchange integral is negative, and that at low temperatures the atomic moments

* C. Zener, *Phys. Rev.*, 1951, [ii], **81**, 440; **82**, 403; **83**, 299.
† L. Pauling, *Proc. Nat. Acad. Sci.* (*U.S.A.*), 1953, **39**, 551.
‡ The electron distribution assumed here is not quite the same as that in the earlier Pauling hypothesis described on p. 362.

of + and − sign are arranged in definite alternating patterns, analogous to the superlattices of binary alloys. In the body-centred cubic structure, for example, the structure may be composed of two interpenetrating simple cubic lattices with opposed spins, as shown in Fig. VII.4. Substances of this type are sometimes called *antiferromagnetics of the first order*, and many examples are known of non-metallic substances in which the two lattices of opposed spin are occupied by atoms of different kinds. In the structure of Fig. VII.4 each atom of one spin is surrounded by 8 close-neighbours of the opposite spin, but in non-metallic substances the body-centred cubic structure may give rise to an *antiferromagnetic of the second order* in which the spins are arranged as in Fig. VII.5, so that

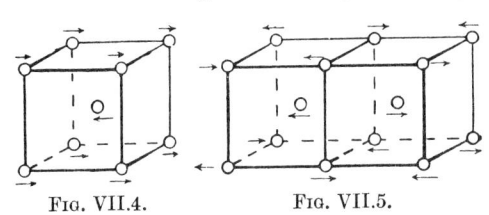

FIG. VII.4.   FIG. VII.5.

each atom is surrounded by four of one spin and four of another, and second-closest neighbours are anti-parallel. The antiferromagnetics are thus superlattices in which the atoms of one spin occupy definite positions relatively to those of the other spin. At the Néel point the ordered arrangement is destroyed, and above this temperature they behave as paramagnetic substances whose susceptibility diminishes with rise of temperature. Below the Néel point the susceptibility diminishes with falling temperature because the balancing of the spins becomes more perfect as the order of the structure increases. An antiferromagnetic is thus characterized by a maximum in the susceptibility/temperature curve, and an interesting series of papers on these phenomena has been published.*

For many years, the idea of antiferromagnetic substances as structures based on interpenetrating lattices of atoms with opposed spins was only a theoretical concept, but the developments of the methods of neutron diffraction have led to direct experimental proof of the original predictions. Neutrons, like all moving particles, are associated with waves of wave-length $\lambda = h/p$ (see Part I, p. 31), and so give rise to diffraction effects analogous to those produced by electrons. Apart from this, when neutrons pass through a crystal, there is additional scattering owing to interaction between the neutron spin moment and

* " Ferromagnétisme et Antiferromagnétisme ", *Colloques Internat. C.N.R.S.*, 1951, **27**, 360.

the moment of paramagnetic ions. It follows, therefore, that an antiferromagnetic structure (which is essentially a magnetic superlattice) will give rise to extra diffraction lines in the same way that an ordered alloy gives additional X-ray superlattice lines in X-ray crystallography. If, for example, a metal had the antiferromagnetic arrangement of paramagnetic ions of Fig. VII.4, the neutron-diffraction spectrum would show a (100) line, just as in the case for the CsCl structure of Fig. III.22 (p. 129), where the ordered arrangement of two kinds of atom produces the (100) line which is not found for a simple body-centred cubic structure.

A further characteristic of neutron diffraction * is that the study of diffuse scattering of neutrons enables the magnetic moments of ions to be estimated. Neutron diffraction would, for example, distinguish between a hypothetical form of tungsten in which ions of moment 5 μ_B were bound together by one conduction electron per atom, and the actual structure in which all the 6 outer electrons per atom take part in the cohesion, and the ions have a zero moment. These characteristics result because a neutron possesses a magnetic moment and it is the interaction between this and the unpaired electron spin of the paramagnetic ion which produces the additional " magnetic " scattering. Neither X-rays nor electrons behave in this way. An electron certainly has a magnetic moment but, unlike the neutron, it is a charged particle and the Coulomb repulsion between its charge and the electrons of the ion prevents the two from coming into close enough contact for the spin–spin interaction to be effective. The neutron succeeds in achieving such an interaction because it is uncharged.

As explained above (p. 392), Zener suggested that in transition metals, the exchange integral was always negative. From this he concluded that, unless the coupling between conduction and core electrons was considerable, these metals would have antiferromagnetic structures. In particular, he suggested that the great stability of the body-centred cubic structures in Group VA and VIA was the result of their having the antiferromagnetic structure of Fig. VII.4. This concept has been disproved by the work of Shull and his collaborators,† which is of great interest. For tungsten, molybdenum, niobium, and vanadium, extensive work has shown no sign of antiferromagnetic order, and the diffuse-scattering data disprove the existence of large moments—only moments of the order of a few tenths μ_B can exist in either ordered or disordered arrangements.

* For a general account of neutron diffraction, the reader may consult " Neutron Diffraction " by G. E. Bacon. **1955**: London (Oxford University Press).
† For a general review, see C. G. Shull, Report on 10th Solvay Congress, **1955**, p. 227.

In metallic chromium weak antiferromagnetism has been observed with a Néel temperature of 310° K and a moment of about $0.4\mu_B$ per atom. It seemed at first that the magnetic structure was that of Fig. VII.4 but careful neutron-diffraction measurements have shown that the periodicity of the spin structure is not exactly commensurate with that of the lattice structure. This is possible because the magnetically active electrons are Bloch electrons, and rather than strictly localized magnetic moments one has to envisage spin-density waves whose periodicity is slightly different from the lattice period.

Neutron-diffraction results have shown * that the abnormal crystal structure of α-manganese is antiferromagnetic at 4·2° K, and contains three kinds of ion. A unique structure has not yet been obtained, but the atomic moments in μ_B are either (1·54, 3·08, and 0) or (2·5, 1·7, and 0). The results suggest that, for small (<2·37 Å) separation of the manganese atoms, there is no spin coupling, whilst the coupling is antiparallel for the range 2·37–2·82 Å, and is parallel for distances greater than 2·96 Å.

The complex structure of β-manganese is not antiferromagnetic, but from data on quenched manganese–copper alloys it is concluded that, in the quenched face-centred tetragonal state, γ-manganese would be antiferromagnetic with a moment of 2·4 μ_B. It is, however, not a normal antiferromagnetic, because its susceptibility increases with rise of temperature above the Néel point.

Many of the rare-earth metals become antiferromagnetic at low temperatures (Section VII.1) and the study of their magnetic structures has revealed a very complicated situation. The orientations of the atomic moments have been found to follow spiral or screw structures of varying pitch rather than lying parallel or antiparallel to a given direction.

Antiferromagnetism has also been found in a number of alloys and intermetallic compounds; some of these also have very complex structures. A useful survey of the methods and results will be found in an article† by Nathans and Pickart.

Ferrimagnetism.

The antiferromagnetic structures referred to above consist of interpenetrating magnetic sub-lattices, whose opposed spins cancel each other at very low temperatures where the magnetic order is perfect.

---

* C. G. Shull, *loc. cit.*
† R. Nathans and S. J. Pickart, " Spin Arrangements in Metals," in Vol. III of " Magnetism " (edited by G. T. Rado and H. Suhl). **1963**: New York and London (Academic Press).

In 1948, Néel proposed a further type of co-operative magnetic phenomenon, to which the name *ferrimagnetism* * is given. At high temperatures, ferrimagnetic substances are paramagnetic, but the curves of $1/\chi$ against T are hyperbolas instead of the straight lines characteristic of normal paramagnetics. At low temperatures, magnetic sub-lattices develop, but the different sub-lattices are not identical, and a resultant permanent magnetic moment is acquired. In some ferrimagnetics, two magnetic sub lattices may be geometrically identical, but may be occupied by atoms of different spins. In other cases the different substructures may contain different kinds of crystallographic site.

One example of ferrimagnetism is provided by the *Ferrites*, which are compounds of the general formula $M\text{Fe}_2\text{O}_4$, where M is a divalent metallic atom. These include the magnetic oxide of iron, magnetite, Fe_3O_4, whose structure is really $\text{Fe}.\text{Fe}_2\text{O}_4$.

Neutron Diffraction and Ferromagnetism.

Neutron-diffraction methods have shown that, in iron, there is no evidence for the existence of a magnetic superlattice, so that the non-integral value of the saturation moment cannot be ascribed to interpenetrating sub-lattices of atoms of different spins. All atoms appear to have the same moment, and at high temperatures the paramagnetic scattering is consistent with a moment equal to that present in the ferromagnetic state. The data suggest that, as the temperature is raised and the Curie point is approached, the long-range order diminishes, as in the case of superlattices, whilst the magnetically coherent regions become smaller, for which there is no parallel in the case of superlattices.

Some of the methods of neutron diffraction enable an estimation to be made of the spin associated with an atom. They will, for example, distinguish between a structure in which metallic vanadium has 3 localized unpaired spins and 2 bonding electrons per atom, and one in which all 5 electrons take part in the bonding and there are no unpaired spins on the atoms; the latter is confirmed.

Paramagnetism and Diamagnetism.

In Section IV.3 we have seen that the simple free-electron theory of metals indicates a small paramagnetism which is almost independent

* For general accounts of ferrimagnetism, see C. Kittel, " Introduction to Solid State Physics ", 2nd edn. **1956**: New York (John Wiley and Son); London (Chapman and Hall), and J. S. Smart, *Amer. J. Physics*, 1955, **23**, 356. Also " Ferromagnétisme et Antiferromagnétisme ", *Colloques Internat. C.N.R.S.*, 1951, **27**, 360.

Some Physical Properties

of the temperature, and a diamagnetism which in the simplest approximation is one-third of the paramagnetism, so that the resultant effect of the free-electron gas is slightly paramagnetic. In the more complete theories the position is very much more complicated, and a large number of factors have to be considered.

In the first place when the results of the calculations are compared with the observed values,* an allowance must always be made for the effect of the ions. Ions with complete outer groups of 8, or 18, electrons are diamagnetic, and their susceptibilities are of the order of 10^{-6}, and hence are roughly of the same magnitude as the temperature-independent paramagnetism predicted by the free-electron theory. The fact that one metal is weakly diamagnetic and another weakly paramagnetic does not therefore necessarily indicate any essential difference between the electronic structures of the two metals, but may simply be the result of the diamagnetic contributions due to the ions. Comparison between the observed results and those calculated by the free-electron theory, and corrected for the effects of the ions, is really justified only for the alkali metals, and the general conclusion is that, except perhaps for lithium, the observed paramagnetism is somewhat greater than that calculated.†

Equation IV (35) (p. 192) is an approximation only when χ is small. In a general way the effect of temperature on $1/\chi$ (the reciprocal of the susceptibility) is similar to that on the energy of the electron gas. At low temperatures $1/\chi$, like the energy (Fig. IV.6, p. 178), is constant, and it then increases asymptotically to a linear variation with the temperature in agreement with the Curie law. In the more general case, in the absence of exchange interaction effects, the low-temperature paramagnetic susceptibility of the collective-band electrons may be written $\chi_e = 2\mu^2 N(\varepsilon^*)$, where μ is the Bohr magneton, and $N(\varepsilon^*)$ the value of $N(E)$ at the Fermi limit. In this way susceptibility measurements at low temperatures can be used to estimate the values of $N(E)$, and the resulting values can be compared with those deduced from the specific heats (p. 259), subject to the assumption of no exchange interaction. For the transition metals, the specific heats and susceptibility data are not in agreement, and the paramagnetic susceptibilities are greater than would be expected from the specific heats. This has been taken to imply that, in general, positive exchange interaction ‡ exists in most

---

* Great care is, of course, necessary to ensure that the observed values are not invalidated by the presence of ferromagnetic impurities.

† Thus rubidium and caesium are slightly paramagnetic, although the calculations indicate slight diamagnetism.

‡ In view of the doubts about the sign of the exchange integral, it is perhaps, more correct to say positive magnetic interaction.

transition metals, even though it is not sufficient to produce ferromagnetism.

When the previous theory (Section IV.3) of weak spin paramagnetism is examined more carefully, it will be seen that it involves the assumption that there are unoccupied energy states which the electrons can enter when the external field is applied. It follows, therefore, that this type of paramagnetism is not to be expected for insulators which are full zone structures with an energy gap between the occupied and the next higher zone.

In the case of a nearly empty zone for which $N(E) = C\sqrt{E - E_0}$, where E_0 is the energy at the bottom of the zone, the equation IV (35) of the free-electron theory still applies, and may be written in the form:

$$\kappa = \frac{3\,N\mu^2}{2\,k\bar{\theta}_0}\left[1 - \frac{\pi^2}{12}\left(\frac{\theta}{\bar{\theta}_0}\right)^2\right] \quad \ldots \quad \text{VII (1)}$$

where the temperature $\bar{\theta}_0$ is defined as before by the relation $k\bar{\theta}_0 = \varepsilon^*$ and N is the number of electrons per unit volume. In this case $\bar{\theta}_0$ is very large, and at room temperatures κ is almost independent of temperature. If we consider the case where $\theta/\bar{\theta}_0$ is large, the variation of κ with temperature is given by:

$$\kappa = \frac{N\mu^2}{k\theta}\left[1 - \frac{2}{3\sqrt{2\pi}}\left(\frac{\bar{\theta}_0}{\theta}\right)^{\frac{3}{2}}\right]. \quad \ldots \quad \text{VII (2)}$$

These equations also apply to the condition of affairs in an almost full band, where if E_ω is the energy at the head of a band, $N(E) \propto \sqrt{E_\omega - E}$. In this case N is the number of unoccupied states or positive holes per unit volume, and $\bar{\theta}_0$ is defined by $k\bar{\theta}_0 = E_\omega - \varepsilon^*$. As defined in this way $\bar{\theta}_0$ for the positive holes is much smaller than $\bar{\theta}_0$ for the degenerate electron gas.

The weak spin paramagnetism of the conduction electrons referred to above is always accompanied by a diamagnetic effect. In the free-electron theory, this diamagnetic susceptibility is one-third of the paramagnetic susceptibility, but this simple relation no longer holds for electrons in a periodic field. The subject is very complicated, but the theory works out in such a way that the diamagnetism becomes greater when the effective electronic mass m is very small. As we have seen before, m^* is given by:

$$m^* = \frac{h^2}{4\pi^2} \cdot \frac{1}{\dfrac{d^2E}{dk^2}}$$

where k is the wave number.†

† The symbol k is again used to distinguish the wave number from the k of Boltzmann's constant.

In general for a given zone, $\frac{d^2E}{dk^2}$ is greatest, and the effective electronic mass is least, at the bottom of a zone, and an abnormally large diamagnetism is to be expected if there is a small overlap into a zone for which $\frac{d^2E}{dk^2}$ is very large at the bottom of the zone into which the overlap occurs. This effect has been worked out in great detail by H. Jones for the case of bismuth, where a number of the magnetic phenomena are explained by assuming that a very small overlap occurs from a zone which, if full, would contain exactly five electrons per atom. This overlap takes place in six directions perpendicular to the axis of the crystal, and the theory explains the diamagnetic anisotropy of single crystals of bismuth.* The phenomena indicate that the number of electrons involved in the overlap is of the order of one in every 10^3 or 10^4 atoms, so that the number of electrons concerned is exceedingly small, though this is counterbalanced by the fact that the effective electronic mass is also very small. It is to be noted that the abnormal diamagnetism is lost when the solid melts, in agreement with the view that the large diamagnetism is a characteristic of the crystal structure, and not of the atom, of bismuth. It is, however, hardly correct to claim this as a conclusive proof of the truth of the zone theories, because a large diamagnetism would also be expected if the crystal structure were regarded as a valency crystal with the bonding electrons moving in relatively large orbits, and here again the effect would be destroyed on melting. As we have already explained, the co-valency and Brillouin-zone theories of the bismuth structure are to be regarded as complementary and not antagonistic.

The relations described above show that the paramagnetic susceptibility of the conduction electrons is small and is only slightly affected by temperature, and is directly proportional to the value of $N(E)$ at the Fermi surface. In contrast to this, the magnetic moment arising from the spin of unpaired localized electrons is generally much larger, and is strongly temperature-dependent. For this reason, measurements of the magnetic susceptibilities of solid solutions may show whether the solute atoms contribute electrons to a common conduction band, or exist with localized moments. Work of this kind † has shown that in dilute solid solutions of Cr, Mn, Fe, Co, and Ni in vanadium the outer electrons enter a common conductivity band, and that the susceptibility

* The diamagnetic susceptibility parallel to the principal axis is due to the electrons which have overlapped from the zone which would contain 5 electrons per atom if completely full. The susceptibility perpendicular to the axis is due to the electrons within the zone.

† B. G. Childs, W. E. Gardner, and J. Penfold, *Phil. Mag.*, 1960, 5, 1267.

depends mainly on the electron concentration of the alloys. Only alloys of iron of high concentration appeared to involve localized moments.

Copper–Nickel and Silver–Palladium Alloys.

The rather large paramagnetism of the elements of Groups VII and VIII has been interpreted in terms of Fig. VII.3, the condition being assumed to be one in which the exchange integral is not quite large enough to produce ferromagnetism. We have already explained that in the case of nickel, the properties of the alloys confirm the view that there are 0·6 positive holes per atom in the d-shell. In the same way the solid solutions of some elements in palladium show properties which vary with the composition in such a way as to suggest that there are about 0·5–0·6 positive holes per atom in the d-shell of palladium. The general electronic structure of palladium is therefore not so very different from that of nickel, and the appearance of ferromagnetism in the latter element is regarded as the result of the interatomic distances being at the critical value necessary to give a relatively large positive exchange integral, and there is a similar resemblance between the other elements of Group VIII in the different Periods.

The work of Coles * has shown that interesting differences exist between the copper–nickel and silver–palladium alloys, in each of which a continuous series of solid solutions is formed. Palladium is paramagnetic, with approximately 0·6 holes per atom in the ($4d$) shell. An atom of silver contains one electron more than an atom of palladium and as the $N(E)$ curve for the ($4d$) is much higher than that for the ($5s$) band (cf. Fig. VI.36, p. 355), we may as a first approximation regard the substitution of silver for palladium as filling up the holes in the d-band, since the proportion of electrons entering the s-band will be very small. In agreement with this, it is found that the paramagnetism of palladium is reduced by the solution of silver and becomes zero at about 60 at.-% silver; at higher silver contents the alloys become diamagnetic. We shall see (p. 408) that the electrical resistance of the transition elements is increased by the possibility of $s \to d$ transitions, which take place in addition to the normal scattering due to the thermal vibrations of the lattice. For the silver–palladium alloys the observed electrical resistance/composition curve can be regarded as the sum of the normal ⌒-shaped curve characteristic of a simple solid solution (p. 409), and of an additional resistance present in the range 0–60 at.-% silver, but disappearing at higher silver contents. These facts clearly fit in with the view that solid palladium contains about 0·6 holes in the d-band, and that these are filled on alloying with silver.

* B. R. Coles, *Proc. Phys. Soc.*, 1952, [B], **65**, 221.

Some Physical Properties 401

In the corresponding copper–nickel alloys, nickel is ferromagnetic with a saturation moment of 0·6 μ_B which indicates the presence of about 0·6 holes per atom in the $(3d)$ band. On alloying with copper, the saturation moment is reduced by 0·6 μ_B for each atomic substitution, and the ferromagnetism is lost at about 60 at.-% copper; over this range of composition the behaviour is analogous to that in silver–palladium alloys. In contrast to this, it is found that copper-rich copper–nickel alloys are paramagnetic and not diamagnetic, and from careful examination this appears to be a real characteristic of the alloys, and probably not due to small "islands" of ferromagnetic atoms remaining in the alloy owing to insufficient homogenization. The electrical resistance/composition curve for copper–nickel alloys cannot be explained as the summation of two curves analogous to those for the silver–palladium alloys referred to above. These and other facts lead to the conclusion that, in dilute solutions of nickel in copper, holes are present in the d-band, in spite of the fact that the solution of copper in nickel produces a rapid filling of the 0·6 holes per atom characteristic of pure nickel. The explanation of this difference between the silver–palladium and copper–nickel alloys probably lies in the argument as follows. In the free atom of nickel, the electron configuration is $(3d)^8(4s)^2$ with 2 holes in the d-shell, whereas in solid nickel there are only 0·6 holes per atom in the $(3d)$ band. The effect of compressing an assembly of free nickel atoms is therefore to diminish the number of holes in the $(3d)$ band, and as explained on p. 356 increasing freedom of a nickel atom in a crystal will increase the number of holes. A nickel atom when dissolved in copper is more nearly free than in metallic nickel, because (a) the lattice spacing of copper is greater, (b) the ion of copper is smaller than that of nickel, with the result that the ionic interaction is reduced, and (c) any effect due to Ni–Ni interaction is less because in a dilute solution the nickel atoms are far apart. We can, therefore, understand why in copper-rich alloys factors may be present which favour the formation of holes in the $(3d)$ band, and counteract the tendency to complete the $(3d)^{10}$ sub-Group. In contrast to this, the electronic configuration of free atoms of palladium is $(4d)^{10}(5s)^0$, whereas in solid palladium there are 0·6 holes in the d-band. Here, therefore, an increase in the distance between palladium atoms tends to keep the d-band filled, and the difference between the silver–palladium and copper–nickel alloys can be understood.

Ferromagnetic Alloys.

Ferromagnetism is shown by some intermetallic compounds of ferromagnetic with non-ferromagnetic metals. Chromium and manganese also give rise to ferromagnetic compounds such as the Heusler alloys and manganese nitride, referred to on p. 389. As explained there, the appearance of ferromagnetism in these alloys has been regarded as due to the alloy structure having produced the correct distance between the atoms of manganese or chromium.

Until recently, it was thought that ferromagnetic compounds of non-magnetic metals were confined to alloys containing elements adjacent to iron, cobalt, and nickel in the Periodic Table or containing rare-earth elements. Recent work has shown* that some Laves phases of quite different elements are ferromagnetic at very low temperatures. Thus, $ZrZn_2$, which has the $MgCu_2$ structure, is ferromagnetic below $35°$ K. The compound Sc_3In has also been reported to be ferromagnetic.

The Mössbauer Effect.

The Mössbauer effect is the recoil-free emission and resonant absorption of nuclear γ-rays in solids, discovered by R. L. Mössbauer† during his work as a graduate student in Heidelberg in 1957.

It provides (amongst other things) a method for studying the magnetic field acting on the nucleus of an atom, and from this conclusions can be drawn regarding the configuration of the extra-nuclear electrons. We may understand this first by considering the behaviour of atoms of the ^{57}Fe isotope. This has a ground state of spin $I = \frac{1}{2}$, and a magnetic moment $\mu = +\,0.903$ nuclear magnetons (see p. 263). There is also an excited state ^{57}Fe$_{ex}$ with an energy 14·4 keV above the ground state, and the excited state has spin $I = 3/2$, magnetic moment $\mu = -\,0.153$ nuclear magnetons, and a half-life of 10^{-7} sec; the decay is accompanied by the emission of a 14·4 keV X-ray.

According to the uncertainty principle, a nuclear half-life of 10^{-7} sec corresponds to an uncertainty in energy of $\sim 10^{-8}$ eV, and the X-ray line width is of this order. In a free atom the recoil of the nucleus can absorb considerable momentum, making the γ-ray line width much greater, but fortunately in solids the momentum is communicated to the lattice as a whole in such a way that a fraction of the transitions takes place with no recoil and the line width is reduced to little more than the ideal value. If, therefore, we were to imagine an assembly of ^{57}Fe atoms to be irradiated by γ-rays of gradually increasing frequency, resonance excitations to the excited ^{57}Fe$_{ex}$ state would occur only when the γ-ray frequency was within about 4×10^{-8} eV of the characteristic

* B. T. Matthias, H. Suhl, and E. Corenzwit, *Phys. Rev. Letters*, 1958, **1**, 449. B. T. Matthias, E. Corenzwit, and W. H. Zachariasen, *Phys. Rev.*, 1958, [ii], **112**, 89.
† R. L. Mössbauer, *Z. Physik*, 1958, **151**, 124.

frequency, and strong absorption* of γ-rays would take place over this very narrow range of frequencies.

Experiments of this kind can be done by preparing radioactive ^{57}Co, whose nuclei have a half-life of 270 days and decay to ^{57}Fe$_{ex}$, which almost instantaneously (half-life 10^{-7} sec) decays to ^{57}Fe with the emission of the characteristic 14·4 keV γ-ray which can then be absorbed by the ^{57}Fe atoms in a foil. The resonance absorption is so sharp that, if the radioactive source is moved relatively to the absorbing atoms with a velocity of only 1 cm/sec, the Doppler Effect† destroys the resonance condition, and the γ-rays are absorbed only weakly.

In practice, the conditions are more complicated, because each nucleus is under a magnetic field resulting partly from neighbouring atoms. This leads to the ground state ^{57}Fe being split into two hyperfine, *i.e.* very slightly different, states with $m = +\frac{1}{2}$ and $m = -\frac{1}{2}$, whilst the excited ^{57}Fe$_{ex}$ state is split into four closely neighbouring states with $m = \pm \frac{1}{2}, \pm 3/2$. The conditions for a possible transition are that m changes by 0 or ± 1, and so six transitions are possible. Most experiments use as γ-ray source a material in which there is no hyperfine field at the ^{57}Fe nucleus, whilst the material under study (enriched in ^{57}Fe, if necessary) in the shape of a foil forms the absorber, whose temperature or external magnetic field can be varied. Transmitted γ-rays are detected by counters and the results are processed by a multi-channel analyser. If the source is then moved with a gradually increasing velocity relatively to the absorber, the condition for resonance will be satisfied and the absorption will rise each time the Doppler velocity of the source makes the emitted γ-ray coincide with an absorption line in the absorber. In this way an absorption spectrum will be obtained (see Fig. VII.6).

Experiments of this kind can often be carried out with ordinary iron or its alloys or compounds since some of the ^{57}Fe isotope is always present. When experiments are made with ionic compounds in which the iron atoms (ions) exist in different valency states, it is found that the Mössbauer absorption peak undergoes slight displacements known as isomer shifts. These are the result of the different electrostatic fields at the nucleus produced by different configurations of extra-nuclear electrons. Since the s electrons have the highest charge density near the nucleus, it is these which are most effective in producing isomer

---

* If a bulb of sodium vapour is illuminated by light from a yellow NaD source, strong resonance absorption takes place, whilst light of slightly different frequency is relatively slightly absorbed.

† If an emitting source is moving away from the observer the wave-length is increased, as in the case of the red shift noted in light received from receding galaxies.

shifts. The ferric^{3+} and ferrous^{2+} ions have outer electronic configurations $(3s)^2(3p)^6(3d)^5$ and $(3s)^2(3p)^6(3d)^6$, respectively. The $(3s)$ electrons penetrate the regions occupied by the $(3d)$ electrons, so that their motion is affected by the number of $(3d)$ electrons present. The change from $(3d)^5$ to $(3d)^6$ thus affects the isomer shift indirectly by its effect on the behaviour of the $(3s)$ electrons, and is not a direct effect of the number of $(3d)$ electrons, since the charge density of these is small near the nucleus. The results obtained show that an increase in the d-electron density decreases the electronic charge density at the nucleus.

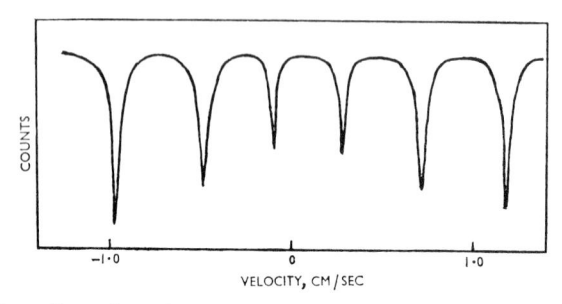

FIG. VII.6. γ-Ray absorption spectrum (counting rate vs. source velocity in cm/sec) showing magnetic hyperfine splitting of ground state and first excited state of ^{57}Fe for a magnetically ordered absorber; asymmetry with respect to zero source activity is caused by isomer shift.

These experiments have been used to throw light on the electron distribution in metallic iron. The difference between the wave-functions in the free atom or ion and those in the metal crystal affects mainly the function at distances far from the nucleus. The isomer shifts depend mainly on differences near the nucleus, and so it is considered justifiable to compare the results for the metal with those from ionic compounds. In metallic iron there are 8 electrons to be divided between the $(3d)$ and $(4s)$ bands. By a combination of simplified theory and interpolation between observations, it is possible to estimate the isomer shifts expected for configurations of the type $(3d)^{8-x}(4s)^x$. The curve connecting the shifts with the value of x is then found to intercept the observed shift at a value near to that corresponding with the configuration $(3d)^7(4s)^1$, and this is regarded as the electronic state of the iron crystal. This conclusion must be accepted with reservation, because the true condition is one of hybrid (spd) bonding orbitals.

Isomer shifts can also provide information about the environment of an iron atom, since this modifies the charge distribution of the electrons.

The Mössbauer technique has also been applied to the study of magnetic effects in alloys. From the splitting of the Mössbauer line it is possible to calculate the hyperfine (HF) field in metallic iron and this work has been extended to ^{57}Fe present in solid solutions in Fe–Co, Fe–Ni, and Ni–Cu alloys.

The Mössbauer technique has also thrown light on the behaviour of small amounts of iron when dissolved in palladium. Here as little as 0·1% Fe produces ferromagnetic behaviour in Pd and the saturation moment per Fe atom is very much higher ($\sim 12 \cdot 6\mu_B$) than in pure iron. Mössbauer methods show that the HF field at ^{57}Fe nuclei is the same in these alloys as in pure iron. This suggests that the saturation moment of the iron atoms is $\sim 2\mu_B$, as in pure iron, and that the remainder of the high observed moment is associated with neighbouring Pd atoms—the data suggest that the moment is associated with a cluster of Fe and Pd atoms and not distributed uniformly through the d-band of palladium.

In other dilute alloy systems the character of the low-temperature magnetic order is less well defined and Mössbauer techniques are a useful means of studying its character. In alloys containing very small concentrations of iron the presence or absence of a magnetic moment on an iron atom can be revealed by measurements in very large (~ 40 k gauss) magnetic field at low temperatures, where the paramagnetic magnetization becomes great enough to produce a large HF field if a moment exists on the iron atoms.

^{57}Fe is not the only nucleus yielding a Mössbauer effect, and some others, including an isotope of tin, have been used in alloy studies.

SUGGESTION FOR FURTHER READING.

C. K. Wertherm, " Mössbauer Effect: Principles and Applications." **1964**: New York and London (Academic Press).

2. THE ELECTRICAL RESISTIVITY OF METAL AND ALLOYS.

In Section IV.3 we described electrical conductivity in terms of a Free-Electron Model introducing the idea of a mean free path between collisions, and in Section V.2 we saw how the zone theory of solids leads to a sharp distinction between two types of solid—metals and insulators—in terms of their response to an electric field; the presence of partly filled zones in the former makes possible electron flow under small applied fields, the conductivity σ being limited only by the departures from crystalline perfection introduced by lattice vibrations, impurities, &c. The electrical resistivity $\rho(\equiv 1/\sigma)$ depends on the extent of these departures from perfection and their efficiency in destroying, by scattering, the net momentum acquired by the electrons from the electric field. (In semiconductors the resistivity and conductivity are dominated by the influence of temperature and purity on the number of electrons available for current carrying, but a detailed account still requires consideration of the mechanisms of electron scattering.)

It has been known for a long time that the contributions to the resistivity of the dilute alloys of many metals can be separated into that dependent on temperature and that dependent on the concentration of foreign atoms, the former being simply the resistivity of the perfectly pure host metal. This is known as Matthiessen's rule and can be formally expressed as

$$\rho(T) = \rho_R + \rho_i(T)$$

where ρ_R is the residual resistivity (the resistivity at $0°$ K, or for practical purposes $4 \cdot 2°$ K) and ρ_i is the temperature-dependence of the resistivity of the ideally pure metal. It follows from the fact that, if changes on alloying of the number and nature of the current carriers can be neglected, the scattering mechanisms are simply additive in their effects on the resistivity, and those arising from impurities and defects are independent of temperature. If alloying causes significant changes in the lattice vibrational spectrum, or if the number and nature of the current carriers or the states to which they make transitions is modified by the temperature, the rule will not be valid.

The sources of scattering mechanisms with which one is normally concerned are:

(a) Lattice vibrations
(b) Impurity atoms
(c) Magnetic disorder
(d) Point defects—vacancies and interstitial atoms
(e) Extended defects—stacking faults and grain boundaries.

Except for chemically very pure metals at low temperatures (d) and (e) are not important sources of resistivity in practice, but they are of great importance as a basis for studying the concentrations and kinetic behaviour of defects.

Lattice Vibrational Scattering.

For simple pure metals the resistivity is dominated by the scattering of electrons by the quantized vibrations* of the crystal lattice (or phonons), and this topic has been the subject of a great deal of theoretical work ever since it was realized that a perfect metallic lattice would have

* See p. 280 for a discussion and the introduction of the characteristic temperature $\underline{\theta}$.

Some Physical Properties

no resistance;* but it is still difficult to make a reliable calculation of the resistivity of any metal from first principles. We shall not, therefore, attempt a comparison of theoretical and experimental results, but simply indicate a few general features.

At high temperatures ($T > \bar{\theta}/2$) the details of the quantization of the lattice vibrational spectrum are unimportant and the probability of electron scattering is proportional to the mean square displacement of an atom from its lattice site, which is in turn proportional to the absolute temperature. (In practice it is found that $d\rho/dT$ is constant to a good accuracy at temperatures greater than about $\bar{\theta}/5$ although this does not mean that $\rho = $ const. T.) At low temperatures the requirement that the scattering process must conserve both energy and momentum modifies this behaviour, since the vibrations available correspond to rather inefficient means of changing the electron state from region A to region B in Fig. IV.8 (Section IV.3). In consequence, a stronger temperature-dependence is found, which for simple metals below $0 \cdot 1 \bar{\theta}$, gives a proportionality of the resistivity to T^5. Thus, the temperature-dependence of the resistivity of a typical pure metal follows a curve like that of Fig. VII.7.

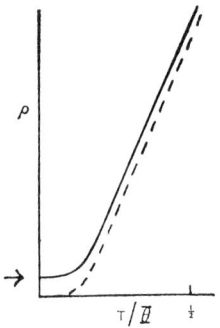

Fig. VII.7. The electrical resistivity of a commercially pure metal specimen; the arrow shows the residual resitivity ρ_R, and the dotted line the ideal lattice vibrational resistivity derived on the assumption that Matthiessen's rule is obeyed.

Electrical Resistivity and the Periodic Table.

From what has been said above and in Section V.2 it will be appreciated that the variation in resistivity at room temperature from metal to metal will depend in part on the electronic structure and in part on the lattice vibrations. In consequence, no simple periodicity is to be expected but a few general points can be made.

* A comprehensive survey of the factors involved will be found in "Electrons and Phonons," by J. M. Ziman. 1962: Oxford (Clarendon Press).

Metals with a Fermi surface of very free-electron-like character will be good conductors, but a low melting point (which correlates with a low value of $\underline{\theta}$) will imply strong lattice vibrational scattering at room temperature. Thus, of the monovalent metals we expect copper, silver, and gold to be better conductors than the alkali metals, but both groups will have higher conductivities than polyvalent metals. Metals whose carriers are a small number of electrons or holes in an otherwise empty or full Brillouin zone will be poor conductors, e.g. bismuth and antimony. Transition metals will transport electric current mainly by the s-electrons, the effective masses of the d-electrons being large, but in spite of their high melting points the scattering probability of these s-electrons will be larger than in, say, copper, since the large number of empty states at the Fermi level in the d-band (see Fig. VI.39) will enhance the scattering of s-electrons to d-states, whereas in copper only $s \to s$ scattering processes are possible. In consequence we find a large increase in resistivity on going from copper to nickel or silver to palladium.

A survey of the experimental data for the electrical resistivities of the various pure metals is given by Meaden.*

Impurity or Alloy Scattering.

Physicists concerned with the conductivity of metals normally refer to the residual resistivity (at $T < 5°$ K) as the impurity resistivity, since it is a valuable measure of the total concentration of impurity atoms and defects, and the residual resistance ratio ($\rho_{4\cdot 2° K}/\rho_{295° K}$) can vary from ~ 0.02 for a commercially pure metal to $\sim 10^{-5}$ for a very carefully refined specimen of the same metal. Metallurgists who are concerned with alloys normally refer to the residual resistivity of a solid solution as the alloy-scattering resistivity.

In dilute solid solutions this resistivity is proportional to the concentration (c) of solute atoms, since these will scatter as separate entities, and the quantity $\Delta\rho/c$ for a particular solvent and solute is a source of useful information. It is not as large as might naively have been assumed, because the impurity potential is efficiently screened (see Fig. VI.46, p. 379). Not surprisingly, it is small (of the order of 0.1×10^{-6} ohm.cm) when solute and solvent are rather similar, as with copper and gold or nickel and cobalt, but large when they are widely separated in the Periodic Table, like copper and arsenic or gold and vanadium. Virtual bound states of the type shown in Fig. VI.46(c) tend to be a source of rather large "resonant" scattering. In the special cases

* G. T. Meaden, "Electrical Resistance of Metals," **1966**: London (Heywood).

of solid solutions in copper, silver, and gold of the metals following them in the Periodic Table it is found that $\Delta\rho/c$ is proportional to $(N-1)^2$ where N is the group number of the solute. In concentrated alloys it is no longer justifiable to regard only one type of atomic site as constituting a disruption of an otherwise perfect lattice, and in simple situations the alloy scattering varies as $c(1-c)$, where c is expressed as an atomic fraction. This expression holds quite well for the silver–gold and disordered copper–gold alloys over the whole concentration range (see Fig. VII.8) but it should not be expected to hold

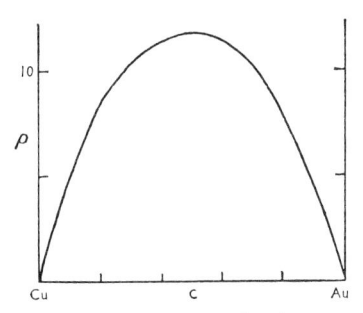

Fig. VII.8. The residual resistivity of the disordered copper–gold alloys (μohm cm).

for any system having a complete range of solid solutions if significant changes in electronic structure occur in this range, as is the case for example in Ag–Pd (p. 400).

Atomic Ordering in Solid Solutions.

When a random solid solution undergoes atomic rearrangement to form a superlattice (p. 121) the increased order is accompanied by a decrease in electrical resistivity, and this is true both for primary solid solutions (*e.g.* Cu_3Au) and intermediate phases (*e.g.* CuZn). If the temperature is changed slowly enough for equilibrium to be reached at each temperature the resistivity/temperature relationship will be as shown in Fig. VII.9, but in many systems quenching from above the critical temperature for ordering will retain the disordered state and yield a much higher room-temperature resistivity. This emphasizes that it is the degree of order that is temperature-dependent, not the alloy scattering produced by a given degree of disorder. In intermediate phases like β-Au–Zn, where ordering and a considerable range of composition coexist, the low-temperature resistivity has a sharp minimum at the equiatomic composition AuZn, for at any other composition some atoms must always be on " wrong " sites.

410   *Atomic Theory*

It was pointed out in Section V.1 that superlattice zone boundaries will modify the Fermi surface in an ordering alloy and hence the effective masses of the conduction electrons. The lattice vibrational resistivity (as measured by $d\rho/dT$) can therefore change on ordering, and is actually larger in ordered Cu_3Au than in the disordered " high-resistance " state.

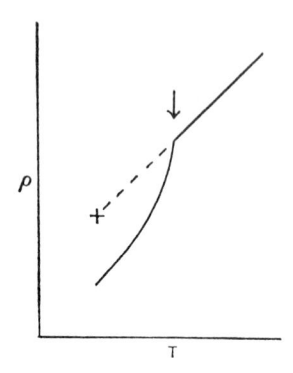

FIG. VII.9. The resistivity/temperature relationship for an ordering alloy brought to equilibrium at each temperature. The arrow indicates the critical temperature and the cross the value produced by quenching to that temperature from above T_c.

Intermetallic Compounds (see p. 122).

Stoichiometric intermetallic compounds correspond to well-defined cusps in isothermal plots of resistivity against composition in alloy systems, but in some cases the compound gives a *maximum* and in others a *minimum* in resistivity. This enables one to distinguish between metallic intermetallic compounds, in which the whole-number ratio of the components is dictated purely by geometric considerations (*e.g.* $MgZn_2$), and quasi-chemical compounds which tend to satisfy normal valency relationships and have strong ionic or covalent character (*e.g.* CsAu or InSb). In the former conduction electrons are present and the ordered nature yields a high conductivity; the latter are semi-conductors or semi-metals and have a very limited carrier concentration.

Magnetic Disorder Scattering.

Electrical resistivity arises from departures of a metal's crystal lattice from perfect periodicity, and we have considered so far departures associated with lattice vibrations (thermal disorder) or impurity atoms (atomic disorder). In metals or alloys possessing localized magnetic moments another form of disorder is present if these are orientated at random, *i.e.* paramagnetically; and this will decrease (and at 0° K

Some Physical Properties

disappear) if magnetic ordering takes place, whether this is ferromagnetic or antiferromagnetic. In recent years it has been realized that magnetic disorder can be an important source of electrical resistivity, since conduction electrons will interact by exchange effects with the local moments and thus be aware of their state of order.* The simplest example is provided by pure gadolinium, which above its Curie temperature of $\sim 300°$ K possesses a large (~ 100 μohm cm) temperature-independent term in its resistivity, and a well-defined change in $d\rho/dT$ at the Curie temperature. The α modification of manganese has a somewhat similar behaviour, although the order there is antiferromagnetic. In the dilute palladium–iron alloys the resistivity provides the clearest indication of the Curie temperature, and in the rare-earth metals the various changes in magnetic character have associated resistance effects.

Resistivity anomalies are found at the Curie temperature in iron and nickel also, and it seems likely that effects of the sort found in gadolinium exist in the former also, but the situation in these d-band metals is much more complicated, since the magnetically active electrons are themselves of Bloch character and contribute to the Fermi surface. In particular, as Mott pointed out many years ago, the $s \to d$ scattering effect mentioned on p. 408 will change at the Curie temperature of nickel in a manner consistent with the observed resistivity behaviour.

3. Superconductivity.

In Section V.2 we remarked that the theoretical description of solids given there, while able to explain the existence of insulators, semiconductors, and metals, did not account for the phenomenon of superconductivity—the complete disappearance of all d.c. resistivity below some characteristic temperature T_c. Values of T_c for the pure metals are shown in Table XVII and for some intermetallic compounds in Table XVIII. A satisfactory theory does now exist (although the lapse of 47 years between discovery† and explanation‡ shows that it did not come easily) and in many ways the theoretical position of superconductivity is now better than that of ferromagnetism. The theory is not easily explained in detail without going outside the main concepts used in this book, but it is sufficiently well established and the phenomenon sufficiently widespread among the metallic elements for an account of the principal experimental facts and theoretical ideas to be justified.

---

* For a now somewhat dated review of these effects see B. R. Coles, *Advances in Physics*, 1958, 7, 40.
† H. Kamerlingh Onnes, *Leiden Commun.*, 1911, Vol. **122b**.
‡ J. Bardeen, L. N. Cooper, and J. R. Schrieffer, *Phys. Rev.*, 1957, **108**, 1175.

TABLE XVII.—*Superconductive Transition Temperatures (upper figure, °K) and Critical Fields (Oersteds) for Normal Stable Modifications of the Elements.*

Li —	Be —											B —	C —	N —	O —	F —
Na —	Mg —											Al 1·19 99	Si —	P —	S —	Cl —
K —	Ca —	Sc —	Ti 0·39 100	V 5·30 1020	Cr —	Mn —	Fe —	Co —	Ni —	Cu —	Zn 0·87 53	Ga 1·09 51	Ge —	As —	Se —	Br —
Rb —	Sr —	Y —	Zr 0·55 47	Nb 9·13 1980	Mo 0·92 98	Tc 8·22	Ru 0·49 66	Rh —	Pd —	Ag —	Cd 0·56 30	In 3·40 293	Sn 3·72 309	Sb —	Te —	I —
Cs —	Ba —	La α 5·0 β 6·3 1600	Hf —	Ta 4·48 830	W 0·015 1·14	Re 1·70 198	Os 0·65 65	Ir 0·14 19	Pt —	Au —	Hg 4·15 412	Tl 2·39 171	Pb 7·19 830	Bi —	Po —	At —
Fr —	Ra —	Ac —	Th 1·37 162	Pa —	U *											

* α-uranium is not superconducting down to 0·1° K at atmospheric pressure but has $T_c = 2·0°$ K if a low-temperature (43° K) phase transition is suppressed by hydrostatic pressure.

Some Physical Properties

TABLE XVIII.—*The Superconducting Transition Temperatures* ($°K$)
of Some Intermetallic Compounds.

Nb_3Sn	18·5	NbC	11	$CeRu_2$	4·9	AuBe	2·64
V_3Ga	16·8	NbN	15·6	$CeCo_2$	0·8	PdSb	2·30
V_3Si	17·0	NbB	8·2	$LaRu_2$	1·6	CuS	1·62
Nb_3Al	17·5	MoN	12·0	$LaOs_2$	6·5	$IrMo_3$	6·8
$Nb_3(Ge,Al)$	20·7	MoC	9·2	KBi_2	3·58	U_6Fe	3·8

General Characteristics of Superconductors.

Although the most striking manifestation of the superconducting state is the disappearance of the resistivity[*]—a disappearance that is as complete in an alloy of large low-temperature normal-state resistance as in a pure metal—it has not proved one of the easiest properties to demonstrate theoretically and it is not always the most suitable experimental parameter for determining the onset of superconductivity. The other striking feature of a superconductor is its refusal to allow a finite value of the magnetic induction, B, within it, for applied fields up to some critical value. This behaviour is found both when a field is applied to a superconductor cooled below T_c in zero field and to one cooled in the presence of a field. The former result would be expected for a material whose characteristic was perfect conductivity; the latter (first discovered

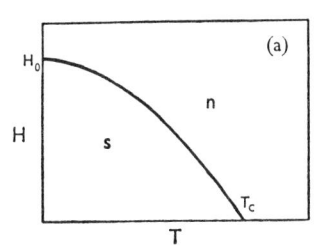

FIG. VII.10. The critical field as a function of temperature for an ideal superconductor.

experimentally by Meissner and Ochsenfeld in 1933, the expulsion of flux from the specimen now being known as the Meissner effect) would not.

In the absence of complications caused by specimen shape (*e.g.*, for a long cylinder whose axis is parallel to the field) the critical value H_c of the magnetic field at which superconductivity is destroyed and flux enters the specimen is well defined for many pure-metal superconductors and its temperature-dependence has the form shown in Fig. VII.10, the

[*] Careful experiments on the persistence of superconducting currents show that the resistivity in the superconducting state cannot exceed 10^{-23} ohm cm.

extrapolation to 0° K yielding the H_0 values included in Table XVIII. Fig.VII.10 is thus a phase diagram for the existence of a superconducting state and, as with other phase changes, the traversal of the phase boundary is accompanied by effects on the specific heat, the specimen length, and other properties. Thermodynamic analyses show that the superconducting phase has a lower entropy than the normal one, and is therefore in some sense a state of greater order.

Factors Affecting the Transition Temperature.

In addition to the effects of an applied magnetic field, changes in the character of the material also modify the T_c, value observed. (Different allotropic modifications of a metal, for example, have different T_c values.) The effects of small amounts of impurities are normally rather slight, in contrast to their effects on the low-temperature normal-state conductivity of pure metals, but the T_c of simple non-transition metals does vary with concentration when primary solid solutions are formed. It has proved possible to separate this variation into a part largely dependent on the change in electron concentration and a part dependent on the change in the normal-state resistivity. The former generally takes the form of an increase with increasing electron : atom ratio and the latter takes the form of a decrease with increasing resistivity.

When solute atoms carry magnetic moments, however, their effects are much more drastic, and the superconductivity of molybdenum went undetected for many years because of the drastic suppression of T_c by a few parts per million of iron.

One change in the character of a superconductor responsible for a change in T_c was independently suggested theoretically and observed experimentally (by two groups of workers) in 1950 and constitutes an important clue to the origin of the phenomenon. That is the change in T_c of a given pure metal with change in its isotopic constitution. For simple metals

$$T_c \propto M^{-\frac{1}{2}}$$

where M is the isotopic mass or average thereof, but smaller effects seem to be found for transition-metal superconductors.

The Theory of Superconductivity.

Although a satisfactory microscopic theory of superconductivity (*i.e.*, one in terms of a quantum-mechanical description of the superconducting state) was long delayed, various phenomenological treatments were extremely successful in providing a macroscopic account of the phenomenon. Many descriptions of these are available in standard works* on superconductivity and it is not appropriate to give details of them here. The Gorter–Casimir two-fluid model successfully linked thermal and magnetic properties. It involved assumptions regarding the condensation of a fraction (temperature-dependent) of the conduction electrons into a condensed state that contributes nothing to the entropy of the system. Phenomenological treatments by Pippard and by Ginzburg and Landau were concerned with the consequences for the magnetic-field behaviour of variations in space of an order parameter specifying the superconductive character.

It is worth, however, outlining the ideas underlying the microscopic theory of the phenomenon, although the mathematical complexities of a complete account are very great.

It was pointed out at the beginning of this Section that a long interval elapsed between the discovery and the explanation of the occurrence of superconductivity, although in the meantime the application of quantum mechanics to solids had given a satisfactory account of most of their other properties. It has often been remarked that the difficulty of constructing a suitable theory is in part due to the very small energy difference ($\sim 10^{-8}$ eV per atom), as compared with Fermi energies of a few volts, between superconducting and normal phases. In this sense the difficulty is similar to the difficulty of predicting the crystal structure of a metal, for the latent heats of transformation between one allotrope and another are very small compared with the total cohesive energy of the metal. On the other hand, even before the very great recent increase in the number of known superconducting elements and compounds, the distribution of superconductors throughout the Periodic Table and the similarity principles that hold (e.g., for H/H_0 as a function of T/T_c) for all of them in spite of differences in electronic and crystal structure, show clearly that the phenomenon must arise from very general features of the metallic state.

The effect referred to above concerning the influence of isotopic mass on the superconducting transition temperature was in accord with the theoretical suggestion of Fröhlich in 1950 that one such feature involved was the interaction of the electrons with the lattice vibrations or phonons. Unfortunately, the first-order interaction of electrons with phonons cannot explain the observed features and it was only in 1957 that Bardeen, Cooper, and Schrieffer succeeded in constructing a

---

* The reader is referred to "Superconductivity" by D. Shoenberg, **1954**: Cambridge (University Press); "Superconductivity" by E. A. Lynton. **1967**: London (Methuen).

EE

satisfactory theory, the B-C-S theory. The final clue had been provided in the proof by Cooper that an attractive interaction, however weak, between a pair of electrons at the Fermi energy rendered the ordinary metallic state unstable with respect to the formation of a bound pair of electrons. The B-C-S theory, therefore, invokes as the interaction leading to superconductivity the interaction via phonons between pairs of electrons.

One electron can give rise to a phonon that interacts with the other and, since the lifetime of the phonon is very short, energy need not be conserved in the process. The phonon is termed " virtual " and can be regarded as a quantization of the oscillatory distortion of the lattice produced by the electron.

The greatest lowering of energy, relative to the normal state, is produced when all electrons at the Fermi surface are bound in pair states of zero total momentum, so that the electrons in a pair have wave-vectors $+k$ and $-k$ (they also have opposite spins); and since the pairing interaction is a phonon one, the electron states so affected lie in an energy shell at the Fermi surface of the order of $\hbar\omega_D$, where ω_D is the Debye characteristic frequency of the lattice vibrations. It should be emphasized that, to be effective, the attractive phonon interaction should dominate the repulsive screened Coulomb interaction between the same pair of electrons, and the net interaction (denoted V) will be more effective the greater the number, $N(E)_F$, of electron states per unit energy at the Fermi surface. In the B-C-S theory, in fact, the critical temperature is given by:

$$k_B T_c = \hbar\omega_D \exp[-1/(N(E)_F V)]$$

where k_B is Boltzmann's constant.

Even when the transition of a metal to a new type of state at low temperatures can be demonstrated, it is not obvious that this will show the properties of zero resistance and perfect diamagnetism (Meissner effect), and a proof that it can do so is beyond the scope of this section. The Meissner effect is associated with the fact that the superconducting state is stabilized by the *correlated* condensation into pairs of *all* the electrons at the Fermi surface, the resultant " rigidity " of the whole system preventing disturbance by a magnetic field. The binding of pairs $+k$ and $-k$ in a superconductor means that scattering of the first to $+k'$ will be correlated with scattering of the second to $-k'$, the net momentum remaining unchanged, and the same situation can be shown to hold for such states when the whole electron distribution is displaced in a way that corresponds to a net current.

Some Physical Properties

T_c and the Electronic Structure.

The B-C-S theory is essentially a free-electron theory and does not take into account the detailed features of the electronic structure. To do so completely is not yet possible, but where other complications are absent a correlation with $N(E)_F$ (as indicated by the electronic specific heat) is found. In alloys of zinc, however, when allowance is made for effects of electron mean free path, effects due to the overlap of the Fermi surface over certain faces of the Brillouin zone seem to be indicated, and in some other alloy systems correlations appear to exist between changes in slope of the T_c vs. composition curve and the type of lattice-spacing anomaly that is suggested to arise from Brillouin-zone overlaps.

Type-II Superconductors and High-Field Superconductivity.

In the above discussion of the effects of magnetic fields it was stated that the critical field at which superconductivity is destroyed is well defined. This is true (apart from geometrical demagnetizing-factor effects) for nearly all pure superconducting metals, and for many years it was assumed that departures from such behaviour in certain alloys were due to metallurgical imperfections and concentration fluctuations. It is now known, however, that in certain types of superconductors, called Type-II superconductors, especially those in which the electronic mean free path in the normal state is rather short, a perfect specimen can show a wide range of applied magnetic fields in which partial penetration of the field occurs and Fig. VII.10 has to be replaced by Fig. VII.11. In this phase diagram, in the region between the upper and lower critical field curves (H_{c2} and H_{c1}), the superconductor is described as being in

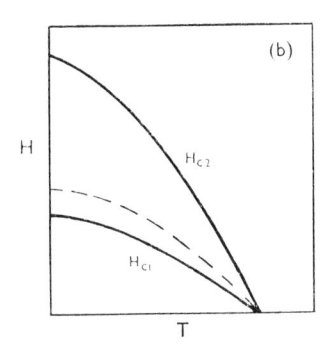

FIG. VII.11. The upper and lower critical fields for a Type-II superconductor; the region of the diagram between the curves H_{c1} and H_{c2} corresponds to the mixed state. The broken curve is the thermodynamic critical field.

the " mixed state." The existence of this state was shown theoretically from arguments concerning the surface energy between normal and superconducting regions, and it has been demonstrated experimentally in very homogeneous solid-solution alloys. In such alloys H_{c2} and H_{c1} separate from the curve of H_c vs. composition at a critical composition dependent on the solvent metal and the normal-state resistivity of the alloy. In the mixed state the superconductor is threaded by flux lines that correspond to normal material and are sustained at a quantized value of $hc/2e$ (or 2×10^{-7} gauss/cm^2) by supercurrent vortices around them. Theory has also shown, and experiment confirmed, that on a rod of Type-II superconductor with an applied field parallel to its axis a surface sheath will remain superconductive to fields even higher than H_{c2}; a field H_{c3} ($\simeq 1{\cdot}69 H_{c2}$) is required to destroy this sheath.

It is of the greatest interest to know whether superconductors with very high values of critical field (100 or more kgauss) can be used to generate such fields in solenoids. For use in such a superconducting solenoid the material needs to be able to carry large currents, and this requirement was not shown by the first alloys found to have large critical fields. Maxwell's equations require a current to be associated with a magnetic field gradient* and in a Type-II superconductor a flow of current through the body of the material is possible, if a gradient in the flux-line density exists within it. The current-carrying paths will, however, be subject to a force proportional to the vector product of the current and the magnetic field (the Lorentz force) and the flux lines will move down the field gradient and out of the superconductor unless restrained.

A metallurgically perfect Type-II superconductor will therefore carry very little current in the mixed state. Fortunately, cold work, internal strains, and other types of imperfection can prove very efficient pinning mechanisms for flux lines, although the exact mechanisms are not completely understood, and superconducting solenoids producing large fields over large volumes are now available. Both solid-solution alloys (Nb–Zr, Nb–Ti) and intermetallic compounds (Nb$_3$Sn) have been used in such magnets.

* In an ideal Type-I superconductor this requirement limits the current to the thin surface layer, called the penetration depth, in which the magnetic field ($< H_c$) at the surface falls to zero.

APPENDIX A

The Metallic Elements and Their Crystal Structures

The elements are arranged below according to the Groups of the Periodic Table, and the following abbreviations are used: B = body-centred cubic; F = face-centred cubic; H = normal close-packed hexagonal structure; D = diamond structure; E = other $(8-N)$ rule structures; C = abnormal or complicated structures.

Where more than one allotropic form exists, the modification stable at the highest temperature is placed first. Thus, BFH means that the body-centred cubic, face-centred cubic, and close-packed hexagonal structures are the stable modifications at high, medium, and low temperatures, respectively.

Group Ia		Group Ib	
Li	BFH	Cu	F
Na	BH	Ag	F
K	B	Au	F
Rb	B		
Cs	B		

Note: the F and H modifications of Li and Na are formed by cold work at very low temperatures.

Group IIa		Group IIb	
Be	BH	Zn	C
Mg	H	Cd	C
Ca	BF	Hg	C
Sr	BHF		
Ba	B		

Note: Zn and Cd have c.p. hex. structures with axial ratios *ca.* 1·9. The structure of Hg is simple rhombohedral.

Group IIIa		Group IIIb	
Al	F	Ga	C
Sc	FH	In	C
Y	H	Tl	BH
La	FH		

Note: It is uncertain which modification of Sc is stable at the higher temperature. Ga crystallizes in a very complicated structure, whilst In has a face-centred tetragonal structure of axial ratio 1·08.

Appendix A

Group IVA		Group IVB	
C	*D*	Ge	*D*
Si	*D*	Sn	*CD*
Ti	*BH*	Pb	*F*
Zr	*BH*		
Hf	*?BH*		
Th	*BF*		

Note: White Sn has a tetragonal structure which may be regarded as a highly distorted modification of the diamond type, but the interatomic distances are much greater than in grey tin with the *D* structure.

Group VA		Group VB	
V	*B*	P	*E*
Nb	*B*	As	*E*
Ta	*B*	Sb	*E*
Pa	*C*	Bi	*E*

Note: Metallic P has an $(8-N)$-rule type of structure which is different from the $(8\text{-}N)$ rule structure of As, Sb, and Bi.

Group VIA		Group VIB	
Cr	*B*	Se	*E*
Mo	*B*	Te	*E*
W	*B*	Po	*CC*
U	*BCC*		

Note: γ-U has the *B* structure and is stable between 775° C and the melting point. The modifications stable at lower temperatures have complicated structures. Se forms $(8-N)$-rule structures of both chain-like and ring-like types, whilst Te forms only a chain-like structure. There are two forms of Po; the high-temperature modification is simple rhombohedral, and the low-temperature form is simple cubic.

Group VIIA	
Mn	*BFCC*
Tc	*H*
Re	*H*

Note: The low-temperature modifications of Mn form very complicated structures.

Group VIIIA		Group VIIIB		Group VIIIC	
Fe	*BFB*	Co	*FH*	Ni	*F*
Ru	*H*	Rh	*F*	Pd	*F*
Os	*H*	Ir	*F*	Pt	*F*

Rare Earth Group

Ce	*?HF*	Eu	*B*	Dy	*H*	Tm	*H*
Pr	*?FH*	Gd	*H*	Ho	*H*	Yb	*F*
Nd	*C*	Tb	*H*	Er	*H*	Lu	*H*
Sm	*C*						

Appendix A

Note: The structures of α-Pr, Nd, and Sm were originally thought to be close-packed hexagonal, but later work showed that a larger unit cell was involved. The structures may be regarded as based on close-packed layers with stacking sequences $ABACABAC$ . . .

Actinide Group
Pu $BCFCCC$

Note: There are 6 modifications of Pu. The modification stable between the B and F structures is f.c. tetragonal.

APPENDIX B

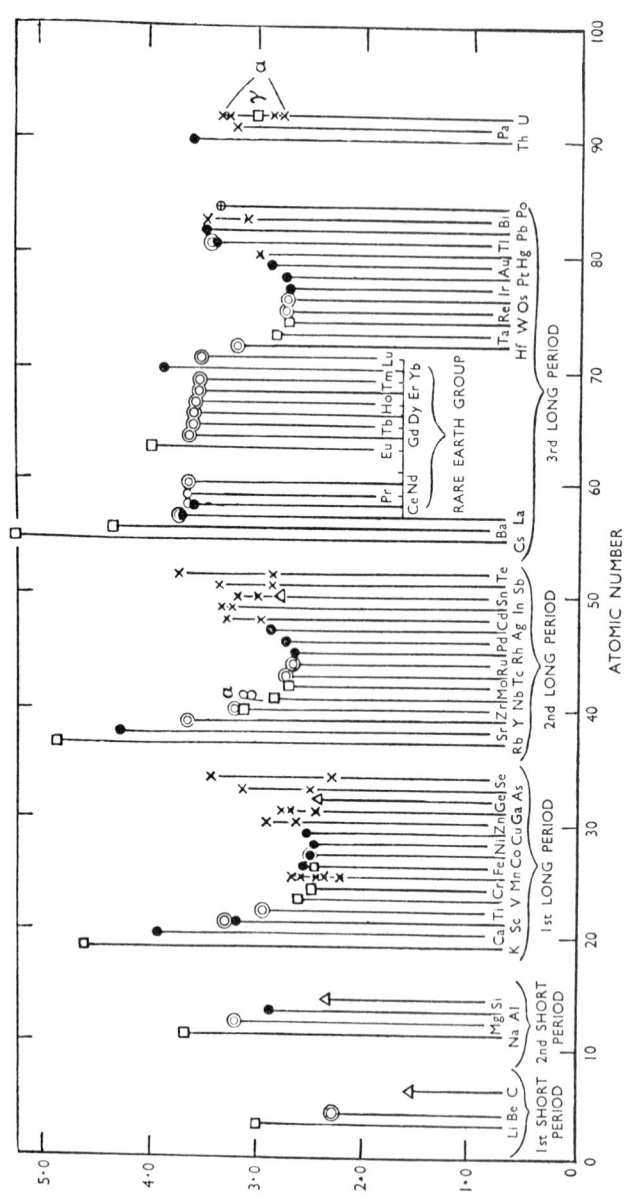

Interatomic distances in the crystals of the elements.

KEY.

Closest distances of approach of atoms in: ⊕ simple cubic structure; □ body-centred cubic structures; ● face-centred cubic structures; ○ close-packed hexagonal structures with axial ratio $c/a = 1.633$; △ diamond-type structures. ◎ Distance between atoms in basal plane for close-packed hexagonal structures with $c/a =$ approximately 1.633. × Interatomic distances in close-packed hexagonal structures with c/a markedly different from 1.633, or in more complex structures.

NAME INDEX

Abarenkov, I., 300.
Altmann, S. L., 367.
Andrews, K. W., 315.
Ashcroft, N., 339.
Aston, F. W., 4.
Axon, H. J., 342.

Bardeen, J., 293, 411, 416.
von Batchelder, F. W., 333.
Baughan, E. C., 366.
Bernal, J. D., 126.
Berry, R. L., 124.
Betterton, J. O., 325.
Birge, R. T., 14.
Blackman, M., 281.
Bloembergen, N., 267.
Bloch, F., 195, 198, 390.
de Boer, F., 102.
Bohm, D., 295, 297.
Bohr, N., 16, 18, 51.
Born, M., 115.
Bradley, A. J., 126, 325.
Bradley, C. C., 193.
Brewer, L., 113.
Brillouin, L., 195.
de Broglie, L., 29.
Brooks, H., 238.
Burns, J., 327.
Butterworth, J., 266.

Callaway, J., 293.
Casimir, H. B. G., 415.
Catterall, J. A., 259.
Chadwick, J., 6.
Channel-Evans, K. M., 120, 126, 310.
Childs, B. G., 399.
Clarebrough, L. M., 127.
Cohen, E. R., 12, 14.
Cohen, M. H., 155, 299, 300.
Coles, B. R., 356, 363, 391, 400.
Compton, A. H., 19.
Cooper, L. N., 411, 416.
Corenzwit, E., 402.
Coulson, C. A., 367.

Dalton, J., 1.
Davisson, C., 23, 31.
Davy, H., 1.
Debye, P., 282.
de Haas, W. J., 286.
Dirac, P. A. M., 21, 72.
Du Mond, J. W. M., 12, 14.

Einstein, A., 282.
Ekman, W., 125.
Engel, K., 113.

Faber, T. E., 193.
Faraday, M., 1.
Fletcher, G. C., 359.

Fowler, R. H., 167.
Frank, F. C., 124.
Frenkel, J. I., 22.
Friedel, J., 305, 378.
Fröhlich, H., 292, 416.
Fuchs, K., 293, 302.

Gardner, W. E., 399.
Germer, L. H., 23, 31.
Ginzburg, V. L., 415.
Gold, A. V., 275.
Goodenough, J. B., 320.
Gorter, C. J., 415.
Goudsmit, S., 63.
Gregory, C. H., 126.
Guggenheim, E. A., 167.
Gunnerson, E. M., 340.

Hamilton, W., 29.
Harrison, W. A., 237, 238, 332, 339, 341.
Hartree, D. R., 67.
Haworth, C. W., 372.
Heine, V., 107, 155, 299, 300, 339, 344.
Heisenberg, W., 18, 241, 388.
Heitler, W., 143, 241.
Hellawell, A., 371.
Herring, C., 266.
von Hippel, A. R., 102.
Hume-Rothery, W., 120, 125, 126, 130, 308, 310, 315, 322, 325, 327, 332, 342, 356, 363, 367, 371, 372.
Huygens, C., 24.

Irving, H. M., 363.

Jones, H., 195, 221, 252, 253, 267, 302, 304, 308, 310, 311, 320, 327, 330, 332, 354, 399.

Kamerlingh Onnes, H., 411.
Kasper, J. S., 124.
Katz, J. J., 54.
King, H. W., 336.
Kleinman, L., 238.
Knight, W. D., 266.
Kohn, W., 239, 267.
Korringa, G., 239.

Landau, L. D., 415.
Laves, F., 124.
Lee, M. F. G., 286, 300.
Lewis, G. N., 99, 117.
van Liempt, 287.
Lipson, H., 325.
London, F., 143, 241.

Mabbott, G. W., 120, 126, 310.
Main-Smith, J. D., 51.
Massalski, T. B., 336.
Matthias, B. T., 402.

Name Index

Meissner, W., 413.
Mendeleev, D. I., 1.
Mooser, E., 118.
Moseley, H., 6.
Mössbauer, R. L., 402.
Mott, N. F., 195, 221, 242, 243, 302, 305, 330, 331, 354.
Myers, H. P., 259.

Nathans, R., 395.
Néel, L., 381, 392, 396.
Newton, Isaac, 24.
Nicholas, J. F., 127, 213.

Ochsenfeld, R., 413.

Pake, G. E., 264.
Pauli, W., 176.
Pauling, L., 51, 111, 113, 156, 162, 337, 360, 392.
Pearson, W. B., 118.
Penfold, J., 399.
Phillips, J. C., 238.
Pickart, S. J., 395.
Pines, D., 295, 297.
Pippard, A. B., 269, 274, 276, 304, 415.
Planck, M., 11.

Raeuchle, R. F., 333.
Raimes, S., 243.
Raynor, G. V., 124, 130, 308, 320, 322, 333.
Reuter, G. E. H., 277.
Reynolds, J., 325.
Reynolds, P. W., 130, 322.
Robertson, W. D., 116.
Rostoker, N., 239.
Rutherford (Lord), 5.

Salkovitz, E. I., 333.
Sato, H., 213.

Schiff, B., 267.
Schindler, A. I., 333.
Schrieffer, J. R., 411, 416.
Schrödinger, E., 21.
Seaborg, G. T., 54.
Seitz, F., 288.
Shoenberg, D., 269, 274, 304.
Shull, C. G., 394, 395.
Skinner, H. W. B., 88, 255.
Slater, J. C., 67, 146, 240.
Smith, A. D. N., 308.
Soddy, F., 6.
Sommerfeld, A., 177, 186, 187.
Sondheimer, E. H., 277.
Spedding, F. H., 376.
Stevens, K. W. H., 243.
Stoner, E. C., 51, 177, 390.
Suhl, H., 402.
Sutton, A. L., 371.

Taylor, A., 325.
Thomson, G. P., 23, 31.
Thomson, J. J., 4.
Toth, R. S., 213.
Townes, C. H., 266.
Trlifaj, M., 331.
Trotter, G., 259.

Uhlenbeck, G. E., 63.
Uhlig, H. H., 116.

Westgren, A., 125.
Wigner, E., 288, 295.
Williams, R. J. P., 363.
Wilson, E. G., 193.
Wohlfarth, E. P., 359, 390, 391.

Zachariasen, W. H., 337.
Zener, C., 322, 392, 394.
Ziman, J. M., 193, 243.

SUBJECT INDEX

Alloys are referred to under their constituent metals arranged in alphabetical order. Thus, all alloys of silver and magnesium will be found under "Magnesium–silver alloys" even though silver may be the chief constituent and the alloy may be mentioned in the text as a "silver–magnesium alloy."

ACCEPTABLE solutions, 37, 66.
Actinides, 55, 60, 377.
Action, 20, 28.
 principle of least, 27, 28.
Alkali metals,
 binding energies, 288.
 elastic constants, 294.
 electronic structure, 284 f.
 Fermi surfaces, 286, 300.
Alkali metal alloys, 286 f.
α-particles, 5, 8.
α-phases,
 lattice spacings, 315.
 solubility limits, 311.
Aluminium,
 electronic structure, 339.
 Fermi surface, 340.
 X-ray spectra of solid, 95.
Aluminium–nickel alloys, 325.
Aluminium–silver alloys, 320.
Amplitude factor, 35.
Ångstrom unit, 10, 14.
Angular momentum, 16, 20, 44, 61.
Anomalous skin effect, 275 f.
Anti-ferromagnetism, 382, 393.
Anti-symmetric function, 139.
Antimony, crystal structure, 101, 102.
Arsenic, crystal structure, 101, 102.
Atomic d-orbitals, 361.
Atomic diameter(s), 120.
 apparent, 342.
Atomic number, 5.
Atomic polyhedra, 239.
Atomic structure,
 poly-electronic atoms, 82.
 table, 52–54.
Atomic unit of length, 10, 14.
Atomic weights,
 scale, 1.
 table, 3.
Auger transitions, 258.
Augmented plane-waves, 240.
Avogadro's number, 4, 14.

Bands. See Electron bands.
Benzene, structure, 158.
Beryllium,
 crystal structure, 330.
 X-ray spectra of solid, 95.
β-particles, 5, 8.
Bismuth,
 crystal structure, 101, 102.
 diamagnetism, 399.
Bloch functions, 198, 235.

Body-centred cubic structure, 104.
 Brillouin zone, 251.
Bohr magneton, 14, 62, 387.
Boltzmann's constant, 12, 14.
Bonding energies, 109 f.
Bragg equation, 200.
Bragg reflection, 200, 204, 217.
Brass,
 α/β equilibrium, 310 f.
 γ-, structure, 323.
 γ-, Jones zone, 324.
Brillouin zones, 195 f, 216.
 construction, 247.
 extended, 252.
 overlaps, 334.
 reduced, 252.

Cadmium,
 crystal structure, 105, 330.
Cadmium–silver alloys, 129.
Calcium-fluoride structure, 116.
Cellular method, 239.
Cerium, 375, 377.
Chromium, anti-ferromagnetism, 395.
Close-packed hexagonal structure, 104.
 axial ratio, 105, 336.
 Brillouin zone, 251, 319, 330.
 electron concentrations, 336.
Collective-electron theories, 235, 354.
Compressibilities, 303, 347.
Compton effect, 19.
Conductivity. See Electrical conductivity and Thermal conductivity.
Copper,
 cohesive energy, 305.
 compressibility, 302.
 diamagnetism, 354.
 electron theory of solid, 134.
 Fermi surface, 304.
Copper alloys, general theory, 308.
Copper–gallium alloys, 127.
Copper–gold alloys, 300.
Copper–nickel alloys, 400.
Copper–silver alloys, 300.
Copper–zinc alloys, 119, 122.
Correlation forces, 134.
Correlation of electrons. See Electron correlation.
Co-valent bonding, 100, 117, 134 f.
Curie point, 381, 385.
Crystal lattices, 243.
Cyclical metal, 182.

d-electrons, 49, 59.
d-states, 79.

425

Subject Index

Debye, (The), 114.
Debye frequency, 282, 416.
Defect structures, 325.
Degeneracy, 49.
δ-states, 138.
Diamagnetism, 381, 396.
  of electron gas, 192.
Diamond,
  bonding, 241.
  crystal structure, 102, 155.
Dilute alloys, theory, 377 f.
Dipole moment, 114.
Directed bonding, 366.
Dissociation, heat of, 109 f.
Domains, magnetic, 382.
Dulong and Petit's law, 283.

Effective magneton number, 387.
Effective mass. *See* Electronic mass.
Electrical conductivity, theory, 184 f, 216, 223.
Electrical resistivity, 405 f.
  impurity scattering, 408.
  lattice vibrational scattering, 406.
  magnetic disorder scattering, 410.
Electrochemical factor, 116, 130.
Electron(s), 4.
  affinity, 58.
  bands, 89, 199.
  charge, 4, 14.
  cloud pattern, 43, 72.
  groups, 48.
  mass, 4, 14.
  mean free path, 186.
  spin, 63.
  velocity, 209.
  velocity in a periodic field, 217.
  wave-length, 23, 26, 31.
Electron compounds, 125, 317.
  relative stabilities, 322.
Electron correlation, 140, 194, 234, 293, 295.
Electron gas, 166.
  degeneracy condition, 175, 176.
Electron volt, 11.
Electronic emission, theory, 187 f.
Electronic mass, effective, 225.
Electronic specific heat, 178, 259.
Energy contours, electronic, 204.
Ether, (The), 24.
Exchange forces, 134 f, 149.
Exchange integral, 388.
Exclusion principle, 50, 138, 169.

f-electrons, 49, 60.
Face-centred cubic structure, 104.
  Brillouin zone, 202, 250.
Faraday, (The), 4, 14.
Fermat's principle, 28.
Fermi–Dirac statistics, 166 f.
Fermi distribution function, 174.
Fermi energy, 172.
Fermi surface, 173, 205, 380.
Ferrimagnetism, 382, 396.
Ferromagnetism, 381, 401.
Free-electron theory, 166 f.
Full metals, 285.

γ-brass. *See* Brass.
γ-rays, 5, 8.
Gas constant, 14.
Gerade. *See* Wave functions.
Germanium, 228.
  crystal structure, 102.
Gold, 300.
Gold–magnesium alloys, 323.
Gold–silver alloys, 300.
Graphite, 228.
Group velocity, 40.
  of electron in a periodic field, 197.
Gyromagnetic ratio, 386.

de Haas–van Alphen effect, 268 f.
Hall coefficient, 227, 229, 260 f, 333.
Hamiltonian, (The), 36.
Hartree–Fock approximation, 295, 298.
Heisenberg principle. *See* Uncertainty principle.
Heitler and London, method of, 143 f.
Heitler–London–Heisenberg approach, 241.
Heusler alloys, 389.
Homopolar bonding. *See* Co-valent bonding.
Hume-Rothery phases. *See* Electron compounds.
Hund's principle. *See* Maximum multiplicity.
Hybrid orbitals, 90, 154.
Hydrogen atom,
  angular momentum, 44, 61.
  Bohr theory, 16.
  classical theory, 15.
  wave-mechanical theory, 66.
Hydrogen ion molecule, 134.
Hydrogen molecule, structure, 141.
Hyperfine field, 405.
Hyperon, 8.

Indistinguishability of electrons, principle, 138.
Indium, crystal structure, 106.
Indium–scandium ferromagnetic alloy, 402.
Ineffectiveness concept, 276.
Insulators, 215 f.
Intermediate phases, 121.
Intermetallic compounds, 123.
  conductivity, 410.
Invar, 389.
Ion core potential, 234.
Ionic bond, 114.
Ionic radii, 336, 338.
Ionization potentials, 56–59.
Isomer shift. *See* Mössbauer effect.
Isotopes, 6, 8.
Isotope effect, superconductivity, 416.

j, 63.
j^*, 63.
Jones zones, 253, 330.

k-space, 196.
kX-unit, 10, 14.
Knight shift, 266.

Subject Index

l, 48, 61.
$*$, 63.
Landé splitting factor, 386.
Lanthanides, 55, 60, 375, 395.
  intermetallic compounds, 376.
  magnetic structures, 376.
Lattice specific heat, 280.
Lattice vibrational spectrum, 281, 406.
Laves phases, 124.
Light, velocity, 11, 14.
Linear combination of atomic orbitals, method of, 236.
Liquid metals, free-electron theory of conductivity, 193.
Lithium,
  anomalous properties, 298.
  Knight shift, 267.
  X-ray spectra of solid, 95, 259.
Lithium–magnesium alloys, 298.
Lithium–silver alloys, 126.

m, 48, 61.
m_s, 48, 63.
Magnesium,
  band structure, 331.
  crystal structure, 330.
  X-ray spectra of solid, 95.
Magnesium alloys, lattice spacing, 333.
Magnesium–silver alloys, 127, 213, 323.
Magnesium–tin alloys, electrical conductivity, 221.
Magnetic Domains. See Domains, magnetic.
Magnetic impurities,
  behaviour, 379.
  electrical resistivity, 410.
  Mössbauer studies, 402.
  nuclear magnetic resonance, 268.
  superconductivity, 414.
Magnetic quantum number, 62.
Magnetic resonance, nuclear. See Nuclear magnetic resonance.
Magnetic saturation moments, 387.
Magnetic susceptibility, 381, 383.
Magneton. See Bohr magneton and Nuclear magneton.
Manganese,
  α-phase, 373.
  α-phase, anti-ferromagnetism, 395.
  β-phase, 318.
  β-phase, Jones zone, 318.
Mass number, 7.
Matthiessen's rule, 406.
Maximum multiplicity, principle, 51.
Meissner effect, superconductivity, 413.
Mercury, crystal structure, 106.
Meson, 8.
Metallic bond, 162.
Metallic orbitals, 362.
Metallic structures, typical, 104.
Mobility, 229.
Molecular orbitals, method of, 141 f.
Molecule, quantum numbers, 138.
Momentum diagram, 168.

Mössbauer effect, 8, 402.
  isomer shift, 403.
Muffin-tin potential, 239.

n, 48, 61.
$n(E)$ curve, 92, 93.
$N(E)$ curve, 92, 97, 172, 205 f.
  α/β structures, 313.
Nearly-free-electron approximation, 237.
Néel point, 382.
Neutron, 6, 7.
Neutron diffraction, 392 f.
Nickel oxide, band structure, 242.
Normalization, 40.
Nuclear magnetic resonance, 262 f.
Nuclear magneton, 263.
Nuclear relaxation, 264.
Nucleon, 8.
Nucleus, atomic, 5.

Orbital(s),
  bonding, 137.
  non-bonding, 137.
Order,
  long-range. See Superlattices.
  short-range, 121.
Orthogonal functions, 254.
Orthogonal plane-wave approximation, 237.
Open metals, 285, 337.
Optical properties, 259.

p-electrons, 49, 59.
p-states, 76, 158.
Palladium–silver alloys, 400.
Paramagnetism, 381, 396.
  free-electron theory, 191.
Paschen–Back effect, 63.
Pauli principle. See Exclusion principle.
Pauling hypothesis, 162, 360 f.
Periodic table of elements, 2.
Phase space, molecular, 168.
Phase velocity, 40.
ϕ-states, 138.
Photoconductivity, 220.
Photoelectric emission, 188.
Photon, 19, 24, 34.
Physical constants, table, 14.
π-states, 138.
Planck's constant, 10, 14, 19, 31.
Plasma frequencies, 296.
Plasma vibrations, 295.
Plutonium, 377.
Polar bond. See Ionic bond.
Polar diagrams, 69.
Positron, 5, 8.
Probability wave-packet, 169.
Proton, 6.
Ψ, 34.
Pseudo-potential method, 107, 238.

Quadrupole moment, 268.
Quantum defect method, 238.
Quantum number, 16, 17, 37, 48, 61 f.
Quantum theory, 11.

Radial factor, 69.
Rare earths. See Lanthanides.
Reciprocal lattice, 245.
Relativity, theory, 17.
Relaxation time, 185.
Residual resistance, 408.
Resonance, 144, 157.
Rydberg constant, 17.
Rydberg unit, 12.

s, 63.
s^*, 63.
s-electrons, 49, 59.
s-states, 72.
sd hybridization, 359, 374.
sp^3 orbitals, 155.
Saturation moments. See Magnetic saturation moment.
Schrödinger equations. See Wave equations.
Screening, in metals, 378.
Screening constant, 58, 83.
Selection rules, 86, 257.
Selenium,
  atom, 152.
  crystal structure, 101.
Self-consistent field, 234.
Self-diffusion, nuclear magnetic resonance effects, 267.
Semi-conductors, 215 f, 227 f.
  impurity, 229.
  intrinsic, 228.
  N-type, 229.
  P-type, 230.
σ-phases, 372.
σ-states, 138.
Silicon, 228.
  crystal structure, 102.
Silver–zinc alloys, 129.
Sodium,
  wave functions, 289.
  X-ray spectra of solid, 95.
Sodium chloride,
  insulating properties, 219.
  structure, 115.
Solid solutions,
  α-phase, liquidus/solidus, 326.
  interstitial, 119.
  primary, 119.
  substitutional, 119.
Space factor, 35.
Specific heat of electrons. See Electronic specific heat.
Spin-lattice relaxation, 265.
Spin wave, 390.
Splitting factor. See Landé splitting factor.
Stationary state, 16, 47.
Sublimation, heat of, 109 f.
Superconductivity, 411 f.
  high-field, 418.
  magnetic impurities, 414.
  theory, 415.
  transition temperatures and critical fields, tables, 412, 413.
  type II, 417.

Superlattices, 121.
Brillouin zones, 213.
Supraconductivity. See Superconductivity.
Surface resistance, 275.
Susceptibility, magnetic. See Magnetic susceptibility.
Symmetric function, 139.

Tellurium, crystal structure, 101.
Tetrahedral bonds, 155.
Thermal conductivity, electronic, theory, 187.
Thermodynamic potential, 174.
Tight-binding approximation, 240.
Tin, grey, crystal structure, 102.
Tin, white, crystal structure, 106.
Transition metals, 55.
  binding energies, 351.
  crystal structures, 345.
  electron theory, 353 f, 374.
  magnetic properties, 352.
  melting points, 350.
  properties, 345 f.
  valencies, 365.
Transition-metal alloys, 370 f.
  liquidus/solidus, 371.
Transuranic elements, 9.
Trouton's rule, 113.
Tungsten, β-phases, 373.

Uncertainty principle, 19, 42, 61, 63, 74, 176.
Ungerade. See Wave functions.

van der Waals forces, 99, 132 f.
Variational method, 239.
Vegard's law, 342.
Vibrational states, 137.
Virtual bound states, 259, 378.

Wave equations, 33–39.
  free-electron theory, 180.
Wave functions,
  gerade, 136, 368.
  ungerade, 136, 368.
Wave group, 40.
  velocity, 40.
Wave mechanics, 21, 23.
  electron motion in a periodic field, 195.
Wave-number diagram, 196.
Wave packet, 22. See also Wave group.
Wiedemann–Franz ratio, 187.
Wurtzite structure, 117.

X-ray spectra, 85, 95, 255.
  absorption spectra, 256.
  satellite lines, 258.
X-unit, 10.

Zeeman effect, 63.
Zinc,
  band structure, 332.
  crystal structure, 105, 330.
Zinc-blende structure, 117.
Zinc–zirconium ferromagnetic alloy, 402.